**Mechanical and Dynamic Properties of Biocomposites**

# Mechanical and Dynamic Properties of Biocomposites

*Edited by*

*Senthilkumar Krishnasamy*
*Rajini Nagarajan*
*Senthil Muthu Kumar Thiagamani*
*Suchart Siengchin*

WILEY-VCH

**Editors**

***Dr. Senthilkumar Krishnasamy***
Kalasalingam Academy of Research and Education
Department of Mechanical Engineering
Anand Nagar
626126 Krishnankoil, Tamil Nadu
India

***Prof. Rajini Nagarajan***
Kalasalingam Academy of Research and Education
Department of Mechanical Engineering
Anand Nagar
626126 Krishnankoil, Tamil Nadu
India

***Dr. Senthil Muthu Kumar Thiagamani***
Kalasalingam Academy of Research and Education
Department of Mechanical Engineering
Anand Nagar
626126 Krishnankoil, Tamil Nadu
India

***Prof. Suchart Siengchin***
King Mongkut's University of Technology North Bangkok
Department of Materials & Production Engineering
1518 Pracharat 1
Wongsawang Road, Bangsue
10800 Bangkok
Thailand

**Cover**
Cement plant © Muhammadfath/Shutterstock, Car manufacturer © xieyuliang/Shutterstock, Cargo Rocket Takes Off © 3D sculptor/Shutterstock, Wood background © Engin_Akyurt/9440 images/Pixabay



**Library of Congress Card No.:**
applied for

**British Library Cataloguing-in-Publication Data**
A catalogue record for this book is available from the British Library.

**Bibliographic information published by the Deutsche Nationalbibliothek**
The Deutsche Nationalbibliothek lists this publication in the Deutsche Nationalbibliografie; detailed bibliographic data are available on the Internet at <http://dnb.d-nb.de>.



**Print ISBN:** 978-3-527-34626-4
**ePDF ISBN:** 978-3-527-82232-4
**ePub ISBN:** 978-3-527-82234-8
**oBook ISBN:** 978-3-527-82233-1

**Typesetting** SPi Global, Chennai, India
**Printing and Binding** CPI Group (UK) Ltd, Croydon, CR0 4YY

Printed on acid-free paper

C094265_260421

# Contents

# 1

# Mechanical Behaviors of Natural Fiber-Reinforced Polymer Hybrid Composites

*Adelani A. Oyeniran[1] and Sikiru O. Ismail[2]*

*[1] Cranfield University, Department of Advanced Mechanical Engineering, School of Water, Energy and Environment, Wharley End, Cranfield, Bedfordshire MK43 0AL, UK*
*[2] University of Hertfordshire, Centre for Engineering Research, School of Engineering and Computer Science, Department of Engineering, College Lane Campus, Hatfield, Hertfordshire AL10 9AB, UK*

## 1.1 Introduction

The use of composites in industrial applications has evolved tremendously over the years, due to the quest for better material performance and cost reduction. They have been found to have exceptional properties in terms of their physical and mechanical properties. Simply put, composites describe a heterogeneous material that comprises two or more different materials that are combined within a single system such that the new material formed now has improved properties, which are suitable for an intended application. The materials that are combined to form a composite material are known as fiber and matrix, reinforcement and binder as commonly called, respectively. The matrix material could be either a natural or synthetic polymer, while fiber material could be glass, boron, or carbon, among others (synthetic type); hemp, jute, flax, among natural type; organic; or ceramic [1]. The increasing use of composite materials in industries has been traced to the fact that they have light weight, and possess high strength as well as exceptional corrosion resistance and acoustic properties, which make them preferred to metallic and alloy materials. Their applications now span into marine, power/energy, automobile, security, aerospace, telecommunications, sport/game, military industries, among others.

Biocomposite has been defined as a composite with at least one of its components derived from biological or natural sources [1]. Their main features that drive research interest are the fact that they are biodegradable, renewable, cheap, and have natural/sustainable resources. These features underscore their environmental friendliness. Some examples of natural fibers frequently used in biocomposites are caraua, sisal, jute, abaca, and kenaf, among others [2]. Other natural fibers used in biocomposites are hemp, agave, and flax, among others [3]. Some natural fibers have been identified in the literature to be used only for craft production and these include kenaf, agave, coir, ramie, and caraua fibers [3].

*Mechanical and Dynamic Properties of Biocomposites*, First Edition. Edited by Senthilkumar Krishnasamy, Rajini Nagarajan, Senthil Muthu Kumar Thiagamani, and Suchart Siengchin.

**Table 1.1** Commonly used natural fibers and their mechanical behaviors.

| Fiber | Density ($g/cm^3$) | Diameter (μm) | Elongation (%) | Tensile strength (MPa) | Young's modulus (GPa) |
|---|---|---|---|---|---|
| *Bast* | | | | | |
| Flax | 1.4–1.5 | 5–38 | 1.2–3.2 | 345–1500 | 27.6–80 |
| Hemp | 1.48 | 10–51 | 1.6 | 550–900 | 70 |
| Jute | 1.3–1.46 | 5–25 | 1.5–1.8 | 393–800 | 10–30 |
| Kenaf | 1.2 | 12–36 | 2.7–6.9 | 295 | — |
| Ramie | 1.5 | 18–80 | 2.0–3.8 | 220–938 | 44–128 |
| *Leaf* | | | | | |
| Abaca | 1.5 | — | 3.0–10 | 400 | 12 |
| Banana | 1.35 | 13.16 | 5.3 | 355 | 33.8 |
| Caraua | 1.4 | — | 3.7–4.3 | 500–1150 | 11.8 |
| Henequen | 1.4 | — | 3.0–4.7 | 430–580 | — |
| PALF | 1.5 | 20–80 | 1–3 | 170–1627 | 82 |
| Sisal | 1.33–1.5 | 7–47 | 2.0–3.0 | 400–700 | 9–38 |
| *Seed* | | | | | |
| Cotton | 1.5–1.6 | 12–35 | 3.0–10.0 | 287–597 | 5.5–12.6 |
| Fruit | | | | | |
| Coir | 1.2 | — | 15.0–30.0 | 175–220 | 4–6 |
| Oil palm EFB | 0.7–1.55 | 19.1–25.0 | 2.5 | 248 | 3.2 |
| *Wood* | | | | | |
| Softwood kraft pulp | 1.5 | 33 | 4.4 | 1000 | 40 |
| *Cane/grass* | | | | | |
| Bagasse | 1.2 | 10–34 | 1.1 | 20–290 | 19.7–27.1 |
| Bamboo | 0.6–1.1 | — | — | 140–230 | 11–17 |

EFB and PALF denote empty-fruit bunches and pineapple leaf fiber, respectively.
Source: Nguyen et al. [4]. © 2017, Elsevier.

Some interesting mechanical behaviors of commonly used natural fibers and many more that are not aforementioned are shown in Table 1.1.

Biocomposites have found application in many different industrial sectors, including packaging, sports articles, and ship building, but most importantly in civil and automotive sectors for nonstructural applications: soundproofing, filling material, and lightening, among others [3]. They favor applications that require low cost and lightness as compared with any other synthetic fiber-reinforced composites. They also demonstrate good thermal and acoustic insulation capacities [3]. Generally, biocomposites are randomly oriented with short fibers that are obtained through the extrusion or molding manufacturing process [3]. Essentially, the low specific weight as well as low cost of biocomposites is a function of the low weight and low cost of most natural fibers, in combination with the low cost of the automated manufacturing processes when mass-producing them [3].

## 1.2 Concept of Natural Fibers and/or Biopolymers: Biocomposites

### 1.2.1 Natural Fiber-Reinforced Polymer Composites or Biocomposites

Natural fiber-reinforced polymer (FRP) composites or biocomposites are gaining widespread interest for many reasons. One such reason is the fact that they have shown a potential for replacement of synthetic fibers at a lower cost. They are also sustainable when compared with their synthetic counterparts [5].

Natural fibers refer to fibers whose origins are natural, that is, they are sourced from plants and animals. These origins give rise to three fundamental natural fiber types, viz:

*Animal fibers*: These contain proteins, such as keratin, fibroin, and collagen. Other classifications in this category are animal wool/hairs (angora wool, alpaca, camel, mohair, lamb's wool, bison, yak wool, cashmere, horse hair, goat hair, and qiviut, among others), keratin fiber (chicken and bird feathers), and silk fibers (spider silk, tussah silkmoths, mulberry silk cocoons).

*Plant fibers*: These are often referred to as cellulosic or lignocellulosic fibers. They are classified in six categories:

- *Seed/fruit fibers*: Coir, coconut, loofah, cotton, oil palm, kapok, sponge gourd, milkweed hairs.
- *Cane, grass, and reed fibers*: Bamboo, corn, albardine, esparto, bagasse, sabai, papyrus, rape, canary.
- *Bast or stem fibers*: Blax, jute, okra, rattan, paper mulberry, hemp, kenaf, isora, urena, ramie, kudzu, roselle hemp, wisteria, mesta and nettle, among others.
- *Wood fibers*: Hardwood and softwood, among others.
- *Leaf fibers*: Caraua, pineapple, abaca, raphia, agave, caroa, banana, fique, piassava, cantala, sansevieria, phormium, Mauritius hemp, sisal, date palm, istle and henequen, to mention but a few.
- *Stalk fibers*: Derivable from barley stalk, rice stalk, maize stalk, wheat stalk, oat stalk as well as other crops. Table 1.2 shows the percentage weight (wt.%) of chemical compositions of the mostly used natural fibers.

*Mineral fibers*: These fibers include fibrous brucite, asbestos group (amosite, chrysotile, anthophyllite, crocidolite, actinolite and tremolite) and wollastonite.

**Table 1.2** Commonly used natural fibers in hybrid composites and their chemical compositions.

| Fiber | Cellulose (wt.%) | Hemicellulose (wt.%) | Lignin (wt.%) | Waxes (wt.%) |
|---|---|---|---|---|
| Abaca | 56–63 | 20–25 | 7–9 | 3 |
| Bagasse | 55.2 | 16.8 | 25.3 | — |
| Bamboo | 26–43 | 30 | 21–31 | — |
| Coir | 32–43 | 0.15–0.25 | 40–45 | — |
| Curaua | 73.6 | 9.9 | 7.5 | — |
| Flax | 71 | 18.6–20.6 | 2.2 | 1.5 |
| Hemp | 68 | 15 | 10 | 0.8 |
| Jute | 61–71 | 14–20 | 12–13 | 0.5 |
| Kenaf | 72 | 20.3 | 9 | — |
| Oil palm | 65 | — | 29 | — |
| Pineapple | 81 | — | 12.7 | — |
| Ramie | 68.6–76.2 | 13–16 | 0.6–0.7 | 0.3 |
| Rice husk | 35–45 | 19–25 | 20 | 14–17 |
| Rice straw | 41–57 | 33 | 8–19 | 8–38 |
| Sisal | 65 | 12 | 9.9 | 2 |
| Wheat straw | 38–45 | 15–31 | 12–20 | — |

Source: Faruk et al. [6]. © 2012, Elsevier.

### 1.2.2 Polymer Matrices

Polymer matrices serve as bonding agents to fibers. They bond the fibers together and help in load transfer to the fibers. Also, the polymer matrices allow for good-quality finish of composite surfaces as well as protection of the reinforcing fibers from chemical attacks. Two common classifications of polymer matrices are thermosetting and thermoplastic resins. They are subsequently elucidated.

*Thermosetting resins*: Curing process (chemical reaction) occurs with this type, thus linking polymer chains and connecting the whole matrix in a three-dimensional (3D) network. It should be noted that once curing occurs, re-melting or reforming becomes impossible. These resins are highly stable in dimension, resist high temperature as well as offer good resistance to solvents, due to their cross-linked 3D structure [4]. Some thermosetting resins that are used frequently in composites are vinylesters, polyesters, phenolics, epoxies, bismaleimides (BMIs), and polyamides (PAs).

*Thermoplastic resins*: These resins differ from thermosetting resins, because their thermoplastic molecules are not cross-linked and can be melted when heated and made into solids and then cooled, thus allowing for reforming and reshaping repeatedly. Apart from being generally ductile, thermoplastic resins have more toughness than their thermosetting counterparts. They are broadly

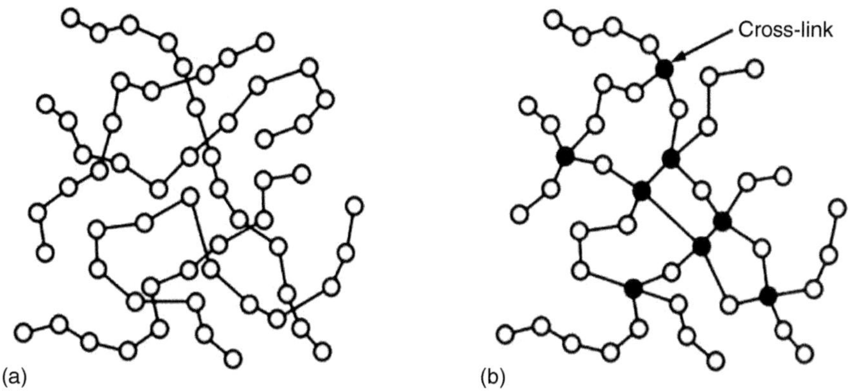


**Figure 1.1** Descriptive molecular structure of both (a) thermoplastic and (b) thermoset polymers. Source: Bergstrom [7]. © 2015, Elsevier

used for nonstructural applications without fillers and reinforcements. Their mechanical properties, which are factors of attraction, include good fatigue and compression strength, excellent tensile strength, excellent stiffness, high dimensional stability, excellent damage tolerance, and excellent durability. Furthermore, their flame-retardant as well as wear-resistant features broaden their applications and make them relevant, especially in an aerospace sector [4]. Common examples of thermoplastic resins include, but are not limited to, polyvinylidene fluoride (PVDF), polypropylene (PP), polyetheretherketone (PEEK), polyphenylene sulfide (PPS), polymethyl methacrylate (PMMA, also called acrylic), polyetherketoneketone (PEKK), and polyetherimide (PEI).

Figure 1.1a,b depicts the molecular structure of thermoplastic and thermosetting resins, respectively. The cross-links in the molecular structure of the thermosetting resins (shaded molecules) are depicted in Figure 1.1b.

There are different categories that exist for the manufacturing process of polymer matrix composites (PMCs). These include squeeze flow methods, short-fiber suspension methods, and porous media methods [4]. Table 1.3 depicts some partial and complete natural and synthetic hybrid FRP composites, their resins/matrices, and manufacturing methods.

It is well known that there is no single engineering material that can be all-encompassing in terms of its applicability to operations and processes. Therefore, natural FRP composites have some limitations, despite their outstanding benefits. Table 1.4 presents some of the benefits as well as disadvantages of natural FRP composites.

The key elements that affect the mechanical response of natural FRP hybrid composites are subsequently identified [5]:

- Fiber selection, which includes the type, method of extraction, time of harvest, natural fiber aspect ratio, content, as well as its treatment
- Interfacial strength
- Matrix choice

**Table 1.3** Manufacturing processes of some hybrid (mainly natural) FRP composites.

| Hybrid fiber | Resin | Curing agent Catalyst | Accelerator | Manufacturing methods |
|---|---|---|---|---|
| Pineapple/sisal/glass | Polyester | MEKP | Cobalt napthenate | Hydraulic press |
| Sisal/silk | Polyester | | | Hand lay-up technique |
| Kenaf/glass | Polyester | | | Hand lay-up and cold press |
| Woven jute/glass | Polyester | | | Hand lay-up |
| Banana/Kenaf | Polyester | | | Hydraulic compression molding process |
| Banana/sisal | Polyester | | | Hand lay-up method followed by compression molding |
| Glass/palmyra | Polyester | | | Hydraulic compression molding process |
| Jute/glass | Polyester | | | Hand lay-up |
| Roselle/sisal | Polyester | | | Hand lay/up technique |
| Silk/sisal | Polyester | | | Hand lay-up technique |
| Banana/sisal | Epoxy | | | Hydraulic compression molding process |
| Glass/glass | Epoxy | HY95 I hardener | | Hand lay-up technique |
| Carbon/glass | Epoxy | HY225 Hardener | | Hand lay-up technique |
| Oil palm/jute | Epoxy | Hardener | | Compression molding process |
| Chicken feather/glass | Epoxy | *n-tert*-Butyl peroxybenzoate | | Hot press |
| Basalt/Hemp | Polypropylene | | | Hot pressing |
| Flax, Hemp, and jute | Polypropylene | | | Hydraulic press |
| Flax/wood fiber | HDPE | | | Twin screw extrusion |
| Banana/glass | Polypropylene | | | Twin screw extrusion |
| Cork/coconut | HDPE | | | Screw extrusion and compression molding |
| Kenaf/pineapple | HDPE | | | Mixing and compression molding |
| Bamboo/glass | Polypropylene | | | Injection molding |
| Cordenka/jute | Polypropylene | | | Injection molding |
| Bamboo/cellulose | Poly lactic acid | | | Injection molding |
| OPEFB/glass | Vinyl ester | | | Resin transfer molding |
| Aramid/sisal | Phenolic | | | Stirring, drying, compression |

HDPE, high-density polyethylene; MEKP, methyl ethyl ketone peroxide; and OPEFB, oil palm empty fruit punch.
Source: Sathishkumar et al. [8]. © 2014, SAGE Publications.

**Table 1.4** Benefits and drawbacks of natural FRP hybrid composites.

| Benefits | Drawbacks |
|---|---|
| • Renewable source of fibers/matrices and sustainability<br>• Low danger/risk during manufacturing processes<br>• Low density, stiffness, and high specific strength<br>• Low process/production energy and environmental friendliness<br>• Lower production cost when compared with synthetic fibers, such as carbon and glass<br>• Low release of harmful fumes when heating and during end of life process (incineration)<br>• Lower abrasive attack on processing tools, when compared with synthetic FRP composites<br>• Possibility of predicting better balanced mechanical behaviors, such as toughness | • Lower responses, especially impact strength in comparison with the synthetic FRP composites<br>• Higher variability of behaviors, due to discrepancies in sources and qualities<br>• Lower durability in comparison with synthetic FRP composites. However, it can be enhanced significantly using treatments<br>• Poor fiber orientation and/or layer stacking sequence, causing weak fiber–matrix interfacial adhesion<br>• High water/moisture absorption, consequently causes swelling effect<br>• Lower processing parameters, such as degradability temperatures. Hence, it causes limiting matrix and fiber options and structural applications |

Source: Modified from Pickering et al. [5]. © 2014, SAGE Publications.

- Fiber distribution
- Composite manufacturing process
- Fiber arrangement [9]
- Void presence/porosity, among others.

## 1.3 Hybrid Natural Fiber-Reinforced Polymeric Biocomposites

Fiber hybridization could offer a further alternative in composites. Hybrid composites are derived by a combination of two or more various fiber types in a common matrix [10]. The fibers can be arranged in different layer pattern/orientations and stacking sequences. Figure 1.2 presents some common arrangements (layering patterns) of natural FRP hybrid composites. This introduces wider spread in their properties than in the regular composite materials, which comprises only one kind of reinforcement. It also enables manufacturing engineers to channel the properties of the composite to the required structural properties. This can possibly be achieved once the hybrid composite behavior can be predicted from the constituent composites.

Operationally, producing hybrid natural FRP composites suggests an intermediate intervention to reduce the negative environmental impact of glass and carbon (synthetic) fibers on the environment, by partially replacing the glass and carbon fibers

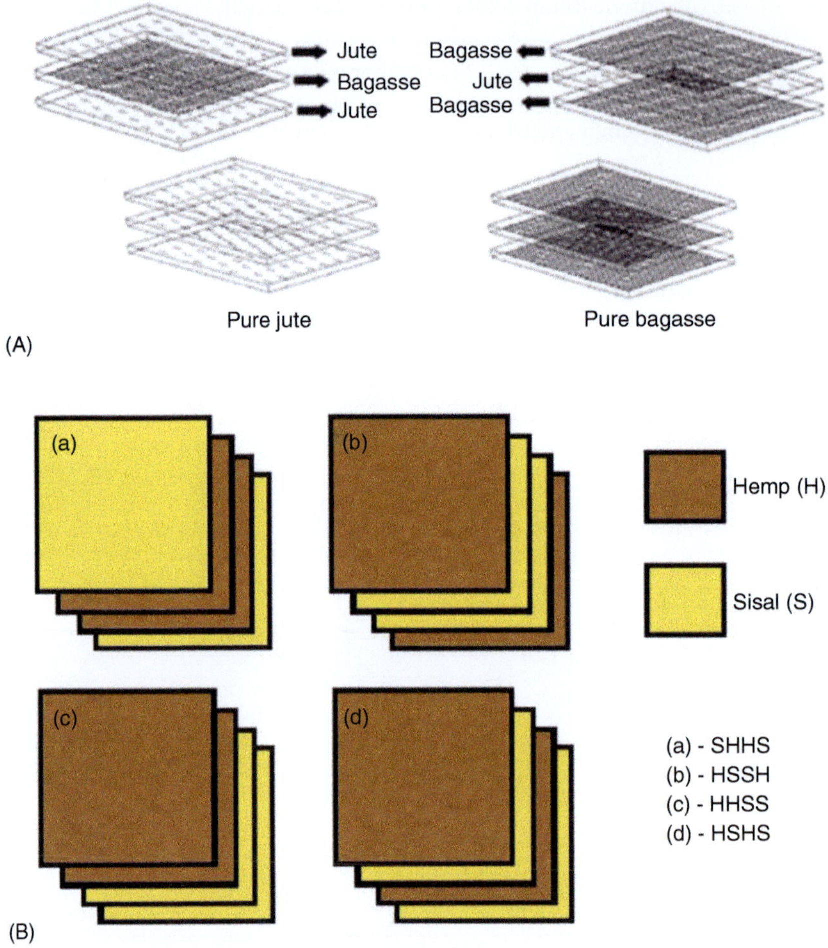


**Figure 1.2** Schematic illustration of different orientations and stacking sequences of natural FRP hybrid composites. Source: Refs. [9, 11]. © 2012; John Wiley and Sons.

with such alternatives as the vegetable fibers jute, flax, hemp, kenaf, sisal, among others [12, 13]. In these substitutions, the limits are revealed by simulating the service performance in such a dynamic testing, including fatigue and impact [14, 15].

Moreover, hybrid biocomposites refer to composites in which two or more different biofibers (natural fibers) are combined in a matrix, or a mixture of natural fibers with synthetic fibers in a matrix [4]. One synthetic fiber commonly used for improving the mechanical response in natural FRP composites is glass or carbon fibers. Several types exist for hybrid composites. These types are dependent on the material constituent mixture [16, 17].

For instance, Figure 1.3 shows higher mechanical properties of unaged hybrid flax/basalt FRP composite sample A, when compared with single or non-hybrid flax FRP composite. However, the impact strength property of the aged hybrid counterpart samples B, C, D, and E changed insignificantly after 15, 30, 45, and 60 aging days of salt-fog environment conditions (Figure 1.3c) of the hybrid types. Additional mechanical behaviors of some FRP hybrid biocomposites are presented later in Table 1.5 by considering their natural/natural fiber combined reinforcements.

In preparing hybrid FRP composites, the rule of mixture comes to play, while the volume fraction can be obtained using Eqs. (1.1)–(1.6) [8].

$$V_{c1} + V_{c2} = 1 \tag{1.1}$$

$$V_{c1} = \frac{V_{f1}}{V_f} \tag{1.2}$$

$$V_{c2} = \frac{V_{f2}}{V_f} \tag{1.3}$$

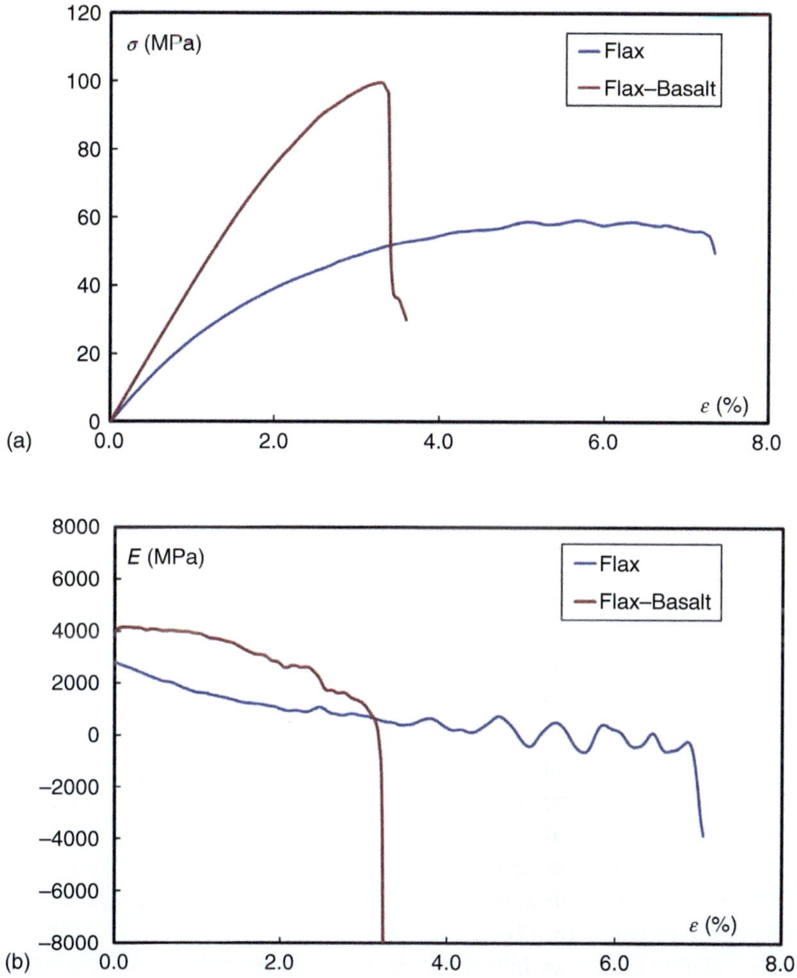


**Figure 1.3** Improved mechanical properties of hybrid flax–basalt fibers FRP composites, depicting (a) stress–strain, (b) modulus–strain curves, and (c) impact strengths of aged and unaged biocomposites. Source: Fiore et al. [18]. © 2016, Elsevier.

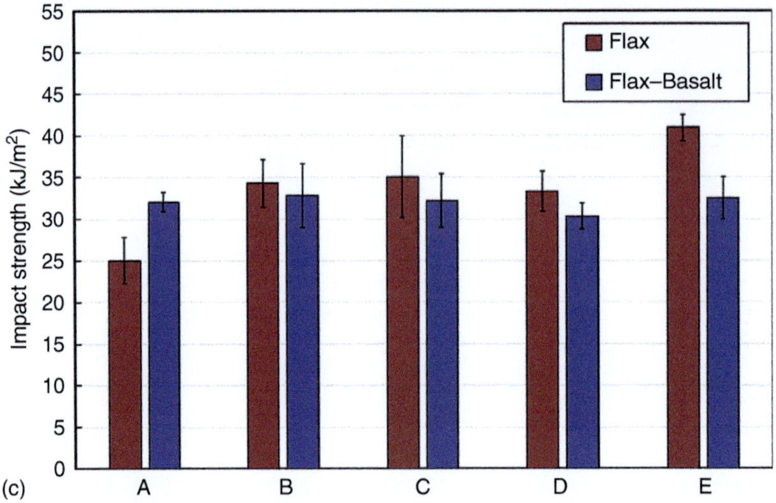

(c)

**Figure 1.3** (*Continued*)

$$V_f = V_{f1} + V_{f2} \tag{1.4}$$

$$W_f = W_{f1} + W_{f2} \tag{1.5}$$

$$V_{f1} = \rho_c \frac{W_{f1}}{\rho_{f1}} \tag{1.6}$$

where $V_f$ denotes total reinforcement volume fraction, $V_{c1}$ and $V_{c2}$ represent the first and second reinforcement relative volume fractions, $V_{f1}$ and $V_{f2}$ stand for the first and second fiber volume fractions, $\rho_c$ and $\rho_f$ designate the densities of the composites and fiber, while $W_f$ indicates the weight of the fiber. The methodology for preparing and characterizing hybrid fiber-reinforced PMCs as well as its applications is presented in Figure 1.4.

However, the present chapter does not cover all the methodologies shown in Figure 1.4 in detail, because the scope of this chapter is not manufacturing processes and techniques of natural FRP composite materials.

## 1.4 Mechanical Behaviors of Natural Fiber-Reinforced Polymer-Based Hybrid Composites

There are many properties of materials that determine where they function or are used in the engineering space. The required characteristics in a proposed design will determine what combinations of materials will be relevant and which of the various mechanical properties are of interest in such instances. Notable among the mechanical properties usually considered in engineering are tensile, compressive, flexural, and impact strengths, among others. These properties are discussed in Section 1.4.1.

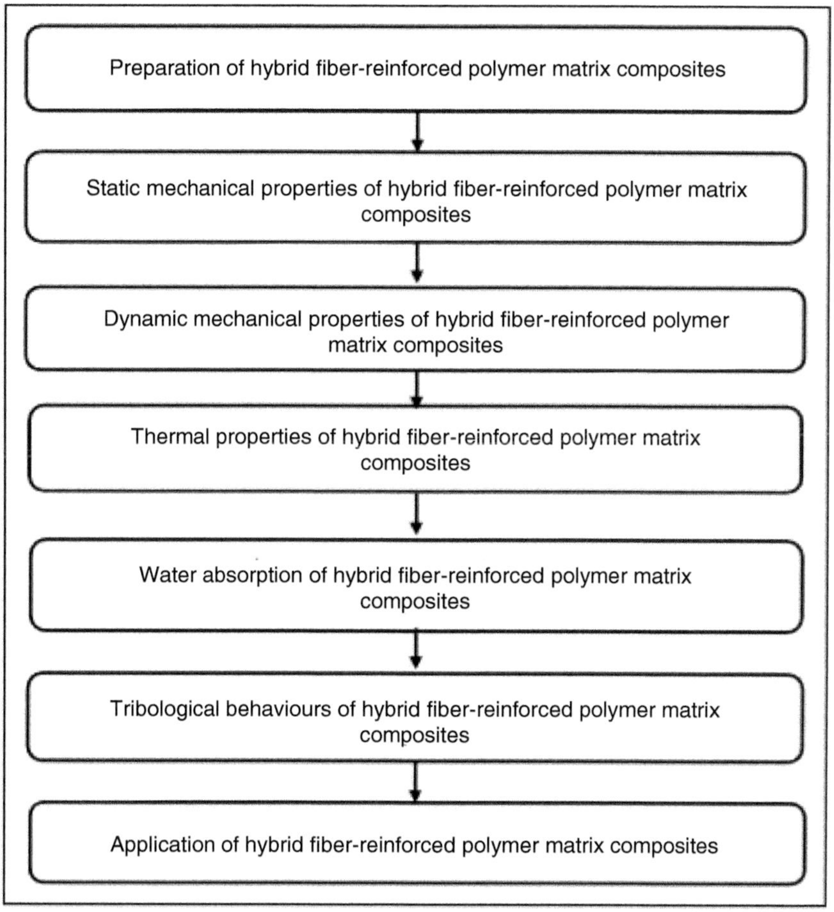


**Figure 1.4** Flowchart of preparation and characterization of the hybrid FRP composites. Source: Sathishkumar et al. [8]. © 2014, SAGE Publications.

### 1.4.1 Hybrid Natural FRP Composites

This section discusses hybrid biocomposites in which their combined fibers are entirely natural (biofibers).

#### 1.4.1.1 Bagasse/Jute FRP Hybrid Composites

Jute is a popular plant-based fiber (vegetable) with dominant presence in tropical countries across the Asian continent, such as China, Brazil, Nepal, Bangladesh, India, and Thailand. They account for about 95% of jute fiber (JF) production worldwide [4]. Jute is considered as a lignocellulosic bast fiber, having comparative advantages with respect to renewability, biodegradability (which makes it eco-friendly), high strength as well as high initial modulus over other fibers [11]. Bagasse, also called sugarcane bagasse, is a lingocellulosic by-product of the sugar industry, mostly utilized as a fuel in boilers and sugar factories. Compared with other residues (by-products), including wheat straw and rice, bagasse is preferred, because its ash content is lower [19].

A study of mechanical behavior of hybrid FRP composites with short JF and short bagasse fiber (BF) bundles reinforcement was carried out by Saw and Datta [20]. They used epoxidized phenolic novolac (EPN) as resin matrix and investigated various fiber surface treatments and fiber ratios. Sodium hydroxide (NaOH) alkali solution was used to treat the JF bundles. The BF bundles were either modified using chlorine dioxide ($ClO_2$) and furfuryl alcohol ($C_5H_6O_2$) or left untreated. The modification of the fiber surface was necessary for quinones creation in the lignin areas of the BF bundles. The created quinones then reacted with the furfuryl alcohol, and thereby improved the BF bundles' (modified) ability for better adhesion. Their result revealed greater mechanical responses (flexural, tensile and impact properties) for hybridized BF (modified) and JF bundles (alkali-treated) in the EPN resin matrix than the BF bundles that were not modified. They obtained an optimum mechanical behavior at a BF/JF ratio of 50 : 50, as depicted in Table 1.5.

#### 1.4.1.2 Bamboo/MFC FRP Hybrid Composites

Asian giants, India and China, are the chief producers of bamboo fiber with more than 80% of global production [21]. This biofiber is highly attractive, due to its renewable nature and low environmental impact. It grows rapidly and has comparative high strength to other biofibers, such as cotton and jute [22].

An unprecedented biocomposite (hybrid) that contained biodegradable poly-lactic acid (PLA) matrix with microfibrillated cellulose (MFC) and bamboo fiber bundles reinforcements was developed by Okubo et al. [23]. Various nomenclatures have been used for describing MFC in the literature, such as microfibril, microfibrillar cellulose, microfibril aggregates, nanofibril, nanofibrillar cellulose, nanofiber, and fibril aggregates [24]. They conducted an investigation on how MFC dispersion influenced the responses of composites reinforced with bamboo fibers by dispersing MFC in a polymer matrix of PLA by a three-roll mill calendering process. This calendering process helps to compress or smoothen a material. They used the PLA (bio-based and biodegradable) polymer matrix for interfacial bonding enhancement with the MFC. The diameter of bamboo fiber bundles was about 200 μm, while that of MFC was just a few microns, which was much smaller. Using gap settings in decreasing order of 70, 50, 35, 25, 15, 10, and 5 μm, they processed the mixture of the MFC and PLA in the three-roll mill. About 200% increase in the fracture energy was realized when they added 1 wt.% of MFC to the PLA matrix and milled the MFC/PLA composite at the smallest gap setting of 5 μm, which was quite significant. This hybrid composite combination of bamboo fiber and the PLA matrix with 1 wt.% MFC reinforcement was observed to prevent an abrupt crack channel through the bamboo fiber effectively, and thus produced a significant improvement in fracture strength. The results of other mechanical behaviors are presented in Table 1.5.

#### 1.4.1.3 Banana/Kenaf and Banana/Sisal FRP Hybrid Composites

A good material for reinforcement in diverse polymer composites is the banana fiber. Its extraction is usually from the bark of banana trees [4]. Banana fiber has such

**Table 1.5** Mechanical behaviors of bagasse/jute, bamboo/MFC, and banana/kenaf FRP hybrid composites.

| Hybrid biocomposites | Fiber ratio (by weight or volume) | Flexural modulus (GPa) | Flexural strength (MPa) | Tensile modulus (GPa) | Tensile strength (MPa) | Impact strength (kJ/m$^2$) |
|---|---|---|---|---|---|---|
| *Natural fibers* | | | | | | |
| | Bagasse fiber bundles (untreated) and jute fiber bundles (treated) | | | | | |
| | 0 : 100 | 0.645 | 31.15 | 0.302 | 11.45 | 6.90 |
| | 20 : 80 | 0.789 | 36.46 | 0.356 | 16.02 | 7.46 |
| | 35 : 65 | 1.101 | 45.32 | 0.420 | 19.45 | 9.53 |
| | 50 : 50 | 1.480 | 55.63 | 0.492 | 23.07 | 10.66 |
| Bagasse/jute | 65 : 35 | 1.311 | 51.19 | 0.399 | 21.15 | 8.33 |
| | 100 : 0 | 0.502 | 26.78 | 0.227 | 9.87 | 6.67 |
| | Bagasse fiber bundles (treated) and jute fiber bundles (treated) | | | | | |
| | 20 : 80 | 1.178 | 42.72 | 0.526 | 18.72 | 10.00 |
| | 35 : 65 | 1.484 | 54.57 | 0.635 | 22.57 | 13.33 |
| | 50 : 50 | 1.748 | 65.22 | 0.753 | 26.77 | 15.93 |
| | 65 : 35 | 1.518 | 60.12 | 0.704 | 23.54 | 10.93 |
| | 100 : 0 | 0.632 | 30.78 | 0.286 | 11.20 | 8.66 |
| | MFC/PLA composites (milled to 5 μm) | | | | | |
| Bamboo/MFC | 1 wt.% of MFC | — | — | 4.61 ± 0.27 | 45.9 ± 4.1 | — |
| | 2 wt.% of MFC | — | — | 3.95 ± 0.14 | 51.7 ± 2.3 | — |
| | 50 : 50, nonwoven hybrid | | | | | |
| Banana/kenaf | 10% NaOH treatment | — | 57.2 | — | 44 | 13 |
| | 10% SLS treatment | | 60.8 | | 50 | 16 |
| | 50 : 50, woven hybrid | | | | | |
| | 10% NaOH treatment | | 62.0 | — | 50 | 18 |
| | 10% SLS treatment | — | 68.0 | | 54 | 21 |

MFC and PLA represent micro-fibrillated cellulose and poly-lactic acid, respectively.
Source: Nguyen et al. [4]. © 2017, Elsevier.

advantaged mechanical properties, including good tensile strength and modulus, due to the high content of cellulose and low microfibrillar angle [25]. Kenaf fiber is a promising element of reinforcement in polymer composites, due to its interesting mechanical features such as eco-friendliness and renewability. Kenaf is usually extracted from bast fiber (kenaf plants) [4]. Sisal, on the other hand, is known to be among the toughest materials for reinforcement. It is also well known for its durability. Sisal FRP composites possess moderate flexural and tensile behaviors and high impact strength, when compared with other composites of natural fiber reinforcements. It has relevant use in some industries, such as agriculture and marine to make twines, ropes, cords, rugs, and bagging, among others [26]. Sisal and kenaf fibers, similar to other natural fibers, have poor interfacial bonding with a polymer matrix, which shows their disadvantage [27].

Moreover, Thiruchitrambalam et al. [28] conducted a study on woven as well as non-woven hybrids of banana/kenaf fiber with unsaturated polyester matrix reinforcement. They kept the fiber contents constant at 40% with equal ratio of banana and kenaf FRP composites (50 : 50 ratio) and treated the fibers with either 10% solution of NaOH or 10% of sodium lauryl sulfate (SLS) for 30 minutes. They observed that the SLS-treated specimen had better improvement with respect to mechanical behavior than the alkali-treated specimen, showing for both woven and non-woven cases of the banana/kenaf hybrid composites enhanced impact, flexural, and tensile strengths, as shown in Table 1.5.

Furthermore, Venkateshwaran et al. [29] evaluated the mechanical properties of banana/sisal FRP epoxy matrix hybrid composite and found out that hybridization increased the flexural, tensile, and impact strengths by 4%, 16%, and 35% respectively. They also reported that the 50 : 50 fiber ratio by weight enhanced the mechanical response of the banana/sisal FRP hybrid composite and decreased the uptake of moisture, as presented in Table 1.6.

#### 1.4.1.4 Coconut/Cork FRP Hybrid Composites

A natural coconut fiber, also known as coir, is usually extracted from coconut trees. These trees are mainly grown in tropical regions of Asian countries, such as Vietnam, India, and Thailand [4]. Cork fiber is usually obtained from cork oak trees (*Quercus suber*). There is a specific species of the tree from whose bark the cork fiber is harvested. The cork oak tree is a renewable resource, as new cork bark regrows naturally [4].

Hybrid composites containing high-density polyethylene (HDPE) with reinforcements of cork powder and short coconut fibers that were randomly distributed was prepared by Fernandes et al. [30]. The interfacial bonding and compatibility between the matrix and fiber was improved by maleic anhydride, a coupling agent (CA). Their results showed 27% and 47% rise in elastic moduli and tensile strengths of the coconut/HDPE/cork hybrid composites, respectively (Table 1.6), when compared with the cork/HDPE composite. Also, they observed that using CA resulted in enhancement of the elongation at break and tensile behaviors of the hybrid composites. As a recommendation for better mechanical responses of

**Table 1.6** Mechanical behaviors of banana/sisal, coconut/cork, coir/silk, corn husk/kenafm and cotton/jute FRP hybrid composites.

| Hybrid biocomposites | Fiber ratio (by weight or volume) | Flexural modulus (GPa) | Flexural strength (MPa) | Tensile modulus (GPa) | Tensile strength (MPa) | Impact strength (kJ/m$^2$) |
|---|---|---|---|---|---|---|
| Banana/sisal | 100 : 0 | 8.920 | 57.33 | 0.642 | 16.12 | 13.25 |
| | 75 : 25 | 9.025 | 58.51 | 0.662 | 17.39 | 15.57 |
| | 50 : 50 | 9.130 | 59.69 | 0.682 | 18.66 | 17.90 |
| | 25 : 75 | 9.235 | 60.87 | 0.703 | 19.93 | 20.22 |
| | 0 : 100 | 9.340 | 62.04 | 0.723 | 21.20 | 22.54 |
| Coconut/cork | 10 : 44 : 44 : 2 (wt.% of coconut/cork/HDPE/coupling agent) | — | — | 0.599 ± 0.02 | 20.4 ± 0.3 | — |
| Coir/silk | Alkali treatment | | | | | |
| | 10 mm fiber | — | 39.53 | — | 15.01 | — |
| | 20 mm fiber | — | 45.07 | — | 17.24 | — |
| | 30 mm fiber | — | 42.02 | — | 16.14 | — |
| Corn husk/kenaf | 0 : 30 (PLA 70 wt.%) | — | — | 2.117 | — | — |
| | 15 : 15 (PLA 70 wt.%) | — | — | 1.547 | — | — |
| | 30 : 0 (PLA 70 wt.%) | — | — | 1.221 | — | — |
| Cotton/jute | 23.7 : 76.3 (jute fabric type III) | | | | | |
| | Test angle, 0° | 9.9 ± 0.8 | 136.7 ± 4.0 | 7.1 ± 0.3 | 59.4 ± 1.7 | 9.3 ± 0.9 |
| | Test angle, 45° | 8.4 ± 0.7 | 84.6 ± 4.7 | 4.6 ± 0.1 | 21.1 ± 1.4 | 7.5 ± 1.0 |
| | Test angle, 90° | 7.2 ± 0.7 | 58.3 ± 5.4 | 4.1 ± 0.1 | 14.6 ± 0.5 | 5.5 ± 1.0 |

HDPE = high-density polyethylene.
Source: Nguyen et al. [4]. © 2017, Elsevier.

the cork-based composites, 10 wt.% of short coconut fibers and 2 wt.% of CA were proposed.

#### 1.4.1.5 Coir/Silk FRP Hybrid Composites

Silk is a continuous protein fiber characterized by its soft, light, and thin nature and produced by different insects. The silkworm and spun synthesize silk fiber, such as silk cocoon. Large quantities of silk proteins (sericin and fibroin) are produced by the silkworm at the last stage of larval development [31]. These silk proteins are key components of the silk cocoons. Silk fiber, with its huge specific strength and stiffness, have wonderful luster and excellent drape. It prides itself as the strongest material in nature. It however has poor resistance when exposed to sunlight [4].

Using unsaturated polyester matrix, Noorunnisa Khanam et al. [32] in an investigation into the coir/silk fiber hybrid composites used various fiber lengths of 10, 20, and 30 mm. Sodium hydroxide (NaOH) solution was used to treat the coir fibers in order to eliminate their lignin and hemicellulose, thereby causing better bonding of the fiber with the matrix. Composites of 20 mm fiber length showed higher tensile and flexural strengths than the 10 and 30 mm counterparts, as shown in Table 1.6. The tensile, flexural, and compressive strengths were improved significantly in the coir/silk hybrid composites, owing to the NaOH treatment that facilitated the bonding at the coir fiber–polyester matrix interface.

#### 1.4.1.6 Corn Husk/Kenaf FRP Hybrid Composites

Several agricultural wastes such as rice straw, corn husk, and rice husk form a huge quantity of raw natural fibers that are used in polymer composites as materials for reinforcement. Corn husks contain fibers that are rich in cellulose. They are the thin and leafy sheaths that surround corn cobs [33]. Kenaf fiber is important in paper as well as other industrial sectors, as a fiber source.

Kwon et al. [34] used PLA matrix and prepared kenaf fiber and corn husk flour hybrid biocomposites, using a constant fiber to matrix ratio of 30 : 70 by weight (Table 1.6). Different kenaf/corn husk flour ratios were examined. Before and after extrusion, the aspect ratio was measured for kenaf fibers and its influence on the mechanical behavior was investigated. The result showed that the aspect ratio post-extrusion had no influence on the values predicted from the Halpin–Tsai equation. Note that the Halpin–Tsai model for predicting elastic response of composites assumes that there is no fiber–matrix interaction and works on the basis of the orientation (geometry) as well as elastic behavior of the matrix and the fibers. They found out that the variation in the Young's modulus of fibers affected the transfer of stress from the matrix to the fiber and reported that a factor of control to optimize mechanical behavior in hybrid biocomposites could be the reinforcements' scale ratio in various aspect ratios.

#### 1.4.1.7 Cotton/Jute and Cotton/Kapok FRP Hybrid Composites

Cotton fibers have been considered as fibers with the greatest importance across the globe. They usually do not have branches and the seed hairs have a single cell (unicellular). They are also rich in cellulose and can elongate up to 30 mm. The wall of cotton fiber does not contain lignin, as distinct from the secondary cell walls of most plants [35]. Cotton fibers are used widely in the textile industry. They possess some advantages, which include excellent drape, high absorbency, as well as good strength.

In the study conducted by De Medeiros et al. [36], the mechanical behavior of woven fabrics of hybrid cotton/jute with phenolic matrix (novolac type) reinforcement was investigated. The results showed strong dependence on the mechanical behaviors on fiber content, fabric characteristics, fiber–matrix adhesion, and fiber orientation. The anisotropy of the composites increased when the test angle was increased and showed dependence on fiber roving/fabric characteristics. There was an inverse proportionality between the mechanical properties and the test

**Table 1.7** Mechanical behaviors of cotton/kapok, cotton/ramie, and jute/oil palm empty-fruit bunches (OPEFB) FRP hybrid composites.

| Hybrid biocomposites | Fiber ratio (by weight or volume) | Flexural modulus (GPa) | Flexural strength (MPa) | Tensile modulus (GPa) | Tensile strength (MPa) | Impact strength ($kJ/m^2$) |
|---|---|---|---|---|---|---|
| Cotton/kapok | 3 : 2 | | | | | |
| | Untreated ($V_f = 60\%$) | — | — | 0.884 | 55.70 | 110.53 |
| | Alkali treatment ($V_f = 43\%$) | — | — | 1.635 | 52.87 | 119.25 |
| | Non-accelerated weather condition ($V_f = 46.6\%$) | 0.709 | 52.40 | — | — | — |
| | Accelerated weather condition ($V_f = 46.6\%$) | 0.703 | 39.55 | — | — | — |
| Cotton/ramie (ramie fibers placed longitudinally to the mold length) | 10.8 : 41.1 (0° composite) | — | — | — | 90.9 ± 12.7 | — |
| | 11.9 : 45.5 (0° composite) | — | — | — | 117.3 ± 13.3 | — |
| | 11.9 : 45.1 (0° composite) | — | — | — | 118.0 ± 6.5 | — |
| Jute/OPEFB | 1 : 4 | | | | | |
| | OPEFB/Jute/OPEFB | — | — | 2.39 | 25.53 | — |
| | Jute/OPEFB/Jute | — | — | 2.59 | 27.41 | — |
| | Pure OPEFB | — | — | 2.23 | 22.61 | |
| | Pure jute | — | — | 3.89 | 45.55 | |

Source: Nguyen et al. [4]. © 2017, Elsevier.

angle as the best performance was obtained in the specimen tested at zero degree (that is, jute roving direction). Brittle failure, though controlled, was displayed in the composites that were tested at angles of 45° and 90° as jute fiber directions (Table 1.6), while at 0° to the longitudinal direction, the tested composites exhibited an uncontrollable catastrophic failure. Jute fiber was identified as a strong material for reinforcement and when combined with cotton could prevent catastrophic failure in the fabric composites. In addition, the results obtained from mechanical behaviors of cotton/kapok FRP hybrid composites are depicted in Table 1.7, showing the effects of accelerated and non-accelerated weather conditions of alkaline-treated and untreated specimens. The results of cotton/ramie FRP hybrid composites, with ramie fibers placed longitudinally to the mold length and various ratio, are also shown in Table 1.7.

#### 1.4.1.8 Jute/OPEFB FRP Hybrid Composites

Oil palm (*Elaeis guineensis*) is a perennial crop known for its high-value fruits from which oil is produced. It mostly grows in tropical regions, such as Southeast Asia and West/Southwest Africa. Oil is extracted by stripping the fruits (nuts) from the bunches, a process that leaves the empty-fruit bunches (EFBs) as waste material [37]. Fibers of oil palm are usually derived from the oil palm empty-fruit bunches (OPEFB) as well as mesocarp. In composite materials, the OPEFB fibers are mostly used, as they contain the highest hemicellulose content in comparison with pineapple, coir, banana, as well as soft and hardwood fibers [11].

A three-ply hybrid sample of jute/OPEFB fibers composites with epoxy resin reinforcement was prepared by Jawaid et al. [38], fixing the jute/OPEFB ratio (by weight) at 1 : 4. They investigated the void content, chemical resistance, as well as tensile behaviors of the hybrid composites. From the results obtained, the OPEFB/jute/OPEFB and jute/OPEFB/jute composites showed great resistance to chemicals: toluene ($C_7H_8$), benzene ($C_6H_6$), water ($H_2O$), 40% of nitric acid ($HNO_3$), carbon tetrachloride ($CCl_4$), hydrochloric acid (HCl), 5% of acetic acid ($CH_3COOH$), 20% sodium carbonate ($Na_2CO_3$), 10% of sodium hydroxide (NaOH), and 10% of ammonium hydroxide ($NH_4OH$). A lower void content was displayed in the jute/OPEFB/jute than pure OPEFB as well as OPEFB/jute/OPEFB composites, because the mats of the jute fiber adhered better to the epoxy resin with higher compatibility. At the outer ply, the jute fibers withstood the tensional stress due to their high strengths, and the core (OPEFB fiber) absorbed and distributed the stresses evenly within the composite sample systems. Also, it was evident from Table 1.7 that the jute/OPEFB hybrid exhibited higher tensile responses (both strength and modulus) as well as improved adhesion bond between the fiber and the matrix, when compared with pure OPEFB composite.

#### 1.4.1.9 Kenaf/PALF FRP Hybrid Composites

A tropical plant, pineapple (*Ananas comosus*) belongs to the family of bromeliad (*Bromeliaceae*). In South America, it is next in line to banana and mango in total production across the globe [39]. Pineapple leaf fibers (PALFs) are waste products when cultivating pineapples and are extracted from pineapple leaves. It has a significant mechanical behavior, because it is high in cellulose (70–82%) as well as in crystallinity (44–60%) [40]. Combining these properties with that of Kenaf fiber, excellent tensile and flexural strengths from FRP composite are obtained, which promises a good material for different applications [4].

Aji et al. [41] studied hybridized Kenaf/PALF specimens with HDPE reinforcement, using 1 : 1 fiber ratio. They investigated into how the size of fiber and its loadings affected the mechanical responses of the hybrid biocomposites (Table 1.8). The four reinforcement lengths considered at a fiber loading range of 10–70% were 0.25, 0.50, 0.75, and 2.00 mm. The smallest of these fiber lengths (0.25 mm) yielded the best result in terms of its flexural and tensile properties, while both 0.75 and 2 mm exhibited enhancement in impact strength. As observed further, an increase in the fiber length reduced some of the mechanical behaviors, which is credited to the entanglement in fibers as against fiber attrition. An inverse proportionality was

**Table 1.8** Mechanical behaviors of kenaf/PALF, roselle/sisal, and silk/sisal FRP hybrid composites.

| Hybrid biocomposites | Fibre ratio (by weight or volume) | Flexural modulus (GPa) | Flexural strength (MPa) | Tensile modulus (GPa) | Tensile strength (MPa) | Impact strength (kJ/m$^2$) |
|---|---|---|---|---|---|---|
| Kenaf/PALF | 1 : 1 (At 0.25 mm fiber length and 60% fiber loading) | 4.114 | 34.01 | 0.874 | 32.24 | 6.167 |
| Sisal/roselle | 1 : 1 | | | | | |
| | Dry condition, fiber length = 15 cm | — | 76.5 | — | 58.7 | 1.30 |
| | Wet condition, fiber length = 15 cm | — | 62.9 | — | 44.9 | 1.28 |
| Sisal/silk | 1 : 1, fiber length = 20 mm | | | | | |
| | Untreated | — | 46.18 | — | 18.95 | — |
| | Alkali treatment | — | 54.74 | — | 23.61 | — |

Source: Nguyen et al. [4]. © 2017, Elsevier.

established between the tensile and impact properties, as the rule of mixture was satisfied by flexural strength. The adhesion between the fiber and the matrix interface was good, as evaluated by scanning electron microscopy (SEM).

#### 1.4.1.10 Sisal/Roselle and Sisal/Silk FRP Hybrid Composites

Sisal (*Agave sisalana)*, from the *Agavaceae* family, is a hard-fiber plant with wide cultivation in the tropical countries of Africa, America, and Asia, though it has its origin in Mexico and Central America. Their fibers are strong and tough, and extracted from sisal plant leaves. Sisal fibers are widely utilized in composites and plastic/paper industries. The nativity of roselle (*Hibiscus sabdariffa*) can be traced to West Africa. It is a species of Hibiscus, whose plant is naturally abundant and majorly used for fruits and bast fibers. Roselle fibers have extensive applications in the textile industry and in composites, because they exhibit greater mechanical behaviors in comparison with some other naturally occurring fibers, such as jute and kenaf.

Moreover, Athijayamani et al. [42] used a fiber ratio of 1 : 1 for the sisal and roselle fibers and investigated, under wet conditions, how absorption of moisture affected the mechanical responses of short hybrid sisal/roselle FRP unsaturated polyester composites. Using different fiber contents and lengths, their results showed an improvement (increase) in the flexural and tensile strengths of the hybrid composites of sisal/roselle fibers at increased lengths and contents of fibers and under dry condition. For the wet condition, the strengths (tensile and flexural) were significantly reduced, while inverse proportionality between the impact strength and fiber length and content was observed for both conditions (wet and dry), as depicted in Table 1.8. In addition, Noorunnisa Khanam et al. [43] also carried out a study on sisal/silk fiber ratio of 1 : 1 and prepared a polyester-based hybrid composite to evaluate various fiber lengths. Their results showed higher mechanical (tensile, flexural, and compressive) responses from the composite sample with fiber length of 20 mm than that of 10 and 30 mm counterpart hybrid composites (Table 1.8). Also, the results obtained after fiber modification depicted that the same mechanical behaviors of the alkali-treated hybrid fiber composites improved significantly.

## 1.5 Other Related Properties that Are Dependent on Mechanical Properties

There are other properties of biocomposites that are relevant for material analysis. Some of these are dynamic mechanical properties, thermal and water absorption behaviors, as well as tribological properties. Only tribological and thermal behaviors are subsequently discussed.

### 1.5.1 Tribological Behavior

Tribological behavior refers to friction-related properties of the materials. The frictional coefficient of hybrid sisal/glass fiber (GF)-reinforced epoxy composites was measured by Ashok Kumar et al. [44], using different sliding speeds of 0.2, 2.0, and 4.0 mm/s, under a constant load of 10 N. At an atmospheric temperature of 22 °C and relative humidity of 45%, both alkali-treated samples of fiber composites and untreated ones were tested. The graph of frictional coefficient against fiber length (Figure 1.5) revealed that the frictional coefficient was lower, up to 2 cm fiber length. However, the frictional coefficient increased with an increasing composite fiber length. Moreover, fiber length addition led to a decrease in the frictional

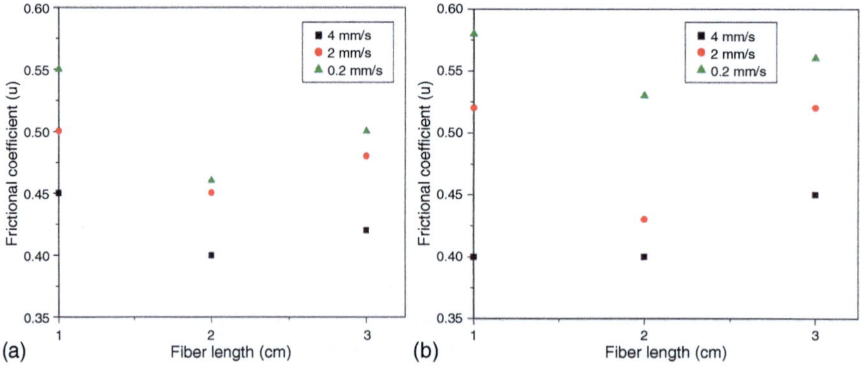


**Figure 1.5** Frictional coefficients of (a) treated and (b) untreated sisal/glass FRP hybrid composites as a function of fiber length, after 50 cycles. Source: Ashok Kumar et al. [44]. © 2010, SAGE Publications.

coefficient when sliding speeds were higher. The treated fiber of the reinforced composites yielded an optimum improvement at 2 cm in comparison with the untreated samples.

Biswas and Xess [45] studied the behavior of short bamboo/E-GF-reinforced epoxy hybrid composites with respect to erosion wear, using different compositions by weight as thus: 65 wt.% of epoxy, 22.5 wt.% of bamboo fiber, 22.5 wt.% of GF; 70 wt.% of epoxy, 15 wt.% of bamboo fiber, 15 wt.% of GF; and 75 wt.% of epoxy, 7.5 wt.% of bamboo fiber, 7.5 wt.% of GF, as well as 100 wt.% epoxy. The graph of the result of erosion rate against impact velocity showed that the 15 wt.% bamboo/GF FRP composites possessed the lowest rate of erosion in comparison with the other composites.

### 1.5.2 Thermal Behavior

The thermal property is concerned with the response of hybrid FRP biocomposites to heat variation. Boopalan et al. [46] worked on hybrid raw jute/banana fiber-reinforced epoxy composites with regard to their thermal analysis by varying the fiber weight ratio. They used ratios of 100/0, 75/25, 50/50, 25/75, and 0/100, while varying the temperature with the use of thermogravimetric analysis (TGA) and heat deflection temperature (HDT) analysis. With the TGA, the curve depicted that the 50/50 jute/banana FRP epoxy hybrid composite demonstrated greater thermal stability. There was a shift in the temperature during degradation from a value of 200 °C to a higher value of 380 °C. For the HDT, thermal property was sustained in the 50% jute with 50% banana FRP epoxy hybrid composite at the highest temperature of 90 °C in comparison with other composite samples.

Also, thermal properties of OPEFB/woven jute FRP epoxy hybrid composites were investigated by Jawaid et al. [47], using various temperatures. It was reported in their work that there was an increase in thermal stability when woven jute fibers were added to the EFB composite in its pure state. That is, hybridizing OPEFB with the woven jute fiber caused the thermal stability to be higher as compared with OPEFB fiber. The temperature of degradation was also reported to have shifted from a value of 292 °C to a higher value of 457 °C, leaving 12.1% char residue.

## 1.6 Progress and Future Outlooks of Mechanical Behaviors of Natural FRP Hybrid Composites

Application of various hybrid natural FRP composites is increasing with their innovative designs and developments through optimized manufacturing techniques. The advent of automation and robots in manufacturing of hybrid natural FRP composite materials has been improving their properties. This will increase in the next century, with synergy of sophisticated processes and techniques. Mechanical properties are most important responses that require serious attention during the design and fabrication of new hybrid natural FRP composites. Other properties of hybrid natural

FRP composites, such as thermal, acoustic, electrical, water absorption, among others, are directly dependent on the mechanical behaviors.

The prospect for FRP composite materials is very high and bright. It is presumed that in the next decade, application of various hybrid natural FRP composite materials would have penetrated all facets of life. This will be made possible through enhanced mechanical behaviors of a new set of composites as fiber selection, extraction, treatment, interfacial adhesion with matrix, and processing of natural FRP composite are improved [6].

In addition, there have been significant developments in natural FRP hybrid composites in the past few decades, due to established advantages in terms of processing, low cost, biodegradability, renewability, high specific strength, sustainability, as well as low relative density. Various natural fiber types have been and are being studied, with the findings forming the basis for replacing synthetic fibers, including both carbon and glass. Primarily, the idea of biocomposites development centers around the generation of novel FRP composites that are environmentally friendly with respect to how they are produced, used, and discarded. Hence, natural FRP composites could be a valid replacement and even superior alternative to synthetic fiber composites. Their biodegradable nature offers a good solution to the problem of waste disposal often experienced with synthetic fiber, petroleum, or non-renewable polymer-based materials. The application of biocomposites is widening continually and is projected to expand more, with more effects in Europe, due to mounting legislative and public pressures.

Till now, adhesion between the natural reinforcements/fiber and matrix interface, as hybridization increases, remains a major object of concern in terms of overall performance of natural FRP hybrid composites. This is a major factor in the ultimate properties, especially mechanical responses of the biocomposites. Further cutting-edge research is therefore necessary in order to overcome this challenge. Also, there is a requirement for more research work in order to get over other challenges such as inadequate toughness, moisture absorption, and stability reduction in long-term outdoor applications. Particularly, various weather conditions, humidity, temperature, and ultraviolet radiation, have significant influence on the product service life of natural FRP hybrid composites. For example, ultraviolet exposure results in discoloring, property deterioration, and deformation.

c Lastly, identifying better extraction of raw materials, sustaining crop growth, product design, and manufacture are activities that will help to reach the goal of better natural FRP hybrid composites. Hence, further research is ongoing across the globe to overcome the aforementioned challenges of biocomposites. Their respective properties should form the basis for generating new applications and create more opportunities for these biocomposites in the present-day green environment and secured future.

## 1.7 Conclusions

Owing to the sustainability, environmental friendliness, low cost of production, biodegradability, as well as enhanced mechanical behaviors of natural FRP composites through hybridization techniques, natural FRP composites are competing with and replacing some synthetic or conventional FRP composites, including both glass and carbon FRP composites. From the studies reported in this chapter, it is evident that mechanical behaviors of some natural FRP hybrid composites, especially impact, tensile, and flexural strengths and moduli, were higher than those of their non-hybrid or single FRP counterparts. Also, the degree of mechanical responses of the natural FRP hybrid composites depends on various areas of their engineering applications, as structural, semi, and/or non-structural composite systems.

## References


1 Elsner, P., Henning, F., and Weidenmann, K.A. (2009). Composite materials. In: *Technology Guide* (ed. H.-J. Bullinger), 24–29. Berlin, Heidelberg: Springer-Verlag.

2 Bispoa, S.J.L., Freire, R.C.S., and De Aquinoa, E.M.F. (2015). Mechanical properties analysis of polypropylene biocomposites reinforced with curaua fibre. *Mater. Res.* 18 (4): 833–837.

3 Zuccarello, B. and Marannano, G. (2018). Random short sisal fibre biocomposites: optimal manufacturing process and reliable theoretical models. *Mater. Des.* 149: 87–100.

4 Nguyen, H., Zatar, W., and Mutsuyoshi, H. (2017). *Mechanical Properties of Hybrid Polymer Composite*. Elsevier Ltd.

5 Pickering, K.L., Efendy, M.G.A., and Le, T.M. (2016). A review of recent developments in natural fibre composites and their mechanical performance. *Composites Part A* 83: 98–112.

6 Faruk, O., Bledzki, A.K., Fink, H.P., and Sain, M. (2012). Biocomposites reinforced with natural fibres: 2000–2010. *Prog. Polym. Sci.* 37: 1552–1596.

7 Bergstrom, J.S. (2015). *Mechanics of Solid Polymers: Theory and Computational Modeling*. William Andrew.

8 Sathishkumar, T.P., Naveen, J., and Satheeshkumar, S. (2014). Hybrid fibre reinforced polymer composites – a review. *J. Reinf. Plast. Compos.* 33 (5): 454–471.

9 Kuma, T.S.M., Senthilkumar, K., Chandrasekar, M. et al. (2019). Investigation into mechanical, absorption and swelling behaviour of hemp/sisal fibre reinforced bioepoxy hybrid composites: effects of stacking sequences. *Int. J. Biol. Macromol.* 140: 637–646.

10 Fragassa, C. (2016). Effect of natural fibres and bio-resins on mechanical properties in hybrid and non-hybrid composites. *AIP Conference Proceedings*, Volume 1736.

11 Saw, S.K., Sarkhel, G., and Choudhury, A. (2012). Effect of layering pattern on the physical, mechanical and thermal properties of Jute/Bagasse hybrid fibre-reinforced epoxy novolac composites. *Soc. Polym. Eng.* 33 (10): 1824–1831.

12 Wambua, P., Ivens, J., and Verpoest, I. (2003). Natural fibres: can they replace glass in fibre reinforced plastics? *Compos. Sci. Technol.* 63 (9): 1259–1264.

13 Santulli, C., Janssen, M., and Jeronimidis, G. (2005). Partial replacement of E-glass fibres with flax fibres in composites and effect on falling weight impact performance. *J. Mater. Sci.* 40 (13): 3581–3585.

14 Almeida Júnior, J.H.S., Ornaghi Júnior, H.L., Amico, S.C., and Amado, F.D.R. (2012). Study of hybrid intralaminate curaua/glass composites. *Mater. Des.* 42: 111–117.

15 Santulli, C. (2007). Impact properties of glass/plant fibre hybrid laminates. *J. Mater. Sci.* 42 (11): 3699–3707.

16 Fukuda, H. (1984). An advanced theory of the strength of hybrid composites. *J. Mater. Sci.* 19 (3): 974–982.

17 Wang, X., Hu, B., Feng, Y. et al. (2008). Low velocity impact properties of 3D woven basalt/aramid hybrid composites. *Compos. Sci. Technol.* 68 (2): 444–450.

18 Fiore, V., Scalici, T., Calabrese, L. et al. (2016). Effect of external basalt layers on durability behaviour of flax reinforced composites. *Composites Part B* 84: 258–265.

19 Pandey, A., Soccol, C.R., Nigam, P., and Soccol, V.T. (2000). Biotechnological potential of agro-industrial residues. I: Sugarcane bagasse. *Bioresour. Technol.* 74 (1): 69–80.

20 Saw, S.K. and Datta, C. (2009). Thermomechanical properties of jute/bagasse hybrid fibre reinforced epoxy thermoset composites. *BioResources* 4 (4): 1455–1476.

21 Han, G., Lei, Y., Wu, Q. et al. (2008). Bamboo-fibre filled high density polyethylene composites: effect of coupling treatment and nanoclay. *J. Polym. Environ.* 16 (2): 123–130.

22 Takagi, H. and Ichihara, Y. (2004). Effect of fibre length on mechanical properties of 'green' composites using a starch-based resin and short bamboo fibres. *JSME Int. J. Series A Solid Mech. Mater. Eng.* 47 (4): 551–555.

23 Okubo, K., Fujii, T., and Thostenson, E.T. (2009). Multi-scale hybrid biocomposite: processing and mechanical characterisation of bamboo fibre reinforced PLA with microfibrillated cellulose. *Composites Part A* 40 (4): 469–475.

24 Siró, I. and Plackett, D. (2010). Microfibrillated cellulose and new nanocomposite materials: a review. *Cellulose* 17 (3): 459–494.

25 Liu, H., Wu, Q., and Zhang, Q. (2009). Preparation and properties of banana fibre-reinforced composites based on high density polyethylene (HDPE)/Nylon-6 blends. *Bioresour. Technol.* 100 (23): 6088–6097.

26 Jacob, M., Thomas, S., and Varughese, K.T. (2004). Mechanical properties of sisal/oil palm hybrid fibre reinforced natural rubber composites. *Compos. Sci. Technol.* 64 (7–8): 955–965.

27 Akil, H.M., Omar, M.F., Mazuki, A.A.M. et al. (2011). Kenaf fibre reinforced composites: a review. *Mater. Des.* 32 (8–9): 4107–4121.

28 Thiruchitrambalam, M., Alavudeen, A., Athijayamani, A. et al. (2009). Improving mechanical properties of banana/kenaf polyester hybrid composites using sodium laulryl sulfate treatment. *Mater. Phys. Mech.* 8 (2): 165–173.

**29** Venkateshwaran, N., ElayaPerumal, A., Alavudeen, A., and Thiruchitrambalam, M. (2011). Mechanical and water absorption behaviour of banana/sisal reinforced hybrid composites. *Mater. Des.* 32 (7): 4017–4021.

**30** Fernandes, E.M., Correlo, V.M., Mano, J.F., and Reis, R.L. (2013). Novel cork-polymer composites reinforced with short natural coconut fibres: effect of fibre loading and coupling agent addition. *Compos. Sci. Technol.* 78: 56–62.

**31** Mondal, M., Trivedy, K., and Nirmal Kumar, S. (2007). The silk proteins, sericin and fibroin in silkworm, Bombyx mori Linn – a review. *Casp. J. Environ. Sci.* 5 (2): 63–76.

**32** Noorunnisa Khanam, P., Ramachandra Reddy, G., Raghu, K., and Venkata Naidu, S. (2010). Tensile, flexural, and compressive properties of coir/silk fibre-reinforced hybrid composites. *J. Reinf. Plast. Compos.* 29 (14): 2124–2127.

**33** Mahalaxmi, Y., Sathish, T., Subba Rao, C., and Prakasham, R.S. (2010). Corn husk as a novel substrate for the production of rifamycin B by isolated *Amycolatopsis* sp. RSP 3 under SSF. *Process Biochem.* 45 (1): 47–53.

**34** Kwon, H.J. et al. (2014). Tensile properties of kenaf fibre and corn husk flour reinforced poly(lactic acid) hybrid bio-composites: role of aspect ratio of natural fibres. *Composites Part B* 56: 232–237.

**35** Kim, H.J. and Triplett, B.A. (2001). Cotton fibre growth in planta and in vitro. Models for plant cell elongation and cell wall biogenesis. *Plant Physiol.* 127 (4): 1361–1366.

**36** De Medeiros, E.S., Agnelli, J.A.M., Joseph, K. et al. (2005). Mechanical properties of phenolic composites reinforced with jute/cotton hybrid fabrics. *Polym. Compos.* 26 (1): 1–11.

**37** Law, K.N., Daud, W.R.W., and Ghazali, A. (2007). Morphological and chemical nature of fibre strands of oil palm empty-fruit-bunch (OPEFB). *BioResources* 2 (3): 351–362.

**38** Jawaid, M., Khalil, H.P.S.A., Bakar, A.A., and Khanam, P.N. (2011). Chemical resistance, void content and tensile properties of oil palm/jute fibre reinforced polymer hybrid composites. *Mater. Des.* 32 (2): 1014–1019.

**39** Davis, S.C., Ming, R., Lebauer, D.S., and Long, S.P. (2015). Toward systems-level analysis of agricultural production from crassulacean acid metabolism (CAM): scaling from cell to commercial production. *New Phytol.* 208 (1): 66–72.

**40** Reddy, N. and Yang, Y. (2005). Biofibres from agricultural byproducts for industrial applications. *Trends Biotechnol.* 23 (1): 22–27.

**41** Aji, I.S., Zainudin, E.S., Khalina, A. et al. (2011). Studying the effect of fibre size and fibre loading on the mechanical properties of hybridised kenaf/PALF-reinforced HDPE composite. *J. Reinf. Plast. Compos.* 30 (6): 546–553.

**42** Athijayamani, A., Thiruchitrambalam, M., Natarajan, U., and Pazhanivel, B. (2009). Effect of moisture absorption on the mechanical properties of randomly oriented natural fibres/polyester hybrid composite. *Mater. Sci. Eng. A* 517 (1–2): 344–353.

**43** Noorunnisa Khanam, P., Mohan Reddy, M., Raghu, K. et al. (2007). Tensile, flexural and compressive properties of sisal/silk hybrid composites. *J. Reinf. Plast. Compos.* 26 (10): 1065–1070.

**44** Ashok Kumar, M., Ramachandra Reddy, G., Siva Bharathi, Y. et al. (2010). Naidu, Frictional coefficient, hardness, impact strength, and chemical resistance of reinforced sisal-glass fibre epoxy hybrid composites. *J. Compos. Mater.* 44 (26): 3195–3202.

**45** Biswas, S. and Xess, P.A. (2012). Erosion wear behaviour of bamboo/glass fibre reinforced epoxy based hybrid composites. *International Conference on Mechanical & Industrial Engineering*, Darjeeling, West Bengal, India.

**46** Boopalan, M., Niranjanaa, M., and Umapathy, M.J. (2013). Study on the mechanical properties and thermal properties of jute and banana fibre reinforced epoxy hybrid composites. *Composites Part B* 51: 54–57.

**47** Jawaid, M., Abdul Khalil, H.P.S., and Alattas, O.S. (2012). Woven hybrid biocomposites: dynamic mechanical and thermal properties. *Composites Part A* 43 (2): 288–293.

# 2

# Mechanical Behavior of Additive Manufactured Porous Biocomposites

*Ramu Murugan and Mohanraj Thangamuthu*

*Department of Mechanical Engineering, Amrita School of Engineering, Amrita Vishwa Vidyapeetham, Coimbatore, India*

## 2.1 Introduction

Tissue engineering is a multidisciplinary process that adopts the features of engineering and life sciences to develop biological substitutes that replace/enhance the functions of a whole organ or tissue [1]. The main application of biomaterials is in the functional restoration of human tissues for better quality of human health and life. The need for bio-implants is also drastically increasing in recent years due to the high incidence of osteoporotic fracture and bone fracture due to road accidents [2, 3]. Numerous researchers have suggested a porous scaffold made of biocompatible materials as a better functional bone substitute. Porous scaffolds are temporary structures that are capable of bearing the load that acts on damaged human bones and supporting the three-dimensional tissue cell. Fabrication of biomaterial into three-dimensional scaffold structures is a challenging process that depends on the nature or type of bone injuries of individual patients, which governs the success of any surgical procedure. The perfect design of the scaffold, the choice of biomaterial, its biocompatibility, precise fabrication to suit the injured bone, and mechanical strength are still under development throughout the world. Under this situation, fabrication of porous scaffolds that mimic the natural bone by using Additive Manufacturing (AM) could be a promising and viable alternative for treating bone-related disorders.

## 2.2 Human Bone

Human bones are rigid and are hard enough to support the frame of the body and have the potential to repair and regenerate. Bone also performs many vital functions in the body such as safeguarding the vital organs and creating an appropriate condition for the marrow for both fat storage and blood forming. It serves as a mineral source for calcium homeostasis, which is a source of growth factors and plays

*Mechanical and Dynamic Properties of Biocomposites,* First Edition. Edited by Senthilkumar Krishnasamy, Rajini Nagarajan, Senthil Muthu Kumar Thiagamani, and Suchart Siengchin.

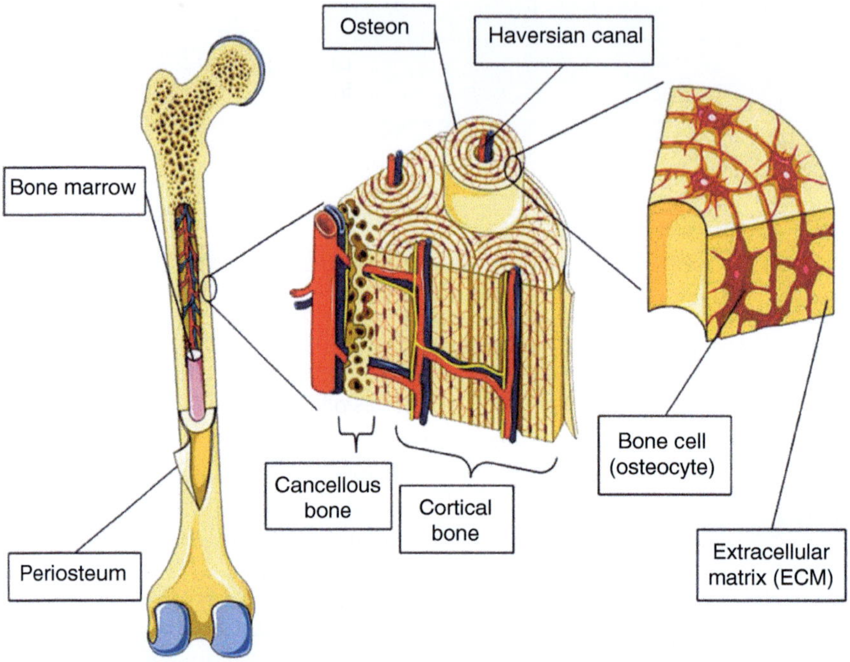


**Figure 2.1** Schematic view of the bone. Source: Bao et al. [5]. Licensed under CC BY 3.0.

a part in acid–base balance [4]. Bone is a gift to the human body and has the ability to continually remodel itself throughout life so as to respond to the ever-changing biomechanical forces. It reconstructs itself in the case of replacing an old bone or any damaged bone or providing a bone that is better in terms of strength and performance. Basically bone has two constituents, the outer cortical bone and the inner cancellous bone (Figure 2.1). Cortical bone is thick and impenetrable and covers the marrow space, whereas the cancellous bone is porous and has the structure of a honeycomb [6].

Bones have two basic forms: the mechanically weak woven form and the stronger lamellar form. New bone formation in the case of fracture healing will be of the woven type and it is remodeled as lamellar. Literally, all healthy matured bones as in adults are of the lamellar type [7–9].

During the bone healing period, the skeletal tissue itself will have the power to self-regenerate, but there is larger scope for getting affected by pathologies, cancers, or congenital defects, which may finally result in the loss of bone mass and strength. Of late, tissue engineering has produced notable results in providing biological alternatives and being the innovative way of advanced surgical procedure for bone fracture and defects [10].

Cancellous bone consists of an interconnected network of pores that are 50–300 μm in size, and the porosity ranges between 75% and 90%, whereas the porosity of cortical bone is less than 10% [11, 12]. Macropores are necessary to facilitate osteogenesis, and interconnected pores are required for body fluid

**Table 2.1** Mechanical properties of human bone [18].

| Property | Cortical bone | Cancellous bone |
|---|---|---|
| Compressive strength (MPa) | 100–230 | 2–12 |
| Tensile strength (MPa) | 50–100 | 10–20 |
| Strain to failure (%) | 1–3 | 5–7 |
| Fracture toughness (MPa $m^{1/2}$) | 2–12 | — |
| Young's modulus (GPa) | 7–30 | 0.05–0.5 |

Source: Henkel et al. [18]. © 2013, Springer Nature.

circulation [13–17]. Typical mechanical properties of the human bone are listed in Table 2.1.

Bone structure is directly proportional to the load-bearing ability. Bone has the tendency to adjust its geometry naturally according to the mechanical conditions prevailing around it. Not only the size and shape of bone but also the cortical bone porosity and characteristics of trabeculae, which is also called the bone micro-architecture, has a significant role to play in defining the mechanical competence of bone [9, 19, 20]. The second most frequent transplanted tissue after blood is bone [21], and roughly 5.3 million orthopedic surgeries are executed yearly worldwide.

## 2.3 Porous Scaffold

The main goal of scaffolds for bone tissue engineering is restoring the functional and structural behavior of degenerated or damaged tissue. A number of alternatives in materials for artificial porous scaffolds are being tried over the years including polymers, ceramics, metals, and composites [22–26]. A porous scaffold should have fully interconnected pores with sufficient porosity so that it aids in cellular infiltration and diffusion of nutrients. The necessary features for a scaffold for biomedical applications are depicted in Figure 2.2.

In biomedical applications, the porosity and pore size of three-dimensional porous scaffolds have direct impacts on their functional behavior. For cell nutrition, proliferation, transfer of nutrients for tissue vascularization, and new tissue formations, it is necessary that the implants must have open pores with interconnectivity. To enhance the mechanical load-bearing strength of the implant, the porous surface works to facilitate mechanical interconnection between the porous scaffold and the nearby tissue. Apart from that, the interlinked structure of scaffold pores helps in developing and paving the way for new tissue growth [28]. High-porosity materials ensure appropriate eliciting of biofactors such as cells and proteins and gives better support for nutrient exchange.

Thus a three-dimensional scaffold environment might enhance cell proliferation or differentiation. Generally, a scaffold with pore size of 100–1500 μm is adopted

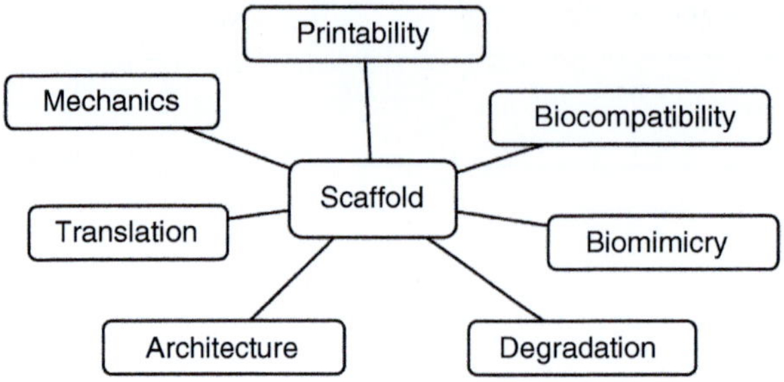


**Figure 2.2** Key requirements of a scaffold for osteochondral tissue engineering. Source: Vyas et al. [27]. © 2017, Elsevier.

in tissue engineering applications. A minimum of 100–150 μm of average pore size is required to facilitate osteogenesis, and the porosity value varies in the range of 10–70% based on the type of bone.

## 2.4 Biomaterials for Scaffolds

The scaffold gives a short-lived biomechanical structure for the growth and multiplication of cells or tissues in order to support damaged body parts. Many factors of the functional design of the scaffold determine the achievement of healthy tissue regeneration. Biomaterials are used to substitute or augment hard or soft tissues, which are injured or spoiled due to infections. Normally the tissues and structures of the body would function for an unlimited period of time in most people, but there may be some sudden, unexpected damages or defects due to some unhealthy conditions such as fracture, deformation, and infection. These may cause functional loss, emotional distress, botheration, and irritation. Based on these conditions, it is advisable to replace the unhealthy tissue with synthetic materials with similar characteristics.

### 2.4.1 Required Properties of Biomaterials

Different materials that are used for bone repair and as an alternative for orthopedic injuries are polymers, ceramics, metals, and composites. The important characteristics of biomaterials are listed below [29]:

i) *Biocompatibility*: No rejection of scaffold by the immune system of the body. The scaffold material must be biocompatible for tissue engineering applications and the cells must be able to attach to it for proper functioning. It should also avoid severe inflammatory reaction, which might reduce healing or cause rejection by the body's immune system.
ii) *Appropriate surface properties*: Permit cells to attach, hold together, multiply, and transit into the scaffold.
iii) *Controlled biodegradability*: The purpose of a bioresorable scaffold is to permit the body's own cells, over time, to replace the implanted scaffold. Scaffold and constructs are not considered to be everlasting implants. So they must be biodegradable and must permit the cells to create their own extracellular

matrix. The supplementary materials of this degradation must be nontoxic, and must leave the body without having much contact with the other organs.

iv) *Adequate mechanical properties*: The mechanical properties of the scaffold should have the proper anatomical site for implantation, and be stronger during implantation, surgical procedures without any clinical difficulties. Although its importance is recognized in all tissues, there are some challenges, particularly for cardiovascular and orthopedic applications. In order to replicate bone, the biggest challenge lies in producing scaffolds with the required load-bearing properties. Although many materials are produced with excellent load-bearing properties, a few have failed when they were implanted. This is because of inadequate ability for vascularization. This happens to the determent of retaining the highly porous structure in many materials. It is to be noted that for the success of any type of scaffold, two aspects must be balanced. They are the mechanical load-bearing properties and porous structure volume for strength, for permitting cell infiltration and vascularization respectively.

### 2.4.2 Types of Biomaterials

#### 2.4.2.1 Metals

Metals are the most commonly used material in load-bearing implants. Implantation of metallic implants is a common feature in orthopedic surgeries. Some of the usages of metals are as simple wires and screws that are used with plates for fractures and artificial joints of shoulders, ankles, hips, knees, etc. Apart from these they are also used for cardiovascular surgery and maxillofacial surgery. Metals and alloys are used for clinical equipment purposes. Stainless steels, and titanium-, magnesium-, and cobalt-based alloys are largely used for orthopedic applications.

#### 2.4.2.2 Polymers

Polymeric materials are also used as implants most commonly either partially or as whole. These are acrylics, polyurethane, polyesters, polysiloxanes, polyamides, polyethylene, and numerous reprocessed biological materials. Polyglycolic acid is used as a resorbable suture material. Also, many polymeric materials are used in pacemakers, artificial liver, kidney, external and internal ear repairs, cardiac assist devices, catheters, bone cement, cornea, eye lens replacements, and contact lens applications. Also, polymeric biomaterials are used in implantable pumps, joint replacements, artificial skin and blood vessels, soft tissue replacement, encapsulations, and sutures. Bioengineers are always in search of the most capable and befitting implants. Polymeric materials, either separately or combined with metals and ceramics, are used largely as implants into the body. When compared to other materials such as metals and ceramics, polymeric biomaterials possess the unique feature of flexibility to produce different shapes such as films, fibers, and latex sheets. There are also polymeric materials that are cost-effective, and can be fabricated with the required mechanical and physical properties. The desired properties of these are related to other biomaterials such as biocompatibility, interconnectivity of the pores, the required mechanical and physical properties, and manufacturability.

#### 2.4.2.3 Ceramics

Bioceramics such as amorphous glasses and crystalline ceramics are mechanically strong (compressive and corrosive behavior) with better bioactivity. The most frequently utilized bioceramics are Hydroxyapatite (HA), Tricalcium Phosphate (TCP), and biphasic calcium phosphate. A few important applications of ceramics are bone grafts, hip prostheses, artificial knees, and various tissue engineering applications. Normally, ceramic biomaterials possess excellent strength but are brittle when used alone. So the usage of ceramics is limited to a few cases such as load-bearing and implant devices that are subjected to tensile stresses.

#### 2.4.2.4 Composites

Composites are used as revitalizing material or dental cement in the field of dentistry. Most successfully, carbon–carbon and carbon-reinforced polymer composites have less flexible modulus and largely employed for bone repair and joint replacement. Composite biomaterials generally refer to only those materials where the various phases are separated on a larger scale than the atomic. The important aspect in a biocomposite is that it must be biocompatible and also the interface between the components should not be devalued by the body environment. The usages of composites in clinical applications are as dental filling composites, orthopedic implants with pores, and bone cement.

Biodegradable materials attract vital attention today. It is natural that if a substance degrades in the body, the natural values of the properties alter and this may result in undesirable or varied performance. The material can be chosen such that the implants degradation rate synchronizes with the rate of tissue formation. For example, suture material holds a wound together. However, it resorbs in the body as the injury is healed. Sometimes, these materials are applied to stimulate the growth of natural tissue. In some cases, the wound dressings and ceramic bone escalation materials enable the tissue to grow into them by giving a scaffold; whether or not the scaffold material resorbs over a period of time, natural tissue grows and restores natural function.

Another important use of biodegradable materials is drug therapy. In this, the drugs are chemically bound to the biodegradable materials and the drug gets released when the materials degrade. This process of releasing the drug is continued for a certain period of time. Biodegradation is a condition that can roughly mean that materials are made ineffective by nature by hydrolytic mechanisms with the help of enzymes. Biodegradation can also be referred as absorbable, restorable, and erodible under in vitro conditions. In the last 10 years, the usage of biodegradable polymeric biomaterials for biomedical engineering has increased drastically, the reason being the two vital merit factors that these classes of biomaterials possess in contrast to nonbiodegradable materials. The fact is that these biomaterials are absorbed in the course of time by the human body. So they do not ever leave any signs of remains in the implantation areas. Hence, there is no scope for any everlasting chronic, foreign body reactions. Thus, for tissue regeneration, operative prosthesis made out of biodegradable biomaterials is applied as a shortening scaffold. This move in the direction of the reconstruction of defective, spoiled, or aged tissues is one of the most encouraging aspects in the twenty-first century [30–34].

## 2.5 Additive Manufacturing of Porous Structures

Additive Manufacturing (AM) is a manufacturing technique where physical objects are built precisely from their three-dimensional CAD models in a layer by layer manner. In contrast to traditional machining methods, most of the AM techniques tend to fabricate parts based on additive (material addition) process rather than subtractive (material removal) method [35].

The most attractive characteristic of AM is that it can build complicated 3D geometries precisely by adding additional cost to the process. Manual intervention is also very limited in the process; thus automated and be part of Industry 4.0. In addition to this, there is no need for part planning, no tools/fixtures, and is done through a generic fabrication machine.

### 2.5.1 Generic Process of AM

AM process is initiated by the creation of a CAD model for the physical object to be built using any conventional CAD modeling package. The CAD model could also be created or reverse engineered from any existing physical model by using 3D laser scanners. For biomedical applications, the Computerized Tomography (CT) scan data can be converted into CAD format and used for prototyping medical models or for patient-specific scaffold applications. All AM systems understand CAD model as Standard Tessellation Language (STL) file format, which is abbreviated from the first AM process developed, which is a stereo-lithography technique. The STL can also be referred as Standard Tesselation Language by the AM community. STL conversion can be done using any software package and during this conversion process all surfaces of the 3D CAD model will be converted into triangular facets. So a typical STL file consists of data pertaining to the vertices of all triangles and the outward normal vector of each triangle to denote its orientation. All other CAD information such as geometrical details, material, rendering features, etc. will be neglected in the STL file and thus the file size will be light compared to a CAD file. The accuracy of the STL file can be increased by mentioning the appropriate resolution/tolerance setting during its conversion from CAD.

The next step in the AM process is to slice the STL file into multiple layers as AM is a layer by layer manufacturing process. AM machines may have their own proprietary slicing software or users can rely on any open-source slicers. Using this slicing software, the different process parameters of AM such as build orientation, layer height, geometrical scaling, part stacking in case of building multiple parts, and support requirements can be set.

This preprocessed data is sent to the AM machine and it builds the first layer based on the contour of first slicing and the subsequent layers are laid over the previous layers. Finally, the model is removed from the bed, cleaned, and post-processed for any finishing requirement.

### 2.5.2 Powder Bed Fusion Process

AM technologies are classified according to the type of raw material used for fabrication and/or by the nature of part consolidation. Powder Bed Fusion (PBF) system is one of the familiar AM systems that uses laser power to build layers of object on a powdered base. The powder can be of different materials ranging from thermoplastics, metals, ceramics, and composites.

PBF systems are further classified into the following categories based on the mechanism of powder consolidation:

- Selective Laser Sintering (SLS) powder consolidation by sintering mechanism; uses less laser power and is not suitable for metal powders.
- Selective Laser Melting (SLM) powder consolidation by melting; uses high power laser source and is suitable for metals.

In the PBF machine, the powders are stored in a powder feed cylinder and distributed evenly on the bed with the help of a roller (Figure 2.3). All powders in the powder feed cylinder, build chamber, and bed are maintained at elevated temperature to minimize the laser energy required for consolidation. Once the laser scans over the powder bed for printing the first layer, the build chamber will move down by one layer thickness and a new layer of the powder is supplied and spread by the powder handling system. The laser beam intensity at a spot is sufficiently controlled so that the temperature profile reaches the previous layer for better fusion. As the entire part is immersed in the powder bed, it eliminates the need for any support structures in the overhanging portions as required by other AM technologies. PBF could be useful for effectively reproducing very complex architectural objects with better resolution [36].

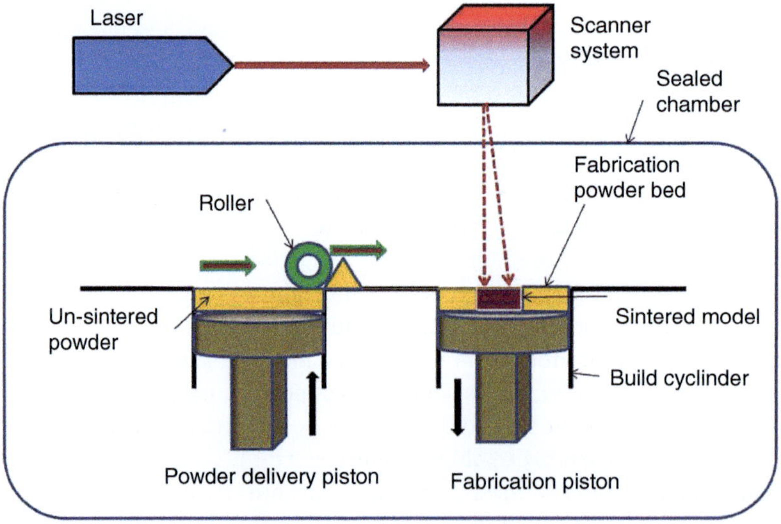


**Figure 2.3** Schematic diagram of Powder Bed Fusion process.

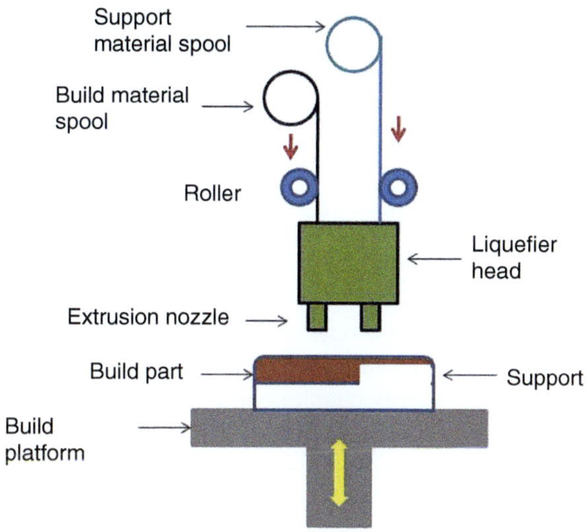


**Figure 2.4** Schematic of Fused Deposition Modeling.

### 2.5.3 Fused Deposition Modeling Process

Fused Deposition Modeling (FDM) is a widely used and the cheapest AM process. The raw material, mostly thermoplastic in the form of filament, is supplied to an extrusion nozzle (Figure 2.4). The extruder head is moved along the surface of the bed for each layer to the required profile based on the slicing code. During its motion over the table, it lays a slender bead of extended plastic to form each layer. The solidification is induced by part cooling fan and/or natural convection, thus sticking to the layer below. Polylactic acid is the commonly used biodegradable thermoplastic.

### 2.5.4 Additive Manufacturing of Porous Biocomposites

To meet the requirements of an individual patient, the architecture and shape of the construct must be evolved. It is also imperative that scaffolds are manufactured in a reproducible, controlled, and cost-effective manner with the flexibility to accommodate the presence of biological components, such as cells and growth factors, in certain applications. AM of composites using the PBF process is a challenging aspect, as it involves cross-disciplinary functionalities. The quality of the product produced in any AM technique depends on the process parameters, which are governed by the material being used.

Major process parameters that govern the PBF process, which are dependent on the material used, are laser power, scan spacing, and bed temperature. Thus in case of AM of composites, optimum process parameters could be achieved by proper experimental design. SLS technique is used to process polymer- and ceramic-based biocomposites with matrix materials such as Polylactic acid (PLA), Poly-L-Lactide (PLLA), Poly-ε-Caprolactone (PCL), polyglycolide, etc.

Metal-based scaffolds are processed using SLM machines with biocompatible materials such as commercially pure Titanium, $Ti_6Al_4V$ alloys, Stainless Steel SS316L, Cobalt Chromium steel, and Magnesium [37]. These metals and their alloys present an important drawback associated with the significant difference between their Young's modulus (~100 GPa for Ti and ~200 GPa for steel) and the corresponding value of the bone (~20 GPa for cortical bone and ~10 GPa for trabecular bone) [38].

The higher Young's modulus value of Titanium entails that implants and prostheses evade the transfer of total load to the bone, producing a stress-shielding phenomenon. Since bone is a dynamic tissue, with its configuration and density adjusted by the load applied to facilitate mismatch between the Young's modulus of Titanium and bone, it produces a defect in bone density, which is commonly called as bone resorption. Numerous failures of titanium implants are associated with bone resorption, usually noticeable in fractures of the surrounding bone and subsequent release of the load-bearing component. Hence, it is advantageous to devise new implants and prostheses with lower stiffness than the present one. This would permit resolving or reducing stress-shielding problems, without any significant unfavorable result on the mechanical strength [39].

Important research and technological efforts have been dedicated to solving stress-shielding problems using AM of customized porous structures. The porous metal scaffolds, with a custom-made porosity, allow the Young's modulus of the implants to be modulated through control of pore size, morphology, distribution, and amount of porosity [38, 40].

## 2.6 Design of Porous Scaffold

The mechanical and biological properties of porous scaffolds are mainly governed by the porosity, pore size, topology of pore, interconnectivity, and the material. Several research works have been described in the field of porous scaffold design; nevertheless, most of the research pays attention toward the material, manufacturing, and in vitro studies on porous scaffolds. The pore pattern used in the scaffold research is a familiar one. The porous 3D scaffold consists of interconnected pores to achieve the set porosity that will aid cell growth. The porosity of the porous scaffold is estimated through the ratio of the void quantity corresponding to the volume of the material. The pore dimension and porosity of the scaffold influence its load-bearing strength in a straightforward manner [41, 42].

### 2.6.1 Pore Size

Mean pore size is a major consideration while designing the porous scaffolds for tissue engineering applications. Porous scaffold should be permeable enough with interconnectivity to aid cell growth and nutrient flow. Too low a pore size limits cell migration and too large pores decrease the surface area, thus reducing the ligand density available for cells to bind [43]. Pores in the range of 20–1500 μm have been fabricated and studied for tissue engineering applications. Earlier studies from researchers demonstrated that bone formation is efficient in scaffolds with pore size of 800 μm, as bone cells prefer to be attached in large pores, which provides adequate space for cell ingrowth, whereas the minimum pore size should be between 75 and 100 μm for a significant amount of cell growth [43–45]. Torres-Sanchez et al. [46] investigated the biological behaviors of porous Ti scaffolds with pore size ranging from 45 to 500 μm.

### 2.6.2 Pore Geometry

A different pore model has been investigated by researchers over the years and a typical porous scaffold with cubical pores is shown in Figures 2.5 and 2.6.

Apart from the basic topology of simple cubic, spherical, and hexagonal patterns of unit cells, research has also been extended to the usage of auxetic and bioinspired patterns [48, 49]. Tissue engineering takes the advantage of AM in the fabrication of porous scaffold with such auxetic patterns with controlled pore geometry.

### 2.6.3 Bioceramics as Reinforcement Material

Bioceramics (HA, TCP) are the common reinforcement materials that are used for the fabrication of scaffold using PBF AM process. HA is one of the main constituents of the human bone and is commonly used as bioactive material in bone implants. HA, when used solely or as a composite as scaffold material, is biologically compatible and stimulates bone growth. But its use is limited due to its brittle nature and thus HA is mixed with other biocompatible material to form a suitable biocomposite. Figure 2.7 shows a typical porous structure made of 85% polyamide:15% HA composite using the SLS process.

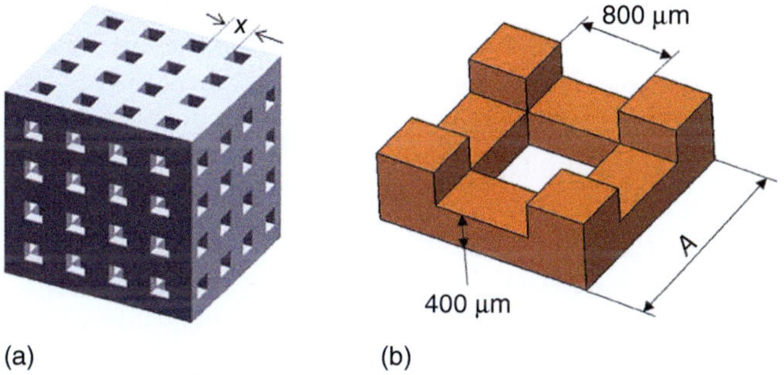


**Figure 2.5** Porous scaffold with cubical pore. (a) Porous scaffold. (b) Details of pore. Source: Kumaresan et al. [47]. © 2019, Inderscience Publishers.

**Figure 2.6** Different topology of pore structure. Source: Maskery et al. [48]. Licensed under CC BY 4.0.

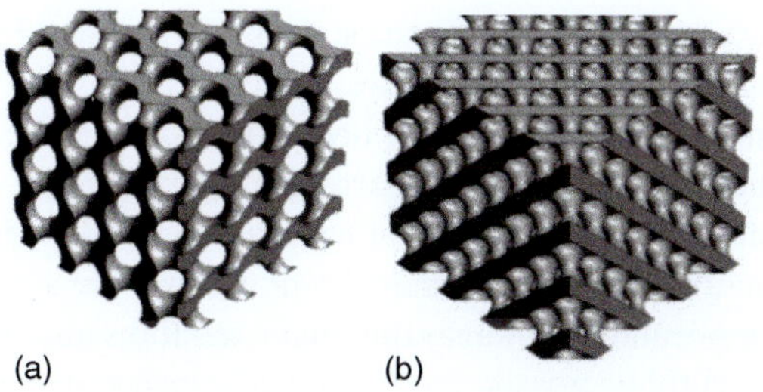

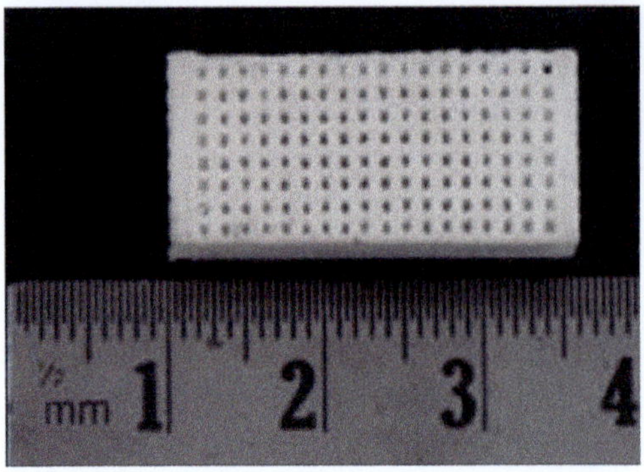


**Figure 2.7** Porous scaffold specimen fabricated using biocomposite (polyamide/hydroxyapatite) by the SLS process.

If powders of matrix material (polymer/metal) and HA are mixed and heated together, various reactions occur. The decomposition of HA has been reported to begin at 800 °C with the appearance of traces of tricalcium phosphate (α-TCP) and tetracalcium phosphate (TTCP). The presence of α-TCP and TCPM will degrade the mechanical property of the sample [50]. This limitation is not applicable with FDM process due to its lower operating temperature (190–240 °C).

## 2.7 Mechanical Characterization of Additive Manufactured Porous Biocomposites

The scaffolds serve as a synthetic alternative for extracellular matrices, which are important for the attachment of cells, cell growth guidance, and cell characteristics maintenance through tissue or organ restoration. Various studies reported that porosity and the kind of materials used for scaffolds extensively control the biological responses [51]. The novel biomaterials used are Polyvinyl alcohol (PVA), HA powders [52], PCL [45], TCP, Tetracalcium Phosphate (TTCP), dicalcium phosphate dihydrate (DCPD; Brushite), dicalcium phosphate anhydrous (DCPA; Monetite), octacalcium phosphate (OCP), calcium pyrophosphate (CPP) [53], poly(ethylene glycol) (EG), PLA, calcium phosphate (CaP), TCP, HA, interleukin (IL), tumor necrosis factor-α (TNF-α), poly(ethylene glycol) dimethacrylate (PEGDMA), and bioactive glass (BG) [54].

Yeo et al. [55] investigated the Polycaprolactone:β-TCP biocomposites by hot extrusion process similar to the FDM process and achieved pore size from 119 to 477 μm with a constant porosity of 60% (approximately). Yan et al. [56] used Gyroid- and Diamond-shaped porous structure with 5–95% porosity with pore size in the range of 560–1600 μm with SLM processing of Ti6Al4V. Ataee et al. [57] investigated the porous scaffold made with commercially pure Ti processed using SLM and have used Gyroid-shaped structure with a pore size of 1660 μm. Li et al. [58] fabricated an Fe-based diamond lattice scaffold by SLM. In a four week immersion test, they found that the mechanical properties of the scaffold were still sufficient for implant applications, whereas the elastic modulus and the yield strength of the samples were slightly decreased by 7% and 5%, respectively. Table 2.2 summarizes the various

**Table 2.2** Various porous biocomposites developed.

| S. No. | Material and its composition | Porosity (%) | Pore size (μm) | References |
|---|---|---|---|---|
| 1 | poly(ε-caprolactone) (PCL) + 0.3% graphene oxide (GO) | 96.9% | — | [59] |
| 2 | PVA | 91 ± 1.1 | 440 | [60] |
| 3 | PVA/n-HA | 90 ± 0.6 | 415 | |
| 4 | PVA/n-HA/CNC-2 | 90 ± 0.2 | 270 | |
| 5 | PVA/n-HA/CNC-4 | 89 ± 0.3 | 220 | |
| 6 | PVA/n-HA/CNC-6 | 87 ± 0.8 | 198 | |
| 7 | PVA/n-HA/CNC-8 | 89 ± 1.1 | 280 | |
| 8 | PVA/n-HA/CNC-10 | 87 ± 1.0 | 180 | |
| 9 | nHAp–chitosan–gelatin–alginate composite scaffold | 30–95% | 200–1000 | [25] |
| 10 | Glycine modified chitosan–gelatin–alginate scaffolds | 80.8 ± 1.20% | — | [61] |
| 11 | Hydroxyapatite | 91.8 ± 1.90% | — | [62] |
| 12 | Hydroxyapatite–high-density polyethylene (HA-HDPE) | 60% and 70% | 400 and 800 | [63] |
| 13 | PVA/n-HA | 91 ± 1.1 | 440 | [60] |
| 14 | PVA/n-HA/CNC (cellulose nanocrystals)-2 | 90 ± 0.6 | 415 | |
| 15 | PVA/n-HA/CNC-4 | 90 ± 0.2 | 270 ± 163 | |
| 16 | PVA/n-HA/CNC-6 | 89 ± 0.3 | 220 ± 146 | |
| 17 | PVA/n-HA/CNC-8 | 87 ± 0.8 | 198 ± 141 | |
| 18 | Ovalbumin (OVA)/polyvinyl alcohol (PVA) | 85 | 240 | [64] |
| 19 | PVA-CNCs/OVA | 87.4 | 170 | |
| 20 | PVA-CNCs-$NH_2$/OVA | 87.7 | 200 | |
| 21 | Hydroxyapatite | 38 and 44 | 366 and 444 | [65] |
| 22 | Tricalcium phosphate cement | 0.2 and 8.7 | 31 and 62 | [66] |
| 23 | Calcium metaphosphate | — | 200 | [67] |
| 24 | Natural coral | 36 | 150–200 | [68] |
| 25 | Hydroxyapatite/tricalcium phosphate | 36 | 100–150 | [69] |
| 26 | Glasses | 5 | 100–200 | [70] |
| 27 | Bioglass | — | 100–600 | [71] |
| 28 | Glass–ceramics | — | 100–600 | [72] |
| 29 | Poly(lactide-*co*-glycolide) | >30 | 72, 164, 101, and 210 | [73] |

(*continued*)

**Table 2.2** (Continued)

| S. No. | Material and its composition | Porosity (%) | Pore size (μm) | References |
|---|---|---|---|---|
| 30 | Polyethylene terephthalate | 93–97 | — | [74] |
| 31 | Melt-blowing | 87 | 150–200 | [75] |
| 32 | Hydroxyapatite/chitosan-gelatin | — | 300–500 | [76] |
| 33 | Hydroxyapatite/β-tricalcium phosphate/chitosan | — | 300–600 | [77] |
| 34 | Collagen/hydroxyapatite | 85 | 30–100 | [78] |
| 35 | Titanium/calcium phosphate | 35 | 50–200 | [79] |
| 36 | Titanium/polyvinyl alcohol | 60 | 170 | [80] |
| 37 | Silica/ceramic | 51, 47, and 43 | 10–300 | [72] |

**Table 2.3** Mechanical properties of additive manufactured porous structure.

| S. No. | Material and its composition | Process | Porosity (%) | Pore size (μm) | Compressive strength (MPa) | References |
|---|---|---|---|---|---|---|
| 1 | Sodium alginate + HA | FDM | 74% | | 9.5 | [81] |
| 2 | 100% PA | | | | 11.2 | |
| 3 | 95% PA:5%HA | | | | 14.3 | |
| 4 | 90% PA:10% HA | SLS | 80% | 800 | 21.5 | [47] |
| 5 | 85% PA:15% HA | | | | 25.2 | |
| 6 | 80% PA:20% HA | | | | 28.1 | |
| 7 | PCL:HA | SLS | 70–78% | — | 1.38–3.17 | [82] |
| 8 | PLA + gelatin-forsterite (10%) | FDM | 36.2 ± 0.6% | 310 ± 3 μm | — | [83] |
| 9 | 80% Ti/20% HA | SLS | — | 50–150 μm | 184.3 | [84] |
| 10 | Polyetheretherketone and hydroxyapatite (PEEK/HA) | SLS | 69.9–76.5% | 20–60 mm | — | [85] |
| 11 | PEEK/40% HA | SLS | — | — | — | [86] |
| 12 | 20 vol.% HA-HDPE | SLS | 40% | 600 mm | — | [87] |
| 13 | PCL | SLS | 23.2–39.3% | >100 μm | — | [88] |
| 14 | PVA | SLS | 55 ± 0.9 | 2000 | — | [45] |

porous biocomposites developed for scaffolds and Table 2.3 summarizes mechanical behavior of porous structures fabricated using the AM process.

## 2.8 Conclusion

This work compiles the methodology of manufacturing porous biocomposites using AM process, mainly focusing on SLS, SLM, and FDM. The fundamental requirements of biocomposites, porous scaffolds, design of porous scaffolds, pore size, and porosity have been reviewed. The main conclusions are as follows:

(1) Biocomposites must satisfy certain criteria based on mechanical behavior and biocompatibility. The biocomposite should be accepted by the human immune system and should be able to withstand the mechanical loads during any kind of physical activities. The composite should have appropriate elastic modulus to match the stiffness behavior of the human bone.
(2) Well-designed porous scaffold is responsible for effective cell attachment, growth, and new bone formation. Porosity also governs the mechanical properties of the structure and strength of the scaffolds being developed.
(3) A well-designed porous configuration is good for cell behavior such as attachment and proliferation, which helps new bone development and regeneration. Additionally, porosity can be utilized to regulate mechanical properties such as strength and Young's modulus. In the case of temporary metal implants, porosity plays a significant role in formatting the corrosion rate, owing to the change in surface area.
(4) One of the advanced AM processes, PBF technique, has enormous potential for metal implant manufacturing, and is used to manufacture high-quality metallic implants. Further variants of the powder bed fusion method such as SLS, SLM, and electron beam melting (EBM) have been employed for implant manufacturing.
(5) Additive manufactured porous biocomposites still remain a great challenge in the field of tissue engineering in terms of understanding the structural, physicomechanical, biological, and immunological properties using a multidisciplinary effort.

## References

**1** Langer, R. and Vacanti, J. (1993). Tissue engineering. *Science* 260: 920–926.
**2** Johnell, O. and Kanis, J.A. (2006). An esitamte of the worldwide prevalience and disability associated with osteoporotic fractures. *Osteopros. Int.* 17: 1726.
**3** Ministry of Health and Family Welfare (MHFW) (2009). *Integrated Disease Surveillance Project – Project Implementation Plan 2004–2009*, 1–18. New Delhi: Government of India.
**4** Russell, S. (2005). Taichman, blood and bone: two tissues whose fates are intertwined to create the hematopoietic stem-cell niche. *Blood* 105 (7): 2631–2639.

5 Bao, C.L.M., Teo, E.Y., Chong, M.S.K. et al. (2013). *Advances in Bone Tissue Engineering*. https://doi.org/10.5772/55916. IntechOpen Limited.

6 Usha, K. and Nandeesh, B.N. (2012). Physiology of bone formation, remodelling and metabolism. In: *Rdionuclide and Hybrid Bone Imaging* (eds. I. Fogelman, G. Gnanesegaran and H. van der Wall), 29–57. Berlin: Springer-Verlag.

7 Rho, J.Y., Kuhn-Spearing, L., and Zioupos, P. (1998). Mechanical properties and the hierarchical structure of bone. *Med. Eng. Phys.* 20 (2): 92–102.

8 Enoka, R. (2002). *The Neuromechanical Basis of Human Movement*, 3e. Champaign, IL: Human Kinetics.

9 Lian, J.B., Gordon, J.A., and Stein, G.S. (2013). Redefining the activity of a bone-specific transcription factor: novel insights for understanding bone formation. *J. Bone Min. Res.* 28 (10): 2060–2063.

10 Stoppato, M., Carletti, E., Sidarovich, V. et al. (2013). Influence of scaffold pore size on collagen I development: a new in vitro evaluation perspective. *J. Bioact. Compat. Polym.* 28 (1): 16–32.

11 Choi, K., Kuhn, J.L., Ciarelli, M.J., and Goldstein, S.A. (1990). The elastic moduli of human subchondral, trabecular, and cortical bone tissue and the size dependency of cortical bone modulus. *J. Biomech.* 23: 1103–1113.

12 Parente, M.A., Geil, M., and Monroe, B. (2006). In the future: prosthetic advances and challenges. In: *Prosthetics and Patient Management: A Comprehensive Clinical Approach* (eds. K. Carroll and J. Edelstein), 215–232. SLACK Incorporated.

13 Yang, S., Leong, K.-F., Zhaohui, D., and Chua, C.-K. (2001). The design of scaffolds for use in tissue engineering. Part I. Traditional factors. *Tissue Eng.* 7 (6): 679–689.

14 Masood, S.H., Singh, J.P., and Morsi, Y. (2005). The design and manufacturing of porous scaffold for tissue engineering using rapid prototyping. *Int. J. Adv. Manuf. Technol.* 27: 415–420.

15 Murphy, C.M., Haugh, M.G., and O'Brien, F.J. (2010). The effect of mean pore size on cell attachment, proliferation and migration in collagen–glycosaminoglycan scaffolds for bone tissue engineering. *Biomaterials* 31 (3): 461–466.

16 Reinwald, Y., Johala, R.K., Ghaemmaghami, A.M. et al. (2014). Interconnectivity and permeability of supercritical fluid-foamed scaffolds and the effect of their structural properties on cell distribution. *Polymer* 55 (1): 435–444.

17 Zhang, K., Fan, Y., Dunne, N., and Li, X. (2018). Effect of microporosity on scaffolds for bone tissue engineering. *Regener. Biomater.* 5: 115–124.

18 Henkel, J., Woodruff, M.A., Epari, D.R. et al. (2013). Bone regeneration based on tissue engineering conceptions – a 21st century perspective. *Bone Res.* 1 (3): 216–248.

19 Seeman, E. and Delmas, P.D. (2006). Bone quality-the material and structural basis of bone strength and fragility. *N. Engl. J. Med.* 354 (21): 2250–2261.

20 Dempster, D.W. (2003). Bone microarchitecture and strength. *Osteoporos. Int.* (Suppl. 5): S54–S56.

**21** Shegarfi, H. and Reikeras, O. (2009). Bone transplantation and immune response. *J. Orthop. Surg.* 17 (2): 206–211.

**22** Li, D., Liu, J., Guo, B. et al. (2016). Osteoclast-derived exosomal miR-214-3p inhibits osteoblastic bone formation. *Nat. Commun.* 7: 1–16.

**23** Fricain, J.C., Schlaubitz, S., and Le Visage, C. (2013). A nano-hydroxyapatite–pullulan/dextran polysaccharide composite macroporous material for bone tissue engineering. *Biomaterials* 34: 2947–2959.

**24** Chan, O., Coathup, M.J., and Nesbitt, A. (2012). The effects of microporosity on osteoinduction of calcium-phosphate bone graft substitute biomaterials. *Acta Biomater.* 8: 2788–2794.

**25** Karageorgiou, V. and Kaplan, D. (2005). Porosity of 3D biomaterial scaffolds and osteogenesis. *Biomaterials* 26: 5474–5491.

**26** Fujibayashi, S., Neo, M., and Kim, H.M. (2004). Osteoinduction of porous bioactive titanium metal. *Biomaterials* 25: 443–450.

**27** Vyas, C., Poologasundarampillai, G., Hoyland, J., and Bartolo, P. (2017). 12-3D printing of biocomposites for osteochondral tissue engineering. In: *Biomedical Composites*, 2e (ed. L. Ambrosio), 261–302. Woodhead Publishing.

**28** García-Gareta, E., Coathup, M.J., and Blunn, G.W. (2015). Osteoinduction of bone grafting materials for bone repair and regeneration. *Bone* 81: 112–121.

**29** Rezaie, H.R., Bakhtiari, L., and Öchsner, A. (2015). *Biomaterials and Their Applications*. Springer Publishers.

**30** Black, J., Shalaby, S.W., and Laberge, M. (1992). Biomaterials education: an academic viewpoint. *J. Appl. Biomater.* 3 (3): 231–236.

**31** Park, J. and Lakes, R.S. (2002). *Biomaterials: An Introduction*. CRC Press.

**32** Bajpai, P.K. and Billotte, W.G. (1995). *The Biomedical Engineering Handbook* (ed. J.D. Bronzino), 552–580. CRC Press.

**33** Beaujot, R. (2003). *Effect of Immigration on the Canadian Population: Replacement Migration*. Halifax: Canadian Population Society.

**34** Parida, P., Behera, A., and Mishra, S.C. (2012). Classification of biomaterials used in medicine. *Int. J. Adv. Appl. Sci.* 1 (3): 31–35.

**35** Pham, D. and Dimov, S.S. (2001). *Rapid Manufacturing: The Technologies and Applications of Rapid Prototyping and Rapid Tooling*. London: Springer Verlag.

**36** Chua, C.K., Lim, C.S., and Leong, K.F. (2010). *Rapid Prototyping: Principles and Applications*. Stallion Press. ISBN: 13-978 981 277 987 0.

**37** Yang, Y., Wang, G., Liang, H. et al. (2019). Additive manufacturing of bone scaffolds. *Int. J. Bioprint.* 5 (1): 148. https://doi.org/10.18063/IJB.v5i1.148.

**38** Torres, Y., Trueba, P., Pavón, J.J. et al. (2016). Design, processing and characterization of titanium with radial graded porosity for bone implants. *Mater. Des.* 110: 179–187.

**39** Torres, Y., Trueba, P., Pavón, J. et al. (2014a). Designing, processing and characterisation of titanium cylinders with graded porosity: an alternative to stress-shielding solutions. *Mater. Des.* 63: 316–324.

**40** Torres, Y., Pavón, J., Trueba, P. et al. (2014b). Design, fabrication and characterization of titanium with graded porosity by using space-holder technique. *Procedia Mater. Sci.* 4: 115–119.

**41** Sanz-Herrera, J.A. and Boccaccini, A.R. (2011). Modelling bioactivity and degradation of bioactive glass based tissue engineering scaffolds. *Int. J. Solids Struct.* 48 (2): 257–268.

**42** Mata, A. (2011). Micro and nanotechnologies for bioengineering regenerative medicine scaffolds. *Int. J. Biomed. Eng. Technol.* 5 (2/3): 266–291.

**43** Murphy, C.M. and O'Brien, F.J. (2010). Understanding the effect of mean pore size on cell activity in collagen-glycosaminoglycan scaffolds. *Cell Adhes. Migr.* 4 (3): 377–381.

**44** Roosa, S.M.M., Kemppainen, J.M., Moffitt, E.N. et al. (2010). The pore size of polycaprolactone scaffolds has limited influence on bone regeneration in an in vivo model. *J. Biomed. Mater. Res. Part A* 92 (1): 359–368.

**45** Williams, J.M., Adewunmi, A., Schek, R.M. et al. (2005). Bone tissue engineering using polycaprolactone scaffolds fabricated via selective laser sintering. *Biomaterials* 26: 4817–4827.

**46** Torres-Sanchez, C., Al Mushref, F.R.A., Norrito, M. et al. (2017). The effect of pore size and porosity on mechanical properties and biological response of poroustitanium scaffolds. *Mater. Sci. Eng., C* 77: 219–228.

**47** Kumaresan, T., Gandhinathan, R., Ramu, M., and Gunaseelan, M. (2019). Biomechanical analysis of implantation of polyamide/hydroxyapatite shifted architecture porous scaffold in an injured femur bone. *Int. J. Biomed. Eng. Technol.* 30 (1): 16–30.

**48** Maskery, I., Sturm, L., Aremu, A.O. et al. (2018). Insights into the mechanical properties of several triply periodic minimal surface lattice structures made by polymer additive manufacturing. *Polymer* 152: 62–71.

**49** Huang, X. and Xie, Y.M. (2011). Topological design of microstructures of cellular materials for maximum bulk or shear modulus. *Comput. Mater. Sci.* 50 (6): 1861–1870.

**50** Ou, S.-F., Chiou, S.-Y., and Ou, K.-L. (2013). Phase transformation on hydroxyapatite decomposition. *Ceram. Int.* 39: 3809–3816.

**51** Griffith, L. (2000). Polymeric biomaterials. *Acta Mater.* 48: 263–277.

**52** Chua, C., Leong, K., Tan, K. et al. (2004). Development of tissue scaffolds using selective laser sintering of polyvinyl alcohol/hydroxyapatite biocomposite for craniofacial and joint defects. *J. Mater. Sci. - Mater. Med.* 15: 1113–1121.

**53** Bose, S. and Tarafder, S. (2012). Calcium phosphate ceramic systems in growth factor and drug delivery for bone tissue engineering: a review. *Acta Biomater.* 8: 1401–1421.

**54** Trombetta, R., Inzana, J.A., Schwarz, E.M. et al. (2017). 3D printing of calcium phosphate ceramics for bone tissue engineering and drug delivery. *Ann. Biomed. Eng.* 45: 23–44.

**55** Yeo, M.G., Simon, C.G. Jr., and Kim, G.H. (2012). Effects of offset values of solid freeform fabricated PCL-β-TCP scaffolds on mechanical properties and cellular activities in bone tissue engineering. *J. Mater. Chem.* 22: 21636.

**56** Yan, C., Hao, L., Hussein, A., and Young, P. (2015). Ti–6Al–4V triply periodic minimal surface structures for bone implants fabricated via selective laser melting. *Mech. Behav. Biomed. Mater.* 51: 61–73.

57 Ataee, A., Li, Y., Brandt, M., and Wen, C. (2018). Ultrahigh-strength titanium gyroid scaffolds manufactured by selective laser melting (SLM) for bone implant applications. *Acta Mater.* 158: 354–368.

58 Li, Y., Jahr, H., Lietaert, K. et al. (2018). Additively manufactured biodegradable porous iron. *Acta Biomater.* 77: 380–393.

59 Wan, C. and Chen, B. (2011). Poly (ε-caprolactone)/graphene oxide biocomposites: mechanical properties and bioactivity. *Biomed. Mater.* 6: 055010.

60 Kumar, A., Negi, Y.S., Choudhary, V., and Bhardwaj, N.K. (2014b). Microstructural and mechanical properties of porous biocomposite scaffolds based on polyvinyl alcohol, nano-hydroxyapatite and cellulose nanocrystals. *Cellulose* 21: 3409–3426.

61 Sharma, C., Dinda, A.K., Potdar, P.D. et al. (2016). Fabrication and characterization of novel nano-biocomposite scaffold of chitosan–gelatin–alginate–hydroxyapatite for bone tissue engineering. *Mater. Sci. Eng., C* 64: 416–427.

62 Sharma, C., Dinda, A.K., and Mishra, N.C. (2013). Fabrication and characterization of natural origin chitosan-gelatin-alginate composite scaffold by foaming method without using surfactant. *J. Appl. Polym. Sci.* 127: 3228–3241.

63 Kruyt, M., de Bruijn, J.D., Wilson, C. et al. (2003). Viable osteogenic cells are obligatory for tissue-engineered ectopic bone formation in goats. *Tissue Eng.* 9: 327–336.

64 Kumar, A., Negi, Y.S., Choudhary, V., and Bhardwaj, N.K. (2014a). Effect of modified cellulose nanocrystals on microstructural and mechanical properties of polyvinyl alcohol/ovalbumin biocomposite scaffolds. *Mater. Lett.* 129: 61–64. https://doi.org/10.1016/j.matlet.2014.05.038.

65 Chu, T.-M.G., Orton, D.G., Hollister, S.J. et al. (2002). Mechanical and in vivo performance of hydroxyapatite implants with controlled architectures. *Biomaterials* 23: 1283–1293.

66 Barralet, J., Grover, L., Gaunt, T. et al. (2002). Preparation of macroporous calcium phosphate cement tissue engineering scaffold. *Biomaterials* 23: 3063–3072.

67 Lee, Y.M., Seol, Y.J., Lim, Y.T. et al. (2001). Tissue-engineered growth of bone by marrow cell transplantation using porous calcium metaphosphate matrices. *J. Biomed. Mater. Res.* 54: 216–223.

68 Chen, F., Mao, T., Tao, K. et al. (2002). Bone graft in the shape of human mandibular condyle reconstruction via seeding marrow-derived osteoblasts into porous coral in a nude mice model. *J. Oral Maxillofac. Surg.* 60: 1155–1159.

69 Zhang, C., Wang, J., Feng, H. et al. (2001). Replacement of segmental bone defects using porous bioceramic cylinders: a biomechanical and X-ray diffraction study. *J. Biomed. Mater. Res.* 54: 407–411.

70 Gong, W., Abdelouas, A., and Lutze, W. (2001). Porous bioactive glass and glass–ceramics made by reaction sintering under pressure. *J. Biomed. Mater. Res.* 54: 320–327.

71 Yuan, H., de Bruijn, J.D., Zhang, X. et al. (2001). Bone induction by porous glass ceramic made from Bioglass®(45S5). *J. Biomed. Mater. Res.* 58: 270–276.

72 El-Ghannam, A.R. (2004). Advanced bioceramic composite for bone tissue engineering: design principles and structure–bioactivity relationship. *J. Biomed. Mater. Res. Part A* 69: 490–501.

**73** Borden, M., El-Amin, S., Attawia, M., and Laurencin, C. (2003). Structural and human cellular assessment of a novel microsphere-based tissue engineered scaffold for bone repair. *Biomaterials* 24: 597–609.

**74** Takahashi, Y. and Tabata, Y. (2004). Effect of the fiber diameter and porosity of non-woven PET fabrics on the osteogenic differentiation of mesenchymal stem cells. *J. Biomater. Sci., Polym. Ed.* 15: 41–57.

**75** Kim, H.-W., Knowles, J.C., and Kim, H.-E. (2004). Hydroxyapatite/poly (ε-caprolactone) composite coatings on hydroxyapatite porous bone scaffold for drug delivery. *Biomaterials* 25: 1279–1287.

**76** Zhao, F., Yin, Y., Lu, W.W. et al. (2002). Preparation and histological evaluation of biomimetic three-dimensional hydroxyapatite/chitosan-gelatin network composite scaffolds. *Biomaterials* 23: 3227–3234.

**77** Zhang, Y. and Zhang, M. (2002). Three-dimensional macroporous calcium phosphate bioceramics with nested chitosan sponges for load-bearing bone implants. *J. Biomed. Mater. Res.* 61: 1–8.

**78** Lickorish, D., Ramshaw, J.A., Werkmeister, J.A. et al. (2004). Collagen–hydroxyapatite composite prepared by biomimetic process. *J. Biomed. Mater. Res. Part A* 68: 19–27.

**79** Taché, A., Gan, L., Deporter, D., and Pilliar, R.M. (2004). Effect of surface chemistry on the rate of osseointegration of sintered porous-surfaced Ti-6Al-4V implants. *Int. J. Oral Maxillofac. Implants* 19: 19–29.

**80** Chang, Y.-S., Gu, H.-O., Kobayashi, M., and Oka, M. (1998). Influence of various structure treatments on histological fixation of titanium implants. *J. Arthroplasty* 13: 816–825.

**81** Kumar, A., Akkineni, A.R., Basu, B., and Gelinsky, M. (2016). Three-dimensional plotted hydroxyapatite scaffolds with predefined architecture: comparison of stabilization by alginate cross-linking versus sintering. *J. Biomater. Appl.* 30: 1168–1181.

**82** Xia, Y., Zhou, P., Cheng, X. et al. (2013). Selective laser sintering fabrication of nano-hydroxyapatite/poly-ε-caprolactone scaffolds for bone tissue engineering applications. *Int. J. Nanomed.* 8: 4197–4213.

**83** Naghieh, S., Foroozmehr, E., Badrossamay, M., and Kharaziha, M. (2017). Combinational processing of 3D printing and electrospinning of hierarchical poly(lactic acid)/gelatin-forsterite scaffolds as a biocomposite: mechanical and biological assessment. *Mater. Des.* 133: 128–135.

**84** Qian, C., Zhang, F., and Sun, J. (2015). Fabrication of Ti/HA composite and functionally graded implant by three-dimensional printing. *Bio-Med. Mater. Eng.* 25: 127–136.

**85** Hao, L., Savalani, M.M., Zhang, Y. et al. (2006). Effects of material morphology and processing conditions on the characteristics of hydroxyapatite and high-density polyethylene biocomposites by selective laser sintering. *Proc. Inst. Mech. Eng., Part L: J. Mater.: Des. Appl.* 220: 125–137.

**86** Tan, K., Chua, C., Leong, K. et al. (2003). Scaffold development using selective laser sintering of polyetheretherketone–hydroxyapatite biocomposite blends. *Biomaterials* 24: 3115–3123.

**87** Tan, K., Chua, C., Leong, K. et al. (2005). Fabrication and characterization of three-dimensional poly(ether-ether-ketone)/-hydroxyapatite biocomposite scaffolds using laser sintering. *Proc. Inst. Mech. Eng., Part H: J. Eng. Med.* 219: 183–194.

**88** Hao, L., Savalani, M., Zhang, Y. et al. (2007). Characterization of selective laser-sintered hydroxyapatite-based biocomposite structures for bone replacement. *Proc. R. Soc. A: Math. Phys. Eng. Sci.* 463: 1857–1869.

# 3

# Mechanical and Dynamic Mechanical Analysis of Bio-based Composites

*R.A. Ilyas[1,2], S.M. Sapuan[3,4], M.R.M. Asyraf[5], M.S.N. Atikah[6], R. Ibrahim[7], Mohd N.F. Norrrahim[8], Tengku A.T. Yasim-Anuar[9], and Liana N. Megashah[9]*

*[1] Universiti Teknologi Malaysia, Sustainable Waste Management Research Group (SWAM), School of Chemical and Energy Engineering, Faculty of Engineering, 81310 UTM Johor Bahru, Johor, Malaysia*
*[2] Universiti Teknologi Malaysia, Centre for Advanced Composite Materials, 81310 UTM Johor Bahru, Johor, Malaysia*
*[3] Universiti Putra Malaysia, Laboratory of Biocomposite Technology, Institute of Tropical Forestry and Forest Products (INTROP), 43400 UPM Serdang, Selangor, Malaysia*
*[4] Universiti Putra Malaysia, Advanced Engineering Materials and Composites Research Centre (AEMC), Department of Mechanical and Manufacturing Engineering, 43400 UPM Serdang, Selangor, Malaysia*
*[5] Universiti Putra Malaysia, Department of Aerospace Engineering, Faculty of Engineering, 43400 UPM Serdang, Selangor, Malaysia*
*[6] Universiti Putra Malaysia, Department of Chemical and Environmental Engineering, Faculty of Engineering, 43400 UPM Serdang, Selangor, Malaysia*
*[7] Forest Research Institute Malaysia, Pulp and Paper Branch, 52109 Kepong, Selangor, Malaysia*
*[8] Universiti Pertahanan Nasional Malaysia, Kem Perdana Sungai Besi, Research Centre for Chemical Defence (CHEMDEF), 57000 Kuala Lumpur, Malaysia*
*[9] Universiti Putra Malaysia, Department of Bioprocess Technology, Faculty of Biotechnology and Biomolecular Sciences, 43400 UPM Serdang, Selangor, Malaysia*

## 3.1 Introduction

Lately, biocomposites materials have been widely implemented in various applications and structures including armors, building, automotive, household, fishing, and sport equipment [1–5]. This happened due to its capability and characteristics with high strength and stiffness along with better corrosion resistance [6, 7]. The additional advantages that can be seen from biocomposites were the material obtained 100% from bio-based resources such as plants and animals, which is highly biodegradable. In order to provide clearer view for biocomposites material, laboratory scale of characterization works should be carried out to examine and analyze the physical, mechanical, thermal, and chemical properties of the material [8–12].

Commonly, the bio-based polymeric composites have been characterized using variety of mechanical test in order to understand the behavior of the material under certain load conditions. The quasi-static mechanical performances of the biocomposites are evaluated using their general testing modes such as tensile, compression, and flexural tests. This experimental works were carried out aids

*Mechanical and Dynamic Properties of Biocomposites*, First Edition. Edited by Senthilkumar Krishnasamy, Rajini Nagarajan, Senthil Muthu Kumar Thiagamani, and Suchart Siengchin.

the material engineers and scientists to comprehend the mechanical behavior of the polymeric composite under tensile, compressive, and shear stress–strain effects [13]. Strain-rate effects on the constitutive behavior of composites, which are necessary for accurate material modeling, are not directly available from the results of these lateral impact tests.

Currently, numerous studies were conducted on natural fiber such as kenaf [6, 14, 15], ginger [16, 17], cogon fiber [18, 19], sugarcane bagasse [20], water hyacinth [21], sugar palm [22–30], and jute [31] incorporated in synthetic-based and bio-based polymer that have shown better mechanical structure than pure polymer. However, due to the synthetic polymer take up thousands of years to be decomposed, a greener technology or bio-based polymer was introduced. A biocomposites material is suggested to create biodegradable composite components that are decomposed naturally throughout their lifetime [32, 33]. The combination of the components usually made up of natural fiber and biopolymer gives rise to the quality properties of the material instead of being a separate entity. The biocomposites material is one of the universal material, which is widely used in siding, pipes, electrical wire, food container, synthetic paper, thermal insulation, food wrapper, plastic bags, and other uses as well [34]. Therefore this chapter overviews the mechanical properties and dynamic mechanical analysis of biocomposites.

## 3.2 Mechanical Properties of Macro-scale Fiber

Many researchers have deliberated a study regarding quasi-static mechanical properties using mechanical testing instrument using ASTM standard depending on whether the material testing fiber or composite and the biopolymer type involved. In this research, the macro fiber was studied and reviewed in term of quasi-static mechanical behavior to understand the mechanical performance under several load effects. Table 3.1 summarizes the quasi-static mechanical behavior of biocomposites.

From Table 3.1, it shows that the mechanical performances of biocomposites are highly dependable on type of surface treatment, concentration of treating agent, and additions of coupling agents. Moreover, it can be seen also that influence of fiber species such as carnauba and flax along with optimum fiber concentration would improve the quasi-static mechanical performances.

## 3.3 Mechanical Properties of Nano-scale Fiber

### 3.3.1 Factors Affecting Mechanical Properties of Bionanocomposites

The superior characteristics of nanocellulose had imparted them to be utilized for various purposes, including as a reinforcement material for bionanocomposites production. Nevertheless, in order to ensure maximum increment of bionanocomposites after the reinforcement of nanocellulose, several factors need to be considered,

**Table 3.1** Mechanical performances of various biocomposites.

| Single fiber | Resin | Optimization concentration of chemical treatment | Coupling agent/ catalyst/ compati-bilizer | Optimum fiber con-centration | Mode | Strength (MPa) | Modulus (GPa) | References |
|---|---|---|---|---|---|---|---|---|
| Sugar palm | Phenolic | 0.5% of NaOH and seawater | – | 30 vol.% | Flex | 32.2 | 3.3 | [35] |
| Bamboo | HDPE | – | – | 30 vol.% | Flex | 27.9 | 29.1 | [36] |
| | PP | – | Maleic anhydride grafted polypropy-lene | 30 vol.% | Flex | 52.3 | 20.7 | [37] |
| Kenaf | PLA | – | Grafting maleic anhydride | 65 vol.% | Flex | 46.7 | 54.9 | [38] |
| Alfa | Starch | NaOH treatment | – | 30 vol.% | Flex | 28.3 | 12.3 | [39] |
| | | | | | Ten | 21.7 | 15.5 | |
| Carnauba | PHB | Peroxide | – | 10 wt.% | Ten | 180.0 | 10.5 | [40] |
| Pineapple leaf fiber (PALF) | PLA | 5% of silane and 5% of NaOH | – | 40 wt.% | Flex | 50.0 | 6.5 | [40] |
| | PLA | 5% dioxide chlorine and 35% of hydrogen peroxide | – | 40 wt.% | Flex | 38.02 | 2.02 | [41] |
| Banana | PLA | 5% of NaOH | Maleic anhydride (MA) | 30 wt.% | Ten | 36.0 | 41.0 | [42] |
| | PLA | NaOH treatment | – | 30 wt.% | Ten | 14.6 | 46.3 | [43] |
| Flax | PLA | – | – | 40 wt.% | Ten | 72.2 | 13.0 | [44] |
| Hemp | PP | – | Maleic anhydride (MAH) | 30 wt.% | Ten | 35.0 | – | [45] |
| | | | | | Flex | 67.0 | – | [46] |
| Waste silk fiber | PBS | – | – | 40 wt.% | Ten | 42.5 | 1.3 | [47] |

Flex = flexural properties and Ten = tensile properties.

as the mechanical properties is highly depended on certain key factors such as fabrication method, nanocellulose loading, dispersion, and orientation [48, 49].

#### 3.3.1.1 Fabrication Method

Various processing methods can be applied to produce bionanocomposites, and the selection of methods is generally depends on three factors, which are the properties of matrices and reinforcement material, targeted shapes and applications. In order to produce bionanocomposites with desirable and greater properties than the individual components, they must be fabricated correctly by taking into consideration certain criteria such as process temperature, pressure, time duration, desired product geometry, performance needed, cost, and the ease of manufacture [50]. All these criteria are important as non-suitable bionanocomposite processing may cause agglomeration, swelling, and dense structure of nanocellulose, which then will disrupt the mechanical properties of bionanocomposite afterwards [51].

The most commonly used manufacturing processes for bionanocomposites are extrusion and internal melt-blending, as nanocellulose was proven to be dispersed well in polymers compounded by these methods. Extrusion process was considered as one of the most versatile, suitable equipment for melt compounding various polymers with numerous types of reinforcements and modifiers at high temperature as shown in Table 3.2 [56]. Extruders generally consist of three main zones, which are feeding, compression, and metering zones. Polymers will firstly be introduced to the heated barrel at the feeding zone, and after being melted, the polymers will be conveyed to the compression zone, in which the polymers will be mixed with nanomaterials. Later, at the metering zone, the desired product cross-section will be extruded out through a die [57, 58]. In polymer industry, extrusion has been widely used for bionanocomposites as it offers numerous advantages: (i) a solvent-free process, (ii) continuous process, and (iii) well-align fibers orientation [59].

Besides extrusion, internal melt-blending is also one of the common fabrication methods used to produce bionanocomposites, as it is a simple, yet multi-purpose method and compatible with current industrial process especially in the plastic industry [60]. An internal mixer is generally composed of screws that allow polymer

**Table 3.2** Bionanocomposites produced by extrusion and internal-melt blending.

| Fabrication method | Polymer matrix | Reinforcement material | Processing condition | References |
|---|---|---|---|---|
| Extrusion | Polylactic acid | Microcrystalline cellulose | Temperature: 170–200 °C<br>Screw rotation speed: 150 rpm | [52] |
| | Thermoplastic starch | Nanocellulose | Temperature: 80–110 °C<br>Screw rotation speed: 200 rpm | [53] |
| Internal melt-blending | High density polyethylene | Carbon nanofiber and nanotube | Temperature: 190 °C<br>Rotor speed: 30–60 rpm<br>Duration: 5 min | [54] |
| | Low density polyethylene | Nanocellulose | Temperature: 160 °C<br>Rotor speed: 50 rpm<br>Duration: 20 min | [55] |

matrix and nanocellulose to mix and blend to each other [61]. In regard to its simple yet effective processing, internal-melt blending method had been applied to produce various bionanocomposites as shown in Table 3.2.

#### 3.3.1.2 Nanocellulose Loading

The main purpose of incorporating nanocellulose is to enhance the properties of neat polymers. Studies had reported that the mechanical properties of composites increased by increasing the nanocellulose loading in the matrix [59, 62]. Nevertheless, the mechanical properties of bionanocomposites may decrease after reaching a threshold limit, due to poor mechanical interlocking and less load transfer between nanocellulose and matrix [23].

Yasim-Anuar et al. [59] studied the mechanical properties of bionanocomposites reinforced 0.5–5 wt.% nanocellulose fiber (NCF) in polyethylene matrix prepared by extrusion. Her studies revealed that the tensile and flexural properties of composites reinforced NCF achieve maximum value after 3% NCF was incorporated in the polyethylene matrix, while the same properties were reduced after 4 wt.% NCF was incorporated in the polymer matrix. Another study by Benhamou et al. [63] also revealed that the maximum increment of tensile strength for polyurethane reinforced NCF was obtained after 7.5 wt.% NCF was incorporated in the polymer matrix. The value was higher compared to the composites with NCF loading at 2.5%, 5%, and 10% NCF.

#### 3.3.1.3 Nanocellulose Dispersion and Distribution

The dispersion and distribution of nanocellulose may also give effects to the properties of bionanocomposites. Challenges usually come from the use of nanocellulose as a reinforcement material, as nanocellulose is hydrophilic and non-compatible with the hydrophobic polymers [53]. The differences in wetting properties lead to the possibility of nanocellulose agglomeration, hence affecting the fiber dispersion and distribution.

#### 3.3.1.4 Nanocellulose Orientation

Nanocellulose orientation is significantly related to the method of processing as well as nanocellulose loading, and it can be classified into continuously aligned, discontinuously aligned, and randomly oriented [64, 65]. Many studies had reported that the strength and stiffness of composites with fiber aligned parallel to the loading direction were higher than the composites with fiber randomly aligned to the loading direction, and this was due to the effective stress transfer between fiber and matrix [55, 66–68]. Yasim-Anuar [55] revealed that bionanocomposites prepared by extrusion recorded higher mechanical properties compared with bionanocomposites prepared by internal melt-blending, and it was contributed by the orientation of nanocellulose in the polymer matrix after compounding. The nanocellulose were found to be aligned to the same direction in bionanocomposites prepared by extrusion, whereas nanocellulose was found to be randomly aligned in the bionanocomposites prepared by internal melt-blending. This findings show that nanocellulose orientation may as well give effects to the mechanical properties of bionanocomposites.

### 3.3.2 The Static Mechanical Properties of Bionanocomposites

Typically, the static mechanical properties of bionanocomposites include elastic modulus, tensile strength, elongation, hardness, and fatigue failure [69]. Table 3.3 demonstrates the various studies using different sources of natural fibers as fillers reinforcement, which result in varied mechanical properties. Examples of such filler for development of bionanocomposites include NCF and nanocellulose crystal NCC. It has been shown that the polymer matrix-reinforced with nanocellulose enhanced the static mechanical properties with outstanding results. Young's modulus within tensile test is significantly higher in following studies that demonstrate a good stiffness of bionanocomposites. On the other hand, the reduction of elongation at break could be the excess amount of fillers resulting a competitive interaction with polymer matrix [73].

**Table 3.3** Static mechanical properties of bionanocomposites reinforced by various types of fillers.

| Sources of natural fibers as fillers | Polymer matrix | Fillers content | Method of processing | Improvement in mechanical properties (%) | References |
|---|---|---|---|---|---|
| NCF from oil palm mesocarp fibers | Polyethylene (PE) | 3 wt.% | One-pot extrusion | Tensile strength – 57.7%, Young's modulus – 92.7%, flexural strength – 198.2%, flexural modulus – 25.0% | [59] |
| NCF from oil palm mesocarp fibers | Polypropylene (PP) | 3 wt.% | One-pot extrusion | Tensile strength – 34.2%, Young's modulus – 175.9%, flexural strength – 27.8%, flexural modulus – 88.9% | [70] |
| NCF from *Gigantochloa scortechinii* bamboo fibers | Polyamine/ epoxy resin | 0.7 wt.% | Acid hydrolysis/solution casting | Tensile strength – 29.9%, Young's modulus – 66.7%, flexural strength – 30.6%, flexural modulus – 21.4% | [71] |
| NCC from purified cellulose | Sorbitol plasticizer | 1.5 wt.% (three layers) | Cross-linking/ coatings | Tensile strength – 56.4%, strain at failure – 8.0%, toughness – 60.0% | [72] |
| NCF from oil palm empty fruit bunch | Polyvinyl alcohol (PVA)/Starch | 10% (v/v) | Acid hydrolysis/solution casting | Tensile strength – 85.2% while elongation at break decreased | [73] |
| NCC from *Luffa cylindrica* fibers | Polycaprolactone (PCL) | 3–12 wt.% | Acid hydrolysis/solution casting | Slight improvement in tensile modulus and strain at break while retaining tensile strength (*No quantitative data reported*) | [74] |
| NCC from Ramie fibers | Poly(ε-caprolactone) | 40 wt.% | Acid hydrolysis/solution casting | Young's modulus – 153% while elongation at break decreased | [75] |
| NCF from *Opuntia ficus-indica* parenchyma cell | Poly(styrene-*co*-butyl acrylate) copolymer | 10 wt.% | Homogenizer/ solution casting | Tensile modulus – 6142%, tensile stress – 104.2% and elongation at break decreased | [76] |

## 3.4 Dynamic Mechanical Analysis (DMA) of Biocomposites

There are various mechanical testing approaches used to characterize biocomposites materials such as static [8], dynamic [77], creep [13], and fatigue [78] modes. In this section, the dynamic mechanical test analyzes the characteristics of a polymer that undergoes viscoelastic behavior (combination of elastic solid and Newtonian fluids) [79]. According to the basic understanding of elasticity, an elastic solid expresses its mechanical properties whereby the applied stress is depending on strain or deformation. During the test, the application of sinusoidal stress estimates thermo-strain and complex modulus of the polymeric material [80]. The test is typically carried out with various temperature of the sample and frequency stress to discover glass transition temperature or ($T_g$) along with to determine the transition for other molecular motion [8].

To be specific, a perfect phase of stress and strain exhibited for a perfectly elastic solid. Meanwhile, a viscous fluid is experienced a 90° phase lag of strain depending on stress applied. Hence, viscoelastic plastics exhibit the characteristics of some phase lag occurred along with the sinusoidal stress [81]. Eq. (3.1) explains the stress and strain behavior along with the thermal-stress effect:

$$\text{Stress: } \sigma = \sigma_0 \sin(t\omega + \delta) \tag{3.1}$$

$$\text{Strain: } \varepsilon = \varepsilon_0 \sin(t\omega) \tag{3.2}$$

In the dynamic mechanical analysis (DMA) test, the storage modulus is the variable used to calculate the stored energy, which represents the elastic region of the plastic. Meanwhile, the measure of the heat energy loss along the viscous region of the plastic is called loss modulus. Eq. (3.2) exhibits the storage modulus and loss modulus:

$$\text{Storage modulus: } E' = \frac{\sigma_0}{\varepsilon_0} \cos\delta$$

$$\text{Loss modulus: } E'' = \frac{\sigma_0}{\varepsilon_0} \sin\delta$$

$$\text{Phase angle: } \delta = \arctan\frac{E''}{E'}$$

For the instrumentation, a dynamic mechanical analyzer (DMA) is composed of displacement sensor (linear variable differential transformer), which measures the voltage shift as a consequence of an instrument probe moving through a magnetic core. Besides that, the machine is also made up of several components, including a temperature control system (furnace) and a drive engine [82]. This machine is operated by driveshaft support with a guidance system. The function of the components is a guide for the force from the motor to sample. The sample clamps to allow placing the sample in a fixed position during the testing. Figure 3.1 displays a schematic diagram of the main components in a DMA instrument.

Numerous factors affect the behavior of composite material in term of thermo-mechanical behaviors such as fiber loading, compatibility, additives,

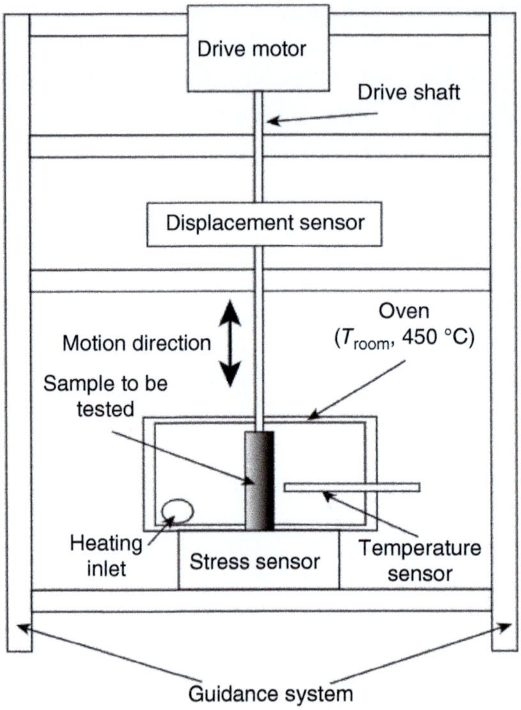


**Figure 3.1** General schematic of a DMA instrumente. Source: Faydi et al. [82]. © 2015, The American Society of Mechanical Engineers.

orientation of fiber, and testing method. These factors play a significant and significant role to have better mechanical properties in the elevated temperature.

Recent studies have shown that the analysis of dynamic mechanical for fiber in biocomposites material behavior into three sizes. The sizes of fiber are categorized into macrofiber, microfiber, and nanofiber. Hence, this work focuses on the dynamic mechanical behavior of various biocomposites with its additives and treatment.

Currently, numerous studies were conducted on natural fiber such as kenaf [6], sugar palm [22], and jute [31] incorporated in synthetic polymer that have shown better mechanical structure than pure polymer. However, due to the synthetic polymer take up thousands of years to be decomposed, a greener technology should be introduced. A biocomposites material is suggested to create biodegradable composites components that are decomposed naturally throughout their lifetime [32, 33]. The combination of the components usually made up of natural fiber and biopolymer gives rise to the quality properties of the material instead of being a separate entity. The biocomposites material is one of the universal material which is widely used in siding, pipes, electrical wire, food container, synthetic paper, thermal insulation, food wrapper, plastic bags, and other uses as well [34].

Macro-scale fiber is evaluated by control the size and the length scale of fiber, which is the original size in order to meet the final requirements [83]. Various researchers have studied changes in dynamic mechanical properties due to chemical treatment and specific additives by subjecting the biocomposites to DMA instrument. The mechanical characterization is done according to the ASTM standard depending on whether the material testing fiber or composite and the biopolymer type involved.

In this research, the macro fiber was studied and reviewed in term of dynamic mechanical behavior to elaborate the polymer composition and glass transition behavior before it is applied in real-life application.

### 3.4.1 Single Fiber

Single fiber composite is made up of one discontinuous phase, which integrated into a continuous matrix to form a single fiber reinforced polymer composite. The single fiber is divided into bio-based fiber and synthetic fiber. Generally, the reinforcement of the single phase of bio-based fiber is derived from natural resources, including plants and animals and reinforced in biopolymer to form a biocomposites [7].

Table 3.4 depicts the parameter and results obtained by various researchers to study the DMA of various biocomposites with their additives and chemical treatment.

#### 3.4.1.1 Sugar Palm

A study was performed by Rashid et al. [84] on the thermo-mechanical behavior of various loading of sugar palm fiber (SPF) composites after undergone chemical treatment in phenolic reinforced biocomposites. The outcomes displayed the maximum storage and loss moduli are at 30 vol.% of SF in phenolic biocomposites. It was shown that the presence of higher SPF content inside phenolic polymer aids the composite system as barriers to limit the molecular chain mobility. On the other hand, the neat phenolic composite is found to be the least storage modulus compared with all other SPF/PF biocomposites. This phenomenon can be explained that the higher the fiber loading, the more the of stress transfer around the composite phenolic matrix by SPF. The improvement also is noticed in glass and rubber regions for all composites since the increase of loss modulus. This happened because adding the SPF contents tends to reduce the damping factor, which indirectly represents as good compatibility of fiber–matrix interphase. Hence, the inclusion of SPF allows better compatibility in the phenolic matrix in elevated temperature, which effectively is used for friction materials.

#### 3.4.1.2 Bamboo

Das and Chakraborty [85] carried out a research on the effect of powdered phenolic resin and chemically modified fiber of bamboo/phenolic biocomposites on thermo-mechanical properties. In the research, varies of concentration of caustic soda solution such as 10%, 15%, and 20% treated the fiber. The results displayed the best stiffness value (storage modulus) and damping parameter ($\tan\delta$) at 20% alkali-treated bamboo-phenolic composites. The observation was noticed since the growing interaction in-between resin and fiber possibly reduced the chain movements at high temperature. Besides that, the interfibrillar region was progressively loosed more hemicellulose and lignin, which results in increasing fiber fibrillation since the effect of mercerization. This aids in better efficiency for the resin to wet the

**Table 3.4** The single fiber reinforced biocomposites in DMA analysis.

| Single fiber | Resin | Optimization concentration of chemical treatment | Coupling agent/ catalyst/ compatibilizer | Optimum fiber concentration | Maximum storage modulus (MPa) | Maximum loss modulus (MPa) | Damping factor (tan $\delta$) | References |
|---|---|---|---|---|---|---|---|---|
| Sugar palm | Phenolic | 0.5% of NaOH and seawater | — | 30 vol.% | 2755 | 178 | 0.275 | [84] |
| Bamboo | Powdered phenolic | 20% of caustic soda solution | — | — | 5.5 | 305 | 0.160 | [85] |
| Kenaf | PLA | | — | 20 vol.% | 3300 | 6039 | 1.83 | [86] |
| | PLA | — | Grafting maleic anhydride | 65 vol.% | 7500 | 900 | 0.96 | [38] |
| Alfa | Starch | NaOH treatment | — | 14 vol.% | 6000 | — | 0.145 (alpha) | [39] |
| | | | | | | | 0.190 (beta) | |
| Carnauba | PHB | Peroxide | — | 10 wt.% | 20 000 | 13 500 | 0.675 | [40] |
| Pineapple leaf fiber (PALF) | PLA | 5% of silane and 5% of NaOH | — | 40 wt.% | 6300 | — | — | [40] |
| Oil palm fiber (OPF) | LLDPE | 5% of NaOH | — | 30 wt.% | 1120 | 690 | 0.061 | [87] |
| Red algae | PBS | 5% dioxide chlorine and 35% of hydrogen peroxide | — | 60 wt.% | 7.5 | 0.30 | 0.045 | [88] |
| | PLA | | — | 50 wt.% | 5.4 | 0.54 | 0.100 | [41] |
| Banana | PLA | 5% of NaOH | Maleic anhydride (MA) | 30 wt.% | 4800 | 755 | 0.370 | [42] |
| | PLA | NaOH treatment | — | 30 wt.% | 5650 | 990 | 0.700 | [43] |
| Flax | PLA | — | — | 40 wt.% | 6950 | 3336 | 0.480 | [44] |
| Jute | PE | Cyano-ethyl | Tertiary butyl perbenzoate | 30 wt.% | 9900 | 460 | 0.100 | [89] |
| | HDPE | Maleic anhydride grafted polyethylene (MAPE) | Maleic anhydride (MA) | 30 wt.% | 5300 | 160 | 0.275 | [90] |
| Hemp | PP | — | Maleic anhydride (MA) – polypropylene | 30 wt.% | 5850 | — | — | [45] |
| | PP | 5% of NaOH | — | 40 wt.% | 9450 | 240 | 0.185 | [46] |
| Waste silk fiber | PBS | — | — | 50 wt.% | 908 | 55 | 0.06 | [47] |

fiber. This tends to improve the interaction of the resin with the alkali-treated fiber that subsequently contributes to escalating the storage modulus values. Other results also showed that pure resin exhibits the higher temperature of glass transition with a reduced loss module value than the whole composite. The molecular movement indicated cured resin sample where cross-linking sites point to the rigidity.

#### 3.4.1.3 Kenaf

Anuar and Zuraida [86] carried out a study on thermal properties of injection molded kenaf fiber (KF)-polylactic acid (PLA) biocomposites. They looked at the effect of fiber contents on the stiffness of PLA-KF biocomposites at elevated temperature has been investigated by using dynamic mechanical analyzer (DMA). The outcome of the research exhibits the increase of storage modulus from unreinforced PLA, 10 wt.% of KF-PLA to 20% of KF-PLA at a higher temperature of 55–75 °C. Thus, the higher the KF content in PLA composite contributes to higher storage modulus after the glass transition temperature ($T_g$) of PLA biopolymer. This shows that the higher the fiber content leads to a stiffer composite. Meanwhile, the damping parameter, which is the sharpness of tan $\delta$ peak reduced with KF content, could suggest an improvement in interfacial interaction occurred.

Another research was conducted by Avella et al. [107] on PLA polymer reinforced with KF to develop biocomposites. The study focused on the effectiveness of compatibilization in composite and fiber content toward the mechanical properties of the material. The results exhibit that the 30 vol.% with compatibilized of kenaf-PLA the highest peak of storage modulus and loss modulus as compared with lower and uncompatibilized composite. Besides, the damping parameter of tan $\delta$ stated the respective composite as the lowest peak value. This shows that reactive compatibility and the higher fiber content tend to promote better fiber–matrix adhesion, which creates a stiffer and stronger composite.

#### 3.4.1.4 Alfa

A scientific study piloted by Belhassen et al. [39] found that biocomposites can be produced by using Alfa reinforced starch polymer. In the DMA results, two relaxations occur at −33 °C ($\beta$) and 30 °C ($\alpha$), which the storage modulus value by a factor between 4 and 5, and 8 and 10, respectively. Two relaxations were observed with fiber incorporation into the polymer with a modest shift at less than 5 °C. At the same time, a significant improvement was noticed in the storage modulus due to the relaxation effect when the higher fiber incorporates in the composite.

#### 3.4.1.5 Carnauba

Melo et al. [91] performed an investigation on the influence of chemically modified carnauba fibers (CFs) incorporated in polyhydroxybutyrate (PHB) and their prospective to create a bio-based composite. Several chemicals were proposed to treat the fiber including alkali, peroxide, potassium permanganate, and acetylation. The results showed that the storage modulus in the dynamic mechanical test was lower at the temperature elevated due to higher activation energy caused vigorous mobility of the polymer chain. The study also found that the addition of the fiber provides better stiffness properties at higher temperatures. This indicates that fiber absorbs energy and reduces crack propagation in PHB matrix. Aside from this, the glass transition temperature of the neat polymer was higher than that of the composites based on observation in the damping parameter (tan $\delta$). It is seen that the glass transition temperature of all fiber-reinforced composites at 20 °C, while the neat polymer the glass transition occurs at a temperature of about 30 °C. This was due to the existence of residual solvent left in the preparation of the prepreg.

#### 3.4.1.6 Pineapple Leaf Fiber (PALF)

A study by Huda et al. [40] studied the chemical modifications of the pineapple leaf fiber (PALF) on PLA biocomposites. They studied the effect of surface treatment on the performance of the PALF/PLA biocomposites. The results indicated that the stiffness properties (storage modulus) were for both treated and untreated PLA/PALF composites as compared with the pure PLA resin at 20–100 °C. In general, it can be said that surface modification improves the fiber adhesion to the PLA biopolymer, increasing when compared with untreated fibers. In addition, alkali-silane treated fiber (PALFNASI) had a significantly high in term of modulus throughout the whole range of temperature studied. This reflects the enhancement of interfacial adhesion of combined attention sodium hydroxide and silane treatment.

#### 3.4.1.7 Oil Palm Fiber (OPF)

Shinoj et al. [87] reported a set of dynamic mechanical test on oil palm fiber (OPF) reinforced linear low-density polyethylene (LLDPE) polymer biocomposites. They worked on the effect of different fiber content, size, and surface modification (sodium hydroxide treatment) on the thermo-mechanical properties of the biocomposites. The results show that the storage and loss moduli are improved with higher fiber content and implementation of alkali-treated fiber in the laminate. Moreover, neat LLDPE glass transition temperature ($T_g$) was increased from −144 °C to −128 °C with 40% of fiber loading. Higher fiber loading, smaller fiber size in composite, and alkali treatment on fiber increased the loss modulus peak and single tan $\delta$. This shows that the higher the fiber loading and smaller the size of fiber when incorporated in biopolymer, the higher the interfacial strength of fiber–matrix due to higher reactive area. Meanwhile, the sodium hydroxide treatment allows the introduction of a reactive group to induce better interaction between fiber–matrix.

#### 3.4.1.8 Red Algae

Lee et al. [88] experimented the influence of bleached fiber along with fiber content on the dynamic mechanical behaviors for the red algae fiber/poly(butylene succinate) (PBS) biocomposites. The observation depicted that a significant increase in storage modulus when the composite contains 50 wt.% of bleached red algae fiber (BRAF). This behavior was noticed due to the higher fiber loading, which permits the transfer of stress from fiber to the matrix and vice versa. Meanwhile, the storage modulus is less significant increment for a 60 wt.% of BRAF compared with others due to lack of filling between fiber–matrix. On the other hand, the inclusion of BRAF inside the biopolymer seems to decrease the damping property (tan $\delta$) of the biocomposites. In general, it can be said that the molecular movement inside the biocomposites and mechanical loss to comprehend inter-friction between molecular chains were reduced. Subsequently, it led to an increase in the intermolecular bonding between BRAF and PBS polymer. However, the fiber loading does not effectively improve the peak glass transition temperature ($T_g$) due to less interaction between the respective fiber and polymer in the biocomposites.

Further comparative study on dynamic mechanical behaviors of the BRAF reinforced with two biopolymers, which were polypropylene (PP), and PLA was continued by Sim et al. [41]. It was found that the BRAF/PLA biocomposites were increased around 74% of storage modulus, which is indicated a more significant improvement of compared with a 37% increase of BRAF/PP biocomposites at 30 °C. This is shown that the dimensional stability of BRAF/PLA biocomposites is higher compared with BRAF/PP. This was observed due to the PP matrix is better adhesion, which leads to higher compatibility to reinforcement fiber as compared with the PLA matrix. At the same time, the implementation of BRAF in PP and PLA matrix seems the reduction of the tan $\delta$ peak, which concludes a lower damping parameter of the respective composite. This indicates that the inclusion of fiber as reinforcement in the biopolymer matrix does improve the damping behavior predominantly at 50 wt.%.

#### 3.4.1.9 Banana

Majhi et al. [42] studied the effect of fiber composition, chemical treatment, and addition of compatibilizer on the dynamic properties banana fiber reinforced poly(lactic acid) (BF/PLA) biocomposites. The compatibilizers used in this study were Maleic anhydride (MA) and glycerol triacetate ester (GTA). The results displayed that the inclusion of MA as a compatibilizer in BF/PLA biocomposites is increased the storage modulus to the optimum value. This improvement is probably because of the formation of ester linkage in the matrix, which subsequently provides better interfacial adhesion. Apart from that, a sufficient improvement seems at the glass transition temperature ($T_g$) up to 65 °C based on the finding in loss modulus peaks. This is due to the combined effect of reinforcement and compatibilizer, which improved the molecular motion in the PLA matrix and the interface of fiber/matrix interaction.

Another similar research replicated by Jandas et al. [43] displayed the influence of chemical treatments of BF on thermo-mechanical properties of BF/PLA biocomposites. The chemical surface treatments of the banana fiber are including sodium hydroxide (Na-BF), 3-aminopropyltriethoxysilane (APS-BF) and bis-(3-triethoxy silylpropyl) tetrasulfane (Si69-BF). The results explained that chemically treated fiber (Na-BF and APS-BF) reinforced biocomposites have higher storage moduli than untreated fiber reinforced composites. This explained that the fiber treatment aids the fiber–matrix interaction in term of better adhesion between them. Si69-BF/PLA biocomposites exhibited the most optimum results in storage modulus due to organofunctional group of silane form a network of polymer change in the matrix structure. In contrast, the maximum shift in the peak of the tan $\delta$ curve in PLA matrix (668–768 °C) in silane treated PLA/BF biocomposites. This was observed due to the presence of unreactive organic group (organosilane), which lowers the cross-linking density of the interfacial network in the polymer. Later, it caused the greater damping magnitude as compared to untreated banana fiber composite or alkali-treated banana fiber composite.

#### 3.4.1.10 Flax

Nassiopoulos have conducted a study and Nijuguna [44] on the dynamic mechanical behavior of PLA reinforced with flax fiber to form biocomposites. In this study, the flax/PLA biocomposites was compared with flax/epoxy synthetic composite. The observation shows that the storage modulus for the flax/PLA biocomposites is higher as compared with flax/epoxy composite. This was attributed because of the flax has better compatibility and adhesion properties with the PLA since the fiber is limited in term of chain mobility in neat PLA. From result also, it shows that the deterioration happened in the damping properties of $\tan \delta$ peak at 74–76 °C. Moreover, additions of fiber inside the matrix tend to escalate the value of stiffness (storage modulus) and damping parameter ($\tan \delta$) of the composites.

#### 3.4.1.11 Jute

Saha et al. [89] conducted a research to study the effect of time immersion (one, two, three, four, and five hours) of jute fiber using cyanoethylation treatment on the thermo-mechanical performance of jute fiber reinforced polyester (JF/PE) biocomposites. The finding presented the storage modulus of the composites shift to higher values when the immersion time in cyanoethylene is in five hours as compared with neat resin, unmodified jute fiber composite, and other time immersion. This could be due to improvement of chemically interfacial interaction between fiber and matrix. Specifically, the jute fiber exhibits hydrophilic properties, which tend to has reduced wettability and adhesion behavior with the resin. This led to the presence of voids interface since the availability of moisture is high in the matrix. Hence, the cyanoethylation of fibers provides better stiffness and strength for the bio-based composite. Moreover, the incorporation of modified fibers reduces the $\tan \delta$ peak height by restricting the movement of polymer molecules due to better compatibility and surface adhesion fiber–matrix interaction.

The previous research has conducted by Mohanty et al. [90] on dynamic mechanical behaviors of maleic anhydride grafted polyethylene (MAPE) treated jute/high-density polyethylene (HDPE) composites. The study is focused on to evaluate the storage modulus and damping parameter ($\tan \delta$) on the fiber chemical treatment and inclusion of fibers. The results show that the storage modulus of pure resin is increased when the jute fiber is incorporated. This was due to the reinforcing effect of the fiber, which increases the stress transfer between polymer and fiber. Besides, the storage modulus of 1% of MAPE treated fiber reinforcement biocomposites improved about 21% as compared with neat resin. This might be happened because of the better interfacial adhesion within the composite laminate. At the same time, the peak of $\tan \delta$ of the unmodified and modified fiber reinforcement biocomposites was lowered comparing the pure HDPE resin. This shows that the energy dissipation in the matrix and at the interface is less since it is incorporated with fiber to have a stronger interface.

#### 3.4.1.12 Hemp

Niu et al. [45] studied on dynamic mechanical properties of hemp/polypropylene (HF/PP) fiber reinforced biocomposites. The study elaborated on the influence of various of compatibilizer (*Maleic anhydride grafted polypropylene* (PP–MAH); *maleic anhydride grafted poly(ethylene octane)* (POE–MAH); *maleic anhydride grafted styrene–(ethylene-co-butylene)–styrene copolymer* (SEBS–MAH)) on thermo-mechanical properties in term of fiber–matrix attraction. Based on the acquired results, the incorporation of POE–MAH, SEBS–MAH, and PP–MAH has increased the storage modulus of the HF/PP biocomposites. This was observed due to the compatibilizer increase the interfacial adhesion between the matrix and natural fiber. Additionally, the composite containing PP–MAH had the highest storage modulus as compared with the other biocomposites. This was acquired due to PP-MAH biocomposites dissipate less energy, which reduces the number of voids on the interface, which tends to have stiffer and stronger characteristics.

Lu and Oza [46] performed a study on the effect of types of chemical modification on thermal stability and thermo-mechanical properties of hemp-high density polyethylene (HF/HDPE) biocomposites. This research was performed to determine the effect of silane and sodium hydroxide (NaOH) treatments of the hemp fiber on thermo-mechanical properties of HF/HDPE biocomposites. According to the observation, the stiffness value of both untreated and treated HF/HDPE biocomposites was increased when the fiber volume fraction was added up until 40%. However, the fiber loading at 50% showed a declined trend due to the maximum benefit of better storage modulus only can be obtained until 40% of fiber loading. Apart from that, the storage modulus was increased when the fiber was treated with NaOH as compared to untreated fiber biocomposites. This phenomenon can be clarified because of chemical treatment improve the roughness surface of the hemp fiber by effectively removed the hemicellulose and lignin from the interface of fiber. Hence, it tended to expose more cellulose to improve interfacial adhesion of fiber–matrix. Meanwhile, the silane treatment allows better storage modulus due to the chemical treatment aids in term of producing more covalent bond and coupling effect. On the other view, the damping parameter of the alkaline and silane treatment biocomposites were significantly reduced, which explained a strong fiber–matrix relation and produced lower energy dissipation in term of viscoelastic behavior.

#### 3.4.1.13 Waste Silk Fiber

A study by Han et al. [47] looked at the dynamic mechanical properties of varies fiber contents and lengths for waste silk fiber reinforced poly(butylene succinate) (WSK/PBS) biocomposites. The results indicated that the storage modulus of all WSF/PBS biocomposites was considerably higher than neat PBS biopolymer at above the glass transition temperature. This was noticed since stress transfer between fiber and matrix is allowed to experience the reinforcing effect. Moreover, the highest value of storage modulus was at WSF/PBS biocomposites with the chopped waste silk fibers of 12.7 mm length. However, the waste silk fiber content affects more significantly on the composite reinforcement than the fiber length. In the meantime, the peak height of the damping parameter was depreciated since the increment of the fiber contents. However, it was not significantly influenced by fiber loading. This acknowledged that the interaction between the WSF and the PBS biopolymer did not seem very strong.

#### 3.4.1.14 Henequen

A study led by Lee and Cho [108] accomplished a finding on the influence of surface fiber modification of henequen fiber (HF) reinforced PP matrix bio-based composite on thermo and static mechanical behavior. The fiber obtained from agave plant leaves was treated with alkaline solution and tap water to analyze the dynamic mechanical properties of the biocomposites. The results found out that 60% of sodium hydroxide treatment was more efficiently treated the HF at 60 minutes soaking. It shows that the fiber treatment increase the stiffness behavior HF in biocomposites. This was proven that the storage modulus is the highest in 60 minutes soaking HF in alkaline solution. However, the glass transition temperature of the biocomposites was not significantly improved with the application of surface modification treatment. This is due to the change of loss modulus of PP biocomposites is greater than storage modulus.

### 3.4.2 Hybrid Fiber

A hybrid fiber composite is composed of more than one discontinuous phases, which embedded in a continuous matrix to form a hybrid composite. The hybrid composite can also exist in biocomposites material [92]. Generally, the hybrid fibers in composites can perform better during higher load as compared with single fiber in the distinct fiber direction. In this case, the fiber aids the surrounding polymer in a fixed position, and orientation subsequently exhibits a higher load stress transfer medium between them [93, 94].

In Section 3.4.2, the dynamic mechanical properties of various hybrid fibers in biocomposites described comprehensively. Table 3.5 depicts the summary of the dynamic mechanical properties of hybrid fiber that was carried out by previous researchers.

#### 3.4.2.1 Sisal/Oil Palm

Scientific research was carried out by Jacob et al. [95] to assess the influence of concentration of chemical treatment of hybrid fibers (sisal and OPF) in natural rubber biocomposites on dynamical mechanical performance. The sisal and OPF is subjected to sodium hydroxide (NaOH) treatment with various concentration including 0.5%, 1%, 2%, and 4%. In the results, the increase of fiber loading does increase the storage and loss modulus, which caused improve the strength and stiffness interface of the composite. Moreover, the observation showed that the gum compound has the highest tan $\delta$ value. This indicated that the composite experienced a higher amount of mobility and better damping behaviors. For the chemical treatment, the higher storage modulus experienced by a higher concentration of NaOH solution, which 4% of NaOH concentration stated the highest storage modulus. This was attributed due to increased cross-linking and formation of a robust fiber–matrix interface. However,

**Table 3.5** Summary of dynamic mechanical properties of hybrid biocomposites.

| Hybrid fiber | Resin | Chemical treatment | Optimum fiber loading ratio | Maximum storage modulus (MPa) | Maximum loss modulus (MPa) | Tan $\delta$ value | References |
|---|---|---|---|---|---|---|---|
| Sisal and oil palm | Natural rubber | 4% of NaOH | 50% of sisal; 50% of oil palm | 840.3 | 755.4 | 0.899 | [95] |
| Coir and PALF | PLA | — | 50% of coir; 50% of PALF | 2400 | 1920 | 0.800 | [96] |
| Kenaf and PALF | Phenolic | 2% of silane | 70% of PALF; 30% kenaf | 3020 | 142.39 | 0.096 | [97] |
| PPLSF and jute | Polyester | — | 25% of PPLSF; 75% of jute | 3.39 | 1.18 | 0.263 | [98] |
| OPEFB and cellulose | Poly-propylene | 8.3% NaOH and 10% of acetic acid | 50% of OPEFB; 50% of cellulose | 7000 | 215 | 0.06 | [99] |

the damping behavior of the biocomposites was depleted when the treated fiber was added in the polymer matrix since better adhesion of fiber to matrix led to reducing the free movement of the macromolecular chain.

#### 3.4.2.2 Coir/PALF

Siakeng et al. [96] operated a study on mechanical, dynamic, and thermo-mechanical properties of coir (CF)/PALF reinforced PLA hybrid biocomposites. The analysis was implemented to evaluate the dynamic mechanical behaviors by comparing the single fiber composites (pineapple and coir) and varies proportions hybrid fiber composite including 1 : 1, 3 : 7, and 7 : 3 of coir/pineapple fiber. The finding showed that the presence hybrid of CF and PALF induced the escalation of storage and loss modulus along with the reduction of the peak height of tan $\delta$ values. It seems that both fibers (CF and PALF) operated as a stiffening agent in the resin matrix and PALF seems to be more intense. The presence of PALF and CF aids to have better stress transfer between fiber–matrix interactions in optimum ratio volume, which is 1 : 1, respectively.

#### 3.4.2.3 Kenaf/PALF

Asim et al. [97] was organized a scientific to characterize the thermal, physical, and flammability after undergone a chemical treatment for kenaf (KF)/PALF phenolic hybrid composites. The study was focused on the effect of hybridization of KF and PALF fibers loading and silane treatment of fiber, which would affect the dynamic mechanical performance of the biocomposites. The work used varies of PALF to KF fiber ratios, including 7 : 3, 1 : 1, and 3 : 7. It was found that the untreated 7 : 3

PALF/KF hybrid composite exhibits the lowest storage modulus value. Meanwhile, the treated 7 : 3 PALF/KF hybrid composite revealed as the highest value of storage and loss modulus. This proved that the PALF enhanced stiffness behavior as well as the silane treatment of fiber and improved the interfacial bonding with KF and matrix. All treated hybrid composites showed similar peaks and the range of highest and lowest peaks was deficient, which indicated that the mobility of polymer chain with fibers are all most constant and showed similar damping characteristics.

#### 3.4.2.4 Palmyra Palm Leaf Stalk Fiber (PPLSF)/Jute

Shanmugam and Thiruchitrambalam [98] carried out a study on both static and dynamic mechanical performance of alkalinization treatment of unidirectional continuous palmyra palm leaf stalk fiber (PPLSF)/jute fiber (JF) hybrid reinforced polyester composites. It is interesting to note that the increase of JF content tends to improve the stiffness behavior of the hybrid composite, which described in term of storage modulus. However, the storage modulus seems to decrease when in higher temperature. Similarly, the loss of modulus is increased when higher JF contents. This was noticed due to the effective stress transfer. In detail, the maximum value of storage and loss moduli when the higher amount of JF, which specifically at 25% of PPLSF and 75% of jute reinforced hybrid composites. Moreover, the peak height of tan $\delta$ is the lowest at 100% jute fiber composite, 25% of PPLSF and 75% of JF reinforced composites compared with 100% PPLSF, 75% of PPLSF and 25% of jute and 50% of PPLSF and 50% of jute reinforced composites. This attribute was adhered since the jute fiber exhibit higher Young's modulus compared with PPLSF, subsequently allows better stress transfer from fiber to the matrix and vice versa. At the same time, the glass transition temperature is shifting toward the right in tan $\delta$, which directs the proof for better interfacial interaction since the sodium hydroxide treatment washed away the lignin and hemicellulose.

#### 3.4.2.5 Oil Palm Empty Fruit Bunch (OPEFB)/Cellulose

Khalid et al. [99] conducted a study regarding thermo-mechanical behavior of hybridization of cellulose, and oil palm empty fruit bunch (OPEFB) reinforced in PP polymer biocomposites. The study found out that the more cellulose content added into the matrix tends to higher the storage modulus. This was noticed due to the fiber imparted by the cellulose allows better transfer of stress along the boundary surface of the PP resin to the cellulose, which tends to have stiffer material. On top of that, the peak height of loss modulus was reduced from neat PP to 50 vol.% of OPEFB and cellulose at $\beta$ transition. This observation was happened because of the higher the filler loading tends to reduce in the glass transition ($T_g$) of the biocomposites.

## 3.5 Dynamic Mechanical Properties of Bionanocomposites

Natural fibers for the nanofillers to biocomposites have numerous studies for advanced applications. Nanofibers reinforced in biocomposites, often termed a bionanocomposites, are relatively new and have attract interest in many advanced applications. Bionanocomposites, on the other hand, contain fillers with nanoscale dimensions of less than 100 nm [100, 101]. For the most part, this review relates to the use of nanocellulose due to its excellent properties such as abundant hydroxyl groups, large surface area, high crystallinity, and high thermal resistance and is known to be compatible with polymer matrix [102]. The good interaction between nanocellulose and polymer matrix makes them the most desirable properties compared to micro sized fiber.

The mechanical properties are the most critical issue in evaluating the behavior of biocomposites in a number of applications. Typically, the static and dynamic mechanical properties are used to evaluate the performance of the bionanocomposites.

### 3.5.1 The Dynamic Mechanical Properties of Bionanocomposites

Dynamic mechanical analysis (DMA) of bionanocomposites is a versatile technique that complements the information provided by the more traditional thermal analysis techniques such as differential scanning calorimetry (DSC), thermogravimetric analysis (TGA), and thermal mechanical analysis (TMA). The dynamic parameters such as storage modulus ($E'$), loss modulus ($E''$), and loss tangent ($\tan \delta$) are

**Table 3.6** Dynamic mechanical properties of bionanocomposites reinforced by various types of fillers.

| Sources of fillers | Polymer matrix | Improvement in dynamic mechanical properties | References |
|---|---|---|---|
| Spruce | Polylactic acid (PLA) | • Storage modulus was increased 14.0% at 20 °C and 25% at −80 °C<br>• Loss modulus was increased | [103] |
| Date palm tree | Polyurethane (PU) | • Significant reduction of the amplitude of two relaxations temperature of $\tan \delta$ | [63] |
| Soy protein isolate | Montmorillonite | • Storage modulus was increased 175.3% at 40 °C<br>• Tan $\delta$ was greatly affected by the montmorillonite content and pH | [104] |
| Wheat straw | Thermoplastic starch | • The storage modulus was increased with increasing nanofiber content However, the modulus was decreased as a function of temperature<br>• The $\tan \delta$ peaks recorded for the nanocomposites were shifted to higher temperatures compared with the $\tan \delta$ peak for pure thermoplastic starch (TPS) | [105] |
| Kenaf pulp | PLA | • The storage modulus of PLA is increased with increased nanofiber content, in glassy as well as in rubbery state<br>• The $\tan \delta$ peak is shifted to higher temperature with increased nanofiber contents | [106] |

temperature dependent and provide information about interfacial bonding between the reinforced nanofiber and polymer matrix of composite material.

Several research works have been conveyed, some of the important research work on DMA studies of nanocomposites are tabulated in Table 3.6. It has been shown that the polymer matrix-reinforced with nanocellulose enhanced the dynamic mechanical properties with outstanding results. The improvement on the DMA properties is usually due to the good interfacial bonding between the nanofiber and the polymer matrix.

## 3.6 Conclusion

Macro- and nano-sized natural fibers have gained huge attention because of their unique properties such as high strength and stiffness, renewability, and biodegradability, as well as their production and application in development of biocomposites. Besides that, this book chapter presents an outline of main results presented on single and hybrid biocomposites focusing the attention in terms of DMA and mechanical properties. Hybrid biocomposites can perform better during higher load as compared with single fiber in the different fiber direction. The fibers aid the surrounding polymer in a fixed position, and orientation subsequently exhibits a higher load stress transfer medium between them. Moreover, the changes in DMA properties are due to the several factors such as chemical treatment or specific additives by subjecting the biocomposites to DMA instrument. It can be seen that the reinforcement of the polymer matrix with natural fibers and nanocellulose enhanced the dynamic mechanical properties with outstanding results. The improvement on the DMA properties is usually due to the good interfacial bonding between the natural fiber and the polymer matrix.

## References


1. Shahroze, R.M., Chandrasekar, M., Senthilkumar, K. et al. (2019). A review on the various fibre treatment techniques used for the fibre surface modification of the sugar palm fibres. *Seminar Enau Kebangsaan*, no. 1, pp. 48–52.
2. Sanyang, M.L., Ilyas, R.A., Sapuan, S.M., and Jumaidin, R. (2018). Sugar palm starch-based composites for packaging applications. In: *Bionanocomposites for Packaging Applications* (eds. M. Jawaid and S.K. Swain), 125–147. Cham: Springer International Publishing.
3. Ilyas, R.A., Sapuan, S.M., Sanyang, M.L. et al. (2018). Nanocrystalline cellulose as reinforcement for polymeric matrix nanocomposites and its potential applications: a review. *Curr. Anal. Chem.* 14 (3): 203–225.
4. Sapuan, S.M., Ishak, M.R., Leman, Z. et al. (2017). Development of products from sugar palm trees (*Arenga pinnata Wurb. Merr*): a community project. *INTROPica*, Serdang, Selangor, pp. 12–13.

**5** Sapuan, S.M. and Ilyas, R.A. (2017). Sugar palm: fibers, biopolymers and biocomposites. *INTROPica*, Serdang, Selangor, pp. 5–7.

**6** Jaafar, C.N.A., Zainol, I., and Rizal, M.A.M. (2018). Preparation and characterisation of epoxy/silica/kenaf composite using hand lay-up method. *27th Scientific Conference of the Microscopy Society Malaysia (27th SCMSM 2018)*, pp. 2–6.

**7** Ilyas, R.A., Sapuan, S.M., Atiqah, A. et al. (2020). Sugar palm (*Arenga pinnata [Wurmb.] Merr*) starch films containing sugar palm nanofibrillated cellulose as reinforcement: water barrier properties. *Polym. Compos.* 41: 1, 459–9, 467.

**8** Jaafar, C.N.A., Rizal, M.A.M., and Zainol, I. (2018). Effect of kenaf alkalization treatment on morphological and mechanical properties of epoxy/silica/kenaf composite. *Int. J. Eng. Technol.* 7: 258–263.

**9** Ilyas, R.A., Sapuan, S.M., and Ishak, M.R. (2018). Isolation and characterization of nanocrystalline cellulose from sugar palm fibres (*Arenga Pinnata*). *Carbohydr. Polym.* 181: 1038–1051.

**10** Ilyas, R.A., Sapuan, S.M., Ishak, M.R., and Zainudin, E.S. (2018). Water transport properties of bio-nanocomposites reinforced by sugar palm (*Arenga Pinnata*) nanofibrillated cellulose. *J. Adv. Res. Fluid Mech. Therm. Sci. J.* 51 (2): 234–246.

**11** Halimatul, M.J., Sapuan, S.M., Jawaid, M. et al. (2019). Water absorption and water solubility properties of sago starch biopolymer composite films filled with sugar palm particles. *Polimery* 64 (9): 27–35.

**12** Halimatul, M.J., Sapuan, S.M., Jawaid, M. et al. (2019). Effect of sago starch and plasticizer content on the properties of thermoplastic films: mechanical testing and cyclic soaking-drying. *Polimery* 64 (6): 32–41.

**13** Asyraf, M.R.M., Ishak, M.R., Razman, M.R., and Chandrasekar, M. (2019). Fundamentals of creep, testing methods and development of test rig for the full-scale crossarm: a review. *J. Teknol.* 81 (4): 155–164.

**14** Mazani, N., Sapuan, S.M., Sanyang, M.L. et al. (2019). Design and fabrication of a shoe shelf from kenaf fiber reinforced unsaturated polyester composites. In: *Lignocellulose for Future Bioeconomy* (eds. S.M. Sapuan, H. Ariffin and M.A. Hassan), 315–332. Elsevier.

**15** Aisyah, H.A. et al. (2019). Thermal properties of woven kenaf/carbon fibre-reinforced epoxy hybrid composite panels. *Int. J. Polym. Sci.* 2019: 1–8.

**16** Abral, H., Ariksa, J. et al. (2020). Transparent and antimicrobial cellulose film from ginger nanofiber. *Food Hydrocolloids* 98: 105266.

**17** Abral, H., Mahardika, M., Handayani, D. et al. (2020). Highly transparent and antimicrobial PVA based bionanocomposites reinforced by ginger nanofiber. *Polym. Test.* 81: 106186.

**18** Jumaidin, R. et al. (2019). Characteristics of cogon grass fibre reinforced thermoplastic cassava starch biocomposite: water absorption and physical properties. *J. Adv. Res. Fluid Mech. Therm. Sci.* 62 (1): 43–52.

**19** Jumaidin, R., Khiruddin, M.A.A., Saidi, Z.A.S. et al. (2020). Effect of cogon grass fibre on the thermal, mechanical and biodegradation properties of thermoplastic cassava starch biocomposite. *Int. J. Biol. Macromol.* 146: 746–755.

**20** Jumaidin, R., Ilyas, R.A., Saiful, M. et al. (2019). Water transport and physical properties of sugarcane bagasse fibre reinforced thermoplastic potato starch biocomposite. *J. Adv. Res. Fluid Mech. Therm. Sci.* 61 (2): 273–281.

**21** Syafri, E. et al. (2019). Effect of sonication time on the thermal stability, moisture absorption, and biodegradation of water hyacinth (*Eichhornia crassipes*) nanocellulose-filled bengkuang (*Pachyrhizus erosus*) starch biocomposites. *J. Mater. Res. Technol.* 8 (6): 6223–6231.

**22** Bachtiar, D., Sapuan, S.M.M., and Hamdan, M.M.M. (2008). The effect of alkaline treatment on tensile properties of sugar palm fibre reinforced epoxy composites. *Mater. Des.* 29 (7): 1285–1290.

**23** Ilyas, R.A. et al. (2019). Effect of sugar palm nanofibrillated cellulose concentrations on morphological, mechanical and physical properties of biodegradable films based on agro-waste sugar palm (*Arenga pinnata [Wurmb.] Merr*) starch. *J. Mater. Res. Technol.* 8 (5): 4819–4830.

**24** Ilyas, R.A. et al. (2020). Thermal, biodegradability and water barrier properties of bio-nanocomposites based on plasticised sugar palm starch and nanofibrillated celluloses from sugar palm fibres. *J. Biobased Mater. Bioenergy* 14: 1–13.

**25** Ilyas, R.A., Sapuan, S.M., Ishak, M.R., and Zainudin, E.S. (2019). Sugar palm nanofibrillated cellulose (*Arenga pinnata [Wurmb.] Merr*): effect of cycles on their yield, physic-chemical, morphological and thermal behavior. *Int. J. Biol. Macromol.* 123: 379–388.

**26** Atikah, M.S.N. et al. (2019). Degradation and physical properties of sugar palm starch/sugar palm nanofibrillated cellulose bionanocomposite. *Polimery* 64 (10): 27–36.

**27** Atiqah, A., Jawaid, M., Sapuan, S.M. et al. (2019). Physical and thermal properties of treated sugar palm/glass fibre reinforced thermoplastic polyurethane hybrid composites. *J. Mater. Res. Technol.* 8 (5): 3726–3732.

**28** Nurazzi, N.M., Khalina, A., Sapuan, S.M., and Ilyas, R.A. (2019). Mechanical properties of sugar palm yarn/woven glass fiber reinforced unsaturated polyester composites: effect of fiber loadings and alkaline treatment. *Polimery* 64 (10): 12–22.

**29** Norizan, M.N., Abdan, K., Ilyas, R.A., and Biofibers, S.P. (2020). Effect of fiber orientation and fiber loading on the mechanical and thermal properties of sugar palm yarn fiber reinforced unsaturated polyester resin composites. *Polimery* 65 (2): 34–43.

**30** Nurazzi, N.M., Khalina, A., Sapuan, S.M. et al. (2019). Thermal properties of treated sugar palm yarn/glass fiber reinforced unsaturated polyester hybrid composites. *J. Mater. Res. Technol.* 9 (2): 1606–1618.

**31** Acharya, S.K. and Dalbehera, S. (2014). Study on mechanical properties of natural fiber reinforced woven jute-glass hybrid epoxy composites. *Adv. Polym. Sci. Technol.* 4 (1): 1–6.

**32** Yu, L., Dean, K., and Li, L. (2006). Polymer blends and composites from renewable resources. *Prog. Polym. Sci.* 31 (6): 576–602.

**33** Curvelo, A.A.S., De Carvalho, A.J.F., and Agnelli, J.A.M. (2001). Thermoplastic starch-cellulosic fibers composites: preliminary results. *Carbohydr. Polym.* 45 (2): 183–188.

**34** Suh, N.P. (2003). Impact of microcellular plastics on industrial practice and academic research. *Macromol. Symp.* 201: 187–201.

**35** Rashid, B., Leman, Z., Jawaid, M. et al. (2016). The mechanical performance of sugar palm fibres (ijuk) reinforced phenolic composites. *Int. J. Precis. Eng. Manuf.* 17 (8): 1001–1008.

**36** Mohanty, S. and Nayak, S.K. (2010). Short bamboo fiber-reinforced HDPE composites: influence of fiber content and modification on strength of the composite. *J. Reinf. Plast. Compos.* 29 (14): 2199–2210.

**37** Samal, S.K., Mohanty, S., and Nayak, S.K. (2009). Polypropylene bamboo/glass fiber hybrid composites: fabrication and analysis of mechanical, morphological, thermal, and dynamic mechanical behavior. *J. Reinf. Plast. Compos.* 28 (22): 2729–2747.

**38** Avella, M., Bogoeva-Gaceva, G., Bužarovska, A. et al. (2008). Poly(lactic acid)-based biocomposites reinforced with kenaf fibers. *J. Appl. Polym. Sci.* 108 (6): 3542–3551.

**39** Belhassen, R. et al. (2009). Biocomposites based on alfa fibers and starch-based biopolymer. *Polym. Adv. Technol.* 20 (12): 1068–1075.

**40** Huda, M.S., Drzal, L.T., Mohanty, A.K., and Misra, M. (2008). Effect of chemical modifications of the pineapple leaf fiber surfaces on the interfacial and mechanical properties of laminated biocomposites. *Compos. Interfaces* 15 (2–3): 169–191.

**41** Sim, K.J., Han, S.O., and Seo, Y.B. (2010). Dynamic mechanical and thermal properties of red algae fiber reinforced poly(lactic acid) biocomposites. *Macromol. Res.* 18 (5): 489–495.

**42** Majhi, S.K., Nayak, S.K., Mohanty, S., and Unnikrishnan, L. (2010). Mechanical and fracture behavior of banana fiber reinforced polylactic acid biocomposites. *Int. J. Plast. Technol.* 14 (1): 57–75.

**43** Jandas, P.J., Mohanty, S., Nayak, S.K., and Srivastava, H. (2011). Effect of surface treatments of banana fiber on mechanical, thermal, and biodegradability properties of PLA/banana fiber biocomposites. *Polym. Compos.* 32 (11): 1689–1700.

**44** Nassiopoulos, E. and Njuguna, J. (2015). Thermo-mechanical performance of poly(lactic acid)/flax fibre-reinforced biocomposites. *Mater. Des.* 66: 473–485.

**45** Niu, P., Liu, B., Wei, X. et al. (2011). Study on mechanical properties and thermal stability of polypropylene/hemp fiber composites. *J. Reinf. Plast. Compos.* 30 (1): 36–44.

**46** Lu, N. and Oza, S. (2013). Thermal stability and thermo-mechanical properties of hemp-high density polyethylene composites: effect of two different chemical modifications. *Composites Part B* 44 (1): 484–490.

**47** Han, S.O., Lee, S.M., Park, W.H., and Cho, D. (2006). Mechanical and thermal properties of waste silk fiber-reinforced poly(butylene succinate) biocomposites. *J. Appl. Polym. Sci.* 100 (6): 4972–4980.

**48** Maia, T.H.S. et al. (2017). Polyethylene cellulose nanofibrils nanocomposites. *Carbohydr. Polym.* 173: 50–56.

**49** Kargarzadeh, H. et al. (2018). Recent developments in nanocellulose-based biodegradable polymers, thermoplastic polymers, and porous nanocomposites. *Prog. Polym. Sci.* 87: 197–227.

**50** Kalia, S., Avérous, L., Njuguna, J. et al. (2011). Natural fibers, bio- and nanocomposites. *Int. J. Polym. Sci.* 2011.

**51** Oksman, K. et al. (2016). Review of the recent developments in cellulose nanocomposite processing. *Composites Part A* 83: 2–18.

**52** Oksman, K., Mathew, A.P., Bondeson, D., and Kvien, I. (2006). Manufacturing process of cellulose whiskers/polylactic acid nanocomposites. *Compos. Sci. Technol.* 66 (15): 2776–2784.

**53** Hietala, M., Rollo, P., Kekäläinen, K., and Oksman, K. (2014). Extrusion processing of green biocomposites: compounding, fibrillation efficiency, and fiber dispersion. *J. Appl. Polym. Sci.* 131 (6): 1–9.

**54** Bonnaud, L., Murariu, O., De Souza Basso, N.R., and Dubois, P. (2013). High viscosity polyethylene-based electroconductive nanocomposites: carbon nanotubes versus carbon nanofibres. *Polym. Bull.* 70 (3): 895–904.

**55** Yasim-Anuar, T.A.T. (2018). *Cellulose Nanofiber from Oil Palm Mesocarp Fiber and Their Utilization as Reinforcement Material in Low Density Polyethylene Composites*. Universiti Putra Malaysia.

**56** Nishida, H., Yamashiro, K., and Tsukegi, T. Biomass composites from bamboo-based micro/nanofibers. In: , vol. 7, 339–362.

**57** Hietala, M., Mathew, A.P., and Oksman, K. (2013). Bionanocomposites of thermoplastic starch and cellulose nanofibers manufactured using twin-screw extrusion. *Eur. Polym. J.* 49 (4): 950–956.

**58** Khanam, P.N. and AlMaadeed, M.A.A. (2015). Processing and characterization of polyethylene-based composites. *Adv. Manuf. Polym. Compos. Sci.* 1 (2): 63–79.

**59** Yasim-Anuar, T.A.T., Ariffin, H., Norrrahim, M.N.F. et al. (2019). Sustainable one-pot process for the production of cellulose nanofiber and polyethylene/cellulose nanofiber composites. *J. Cleaner Prod.* 207: 590–599.

**60** Alves, T.S., Silva Neto, J.E., Silva, S.M.L. et al. (2016). Process simulation of laboratory internal mixers. *Polym. Test.* 50: 94–100.

**61** Salahudeen, S.A., Elleithy, R.H., AlOthmanb, O., and AlZahrani, S.M. (2011). Comparative study of internal batch mixer such as cam, banbury and roller: numerical simulation and experimental verification. *Chem. Eng. Sci.* 66 (12): 2502–2511.

**62** Ariffin, H. et al. (2017). Oil palm biomass cellulose-fabricated polylactic acid composites for packaging applications. In: *Bionanocomposites for Packaging Applications* (eds. M. Jawaid and S.K. Swain), 95–105.

**63** Benhamou, K., Kaddami, H., Magnin, A. et al. (2015). Bio-based polyurethane reinforced with cellulose nanofibers: a comprehensive investigation on the effect of interface. *Carbohydr. Polym.* 122: 202–211.

**64** Ashik, K.P. and Sharma, R.S. (2015). A review on mechanical properties of natural fiber reinforced hybrid polymer composites. *J. Miner. Mater. Charact. Eng.* 3: 420–426.

**65** Mittal, G., Rhee, K.Y., Mišković-Stanković, V., and Hui, D. (2018). Reinforcements in multi-scale polymer composites: processing, properties, and applications. *Composites Part B* 138: 122–139.

**66** Efendy, M.G.A. and Pickering, K.L. (2016). Fibre orientation of novel dynamically sheet formed discontinuous natural fibre PLA composites. *Composites Part A* 90: 82–89.

**67** Hirokazu Kimura, A.S. (2015). Fiber orientation and aspect ratio calculation using image analysis. *J. Textile Sci. Eng.* 05 (01): 1–3.

**68** Tian, W., Qi, L., Zhou, J., and Guan, J. (2014). Effects of the fiber orientation and fiber aspect ratio on the tensile strength of Csf/Mg composites. *Comput. Mater. Sci.* 89: 6–11.

**69** Allemang, R., De Clerck, J., Niezrecki, C., and Wicks, A. (2014). Preface. In: *Conference Proceedings of the Society for Experimental Mechanics Series*, vol. 45, no. 7 (ed. T. Proulx), 577–617. Bethel, CT, USA: Springer.

**70** Norrrahim, M.N.F., Ariffin, H., Yasim-Anuar, T.A.T. et al. (2018). One-pot nanofibrillation of cellulose and nanocomposite production in a twin-screw extruder. *IOP Conf. Ser.: Mater. Sci. Eng.* 368 (1): 1–9.

**71** Saurabh, C.K. et al. (2016). Isolation and characterization of cellulose nanofibers from gigantochloa scortechinii as a reinforcement material. *J. Nanomater.* 2016: 1–8.

**72** M. Herrera Rodriguez (2015). Preparation and characterization of nanocellulose films and coatings from industrial bio-residues (Doctoral dissertation, Luleå tekniska universitet).

**73** Lani, N.S., Ngadi, N., Johari, A., and Jusoh, M. (2014). Isolation, characterization, and application of nanocellulose from oil palm empty fruit bunch fiber as nanocomposites. *J. Nanomater.*: 1–9.

**74** Siqueira, G., Bras, J., Follain, N. et al. (2013). Thermal and mechanical properties of bio-nanocomposites reinforced by Luffa cylindrica cellulose nanocrystals. *Carbohydr. Polym.* 91 (2): 711–717.

**75** Habibi, Y., Goffin, A.-L., Schiltz, N. et al. (2008). Bionanocomposites based on poly(ε-caprolactone)-grafted cellulose nanocrystals by ring-opening polymerization. *J. Mater. Chem.* 18 (41): 5002.

**76** Malainine, M.E., Mahrouz, M., and Dufresne, A. (2005). Thermoplastic nanocomposites based on cellulose microfibrils from Opuntia ficus-indica parenchyma cell. *Compos. Sci. Technol.* 65 (10): 1520–1526.

**77** Ilyas, R.A., Sapuan, S.M., Ishak, M.R., and Zainudin, E.S. (2017). Effect of delignification on the physical, thermal, chemical, and structural properties of sugar palm fibre. *BioResources* 12 (4): 8734–8754.

**78** Zhuang, W.Z. and Halford, G.R. (2001). Investigation of residual stress relaxation under cyclic load. *Int. J. Fatigue* 23: 31–37.

**79** Ferry, J.D. and Rice, S.A. (1962). Viscoelastic properties of polymers. *Phys. Today* 15 (1): 76–78.

**80** Ferry, J.D. (1991). Some reflections on the early development of polymer dynamics: viscoelasticity, dielectric dispersion, and self-diffusion. *Macromolecules* 24 (19): 5237–5245.

**81** Meyers, M. and Chawla, K. (2017). *Mechanical Behavior of Materials*. Michigan: Prentice Hall.

**82** Faydi, Y., Lachat, R., Lesage, P., and Meyer, Y. (2015). Experimental characterization method of the gas diffusion layers compression modulus for high compressive loads and based on a dynamic mechanical analysis. *J. Fuel Cell Sci. Technol.* 12 (5): 1–6.

**83** Thomas, S., Pothan, L.A., and Cherian, B.M. (2009). Advances in natural fibre reinforced polymer composites: macro to nanoscales. *Int. J. Mater. Prod. Technol.* 36 (1/2/3/4): 317.

**84** Rashid, B., Leman, Z., Jawaid, M. et al. (2017). Dynamic mechanical analysis of treated and untreated. *Bioresources* 12: 3448–3462.

**85** Das, M. and Chakraborty, D. (2009). Processing of the uni-directional powdered phenolic resin--bamboo fiber composites and resulting dynamic mechanical properties. *J. Reinf. Plast. Compos.* 28 (11): 1339–1348.

**86** Anuar, H. and Zuraida, A. (2011). Thermal properties of injection moulded polylactic acid – kenaf fibre biocomposite. *Malaysian Polym. J.* 6 (1): 51–57.

**87** Shinoj, S., Visvanathan, R., Panigrahi, S., and Varadharaju, N. (2011). Dynamic mechanical properties of oil palm fibre (OPF)-linear low density polyethylene (LLDPE) biocomposites and study of fibre-matrix interactions. *Biosystems Eng.* 109 (2): 99–107.

**88** Lee, M.W., Han, S.O., and Seo, Y.B. (2008). Red algae fibre/poly(butylene succinate) biocomposites: the effect of fibre content on their mechanical and thermal properties. *Compos. Sci. Technol.* 68 (6): 1266–1272.

**89** Saha, A.K., Das, S., Bhatta, D., and Mitra, B.C. (1999). Study of jute fiber reinforced polyester composites by dynamic mechanical analysis. *J. Appl. Polym. Sci.* 71 (9): 1505–1513.

**90** Mohanty, S., Verma, S.K., and Nayak, S.K. (2006). Dynamic mechanical and thermal properties of MAPE treated jute/HDPE composites. *Compos. Sci. Technol.* 66 (3–4): 538–547.

**91** Melo, J.D.D., Carvalho, L.F.M., Medeiros, A.M. et al. (2012). A biodegradable composite material based on polyhydroxybutyrate (PHB) and carnauba fibers. *Composites Part B* 43 (7): 2827–2835.

**92** Sathishkumar, T.P., Navaneethakrishnan, P., Shankar, S. et al. (2013). Characterization of natural fiber and composites - a review. *J. Reinf. Plast. Compos.* 32 (19): 1457–1476.

**93** Jawaid, M. and Abdul Khalil, H.P.S. (2011). Cellulosic/synthetic fiber reinforced polymer hybrid composites. *Carbohydr. Polym.* 86: 1–18.

**94** Muthuvel, M., Ranganath, G., Janarthanan, K., and Sarinivasan, K. (2013). Characterization study of jute and glass fiber reinforced hybrid composite material. *Int. J. Eng. Sci. Technol.* 2 (4): 335–344.

**95** Jacob, M., Francis, B., and Thomas, S. (2006). Dynamical mechanical analysis of sisal/oil palm hybrid fiber-reinforced natural rubber composites. *Polym. Compos.* 27 (6): 671–680.

**96** Siakeng, R., Jawaid, M., Ariffin, H., and Sapuan, S.M. (2011). Mechanical, dynamic, and thermomechanical properties of coir/pineapple leaf fiber reinforced polylactic acid hybrid biocomposites. *Polym. Compos.* 40 (5): 1, 671–12, 2000.

**97** Asim, M. et al. (2018). Thermal, physical properties and flammability of silane treated kenaf/pineapple leaf fibres phenolic hybrid composites. *Compos. Struct.* 202: 1330–1338.

**98** Shanmugam, D. and Thiruchitrambalam, M. (2013). Static and dynamic mechanical properties of alkali treated unidirectional continuous Palmyra Palm Leaf Stalk Fiber/jute fiber reinforced hybrid polyester composites. *J. Mater. Des.* 50: 533–542.

**99** Khalid, M., Ratnam, C.T., Luqman, C.A. et al. (2009). Thermal and dynamic mechanical behavior of cellulose-and Oil Palm Empty Fruit Bunch (OPEFB)-filled polypropylene biocomposites. *Polym. Plast. Technol. Eng.* 48 (12): 1244–1251.

**100** Pande, V. and Sanklecha, V. (2017). Bionanocomposite: a review. *Austin J. Nanomed. Nanotechnol.* 5: 1–3.

**101** Siqueira, G., Bras, J., and Dufresne, A. (2010). Cellulosic bionanocomposites: a review of preparation, properties and applications. *Polymers* 2 (4): 728–765.

**102** Azammi, A.M.N. et al. (2019). Characterization studies of biopolymeric matrix and cellulose fibres based composites related to functionalized fibre-matrix interface. In: *Interfaces in Particle and Fibre Reinforced Composites- From Macro to Nano Scales*, 1e (eds. K.L. Goh, S. Thomas, R.T. De Silva and M.K. Aswathi), 1–68. London: Woodhead Publishing.

**103** Kowalczyk, M., Piorkowska, E., Kulpinski, P., and Pracella, M. (2011). Composites: Part A Mechanical and thermal properties of PLA composites with cellulose nanofibers and standard size fibers. *Composites Part A* 42 (10): 1509–1514.

**104** Kumar, P., Sandeep, K.P., Alavi, S. et al. (2010). Preparation and characterization of bio-nanocomposite films based on soy protein isolate and montmorillonite using melt extrusion q. *J. Food Eng.* 100 (3): 480–489.

**105** Alemdar, A. and Sain, M. (2008). Biocomposites from wheat straw nanofibers: morphology, thermal and mechanical properties. *Compos. Sci. Technol.* 68 (2): 557–565.

**106** Jonoobi, M., Harun, J., Mathew, A.P., and Oksman, K. (2010). Mechanical properties of cellulose nanofiber (CNF) reinforced polylactic acid (PLA) prepared by twin screw extrusion. *Compos. Sci. Technol.* 70 (12): 1742–1747.

**107** Avella, M., Bogoeva-Gaceva, G., Bužarovska, A. et al. (2008). Poly(lactic acid)-based biocomposites reinforced with kenaf fibers. *J. Appl. Polym. Sci.* 108: 3542–3551. https://doi.org/10.1002/app.28004.

**108** Lee, H.S., Cho, D., and Han, S.O. (2008). Effect of natural fiber surface treatments on the interfacial and mechanical properties of henequen/polypropylene biocomposites. *Macromol. Res.* 16: 411–417. https://doi.org/10.1007/BF03218538.

# 4

# Physical and Mechanical Properties of Biocomposites Based on Lignocellulosic Fibers

*Nadir Ayrilmis[1], Sarawut Rimdusit[2], Rajini Nagarajan[3], and M. P. Indira Devi[4]*

[1] *Istanbul University-Cerrahpasa, Department of Wood Mechanics and Technology, Faculty of Forestry, Bahcekoy, Sariyer, Istanbul, 34473, Turkey*
[2] *Chulalongkorn University, Department of Chemical Engineering, Research Unit on Polymeric Materials for Medical Practice Devices, Faculty of Engineering, Pathumwan, Bangkok, 10330, Thailand*
[3] *Kalasalingam University, Department of Mechanical Engineering, Anand Nagar, Krishnankoil, Tamil Nadu 626126, India*
[4] *Kalasalingam College of Arts and Science, Department of Physics, Anand Nagar, Krishnankoil, Tamil Nadu 626126, India*

## 4.1 Introduction

A biocomposite is defined as a material formed by a polymer matrix and a reinforcing natural fibers derived from wood or nonwood plants. Some property enhancers, such as compatibilizers, biocides, inorganic fillers, fire retardants, UV stabilizers, colorants, and process aid additives such as lubricants and antioxidants, can be also used in the manufacture of biocomposites [1]. According to their sources, natural fibers can be classified as plant fibers, animal fibers, and inorganic (mineral) fibers (Figure 4.1). Other main groups of artificial fibers are natural organic polymers, inorganic materials, and artificial polymers. human-made fibers are subdivided into two subgroups: artificial fibers and semi-artificial fibers (regenerated fibers). Natural fibers are obtained directly from nature. There is no human labor in the synthesis of the chemical substance that forms their structure or in the formation of this substance into fiber shape. They are used for direct weaving or for other purposes after the pre-cleaning and preparation processes. Before man-made fibers were produced, people have been able to meet their need for fibers for many years from natural fibers. Man-made fibers – synthetic fibers – have advantages such as being economical, being produced for specific purposes, product diversity, good control of fiber properties, and product variety, compared to natural fibers although some of them may not be environmentally friendly materials. Today, man-made fibers have been replaced by natural fibers in many areas. Natural fibers are classified based on their origins such as lignocellulosics, animals, and minerals (Figure 4.1) [2].

The examples of flax, hemp, and wood fibers are given in Figure 4.2.

As compared to the wood fibers, the cellulose content of bast fibers is considerably higher than that of wood fibers (Figure 4.3). In addition, natural fibers are quite long

*Mechanical and Dynamic Properties of Biocomposites,* First Edition. Edited by Senthilkumar Krishnasamy, Rajini Nagarajan, Senthil Muthu Kumar Thiagamani, and Suchart Siengchin.

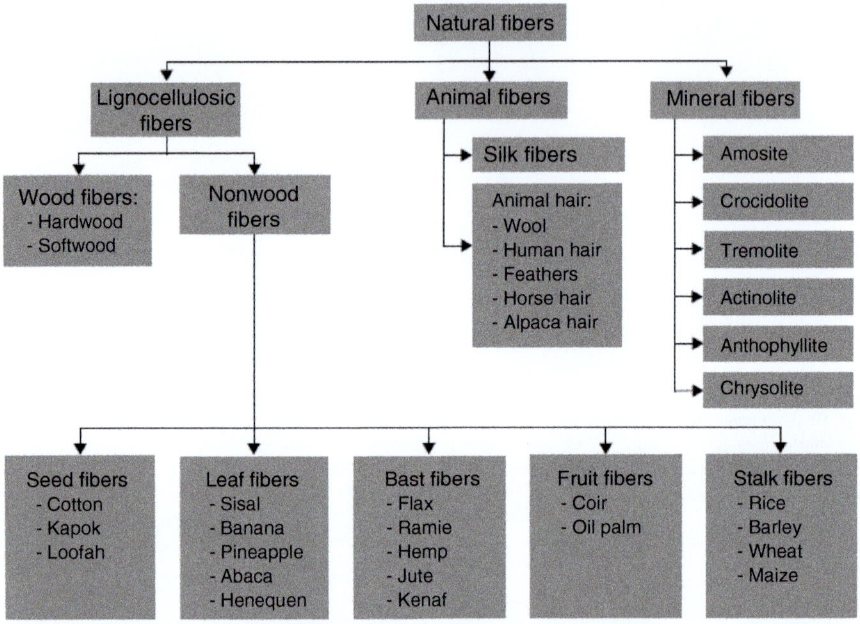


**Figure 4.1** Classification of natural fibers. Source: Campilho [2].

**Figure 4.2** (a) Flax sliver, (b) hemp sliver, (c) flax hackling tow, and (d) thermomechanical pulp (TMP) fiber. Source: Ansell [3].

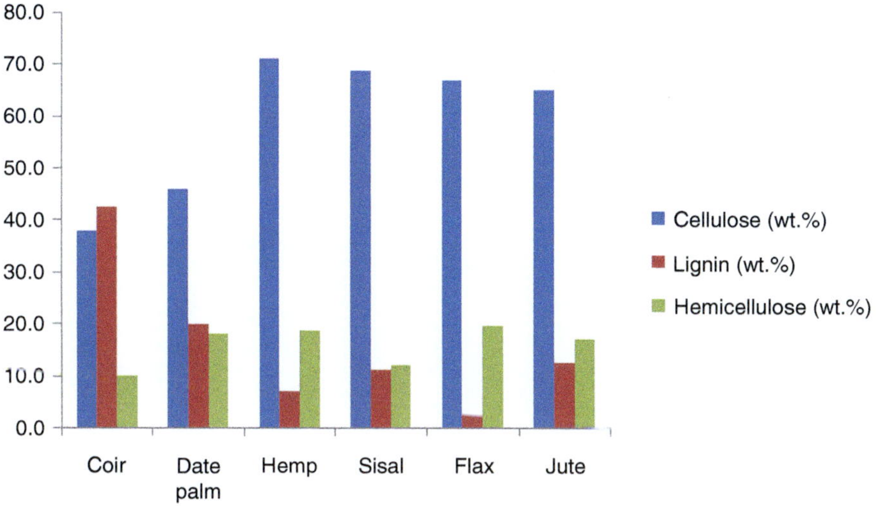


**Figure 4.3** Comparison of chemical composition of natural fibers. Source: AL-Oqla et al. [4].

compared to wood fibers. For this reason, generally, the tensile strength and modulus of thermoplastic composites produced from bast fibers are higher than that of wood fibers.

The sale prices of natural fibers are lower than those of many types of reinforcing synthetic fibers and even synthetic glass fibers. In particular, the low amount of energy required for the production of natural fibers is of great importance in terms of both cost and environment. For example, while the production of 1 kg of kenaf fiber consumes approximately 15 MJ of energy, the same quantity of glass fiber requires 54 MJ of energy [5]. Plant fibers can be directly used in the polymer composites after cleaning, washing, chopping, and drying. Plant fibers also offer significant benefits for workers and consumers. Glass fibers cause skin irritation and respiratory disorders while plant fibers do not show these problems, i.e. many plant fibers can be used without causing health problems [6]. Specific modulus of commonly used plant fibers and E-glass synthetic fibers versus their cost is given in Figure 4.4.

Physical and mechanical properties of some commercial plant fibers are given in Table 4.1.

The advantages of plant fibers used in the production of biocomposites are as follows [8–17]:

1. Environmentally friendly raw material.
2. Lower hazard manufacturing process as compared to glass fiber production. Unlike petroleum-derived polymers when burned, they have less environmental impact.
3. Compared to synthetic fibers and polymer matrices, it is relatively inexpensive, thus reducing the cost of the biocomposite and the price of the final product.
4. Renewable resource and abundant in nature.

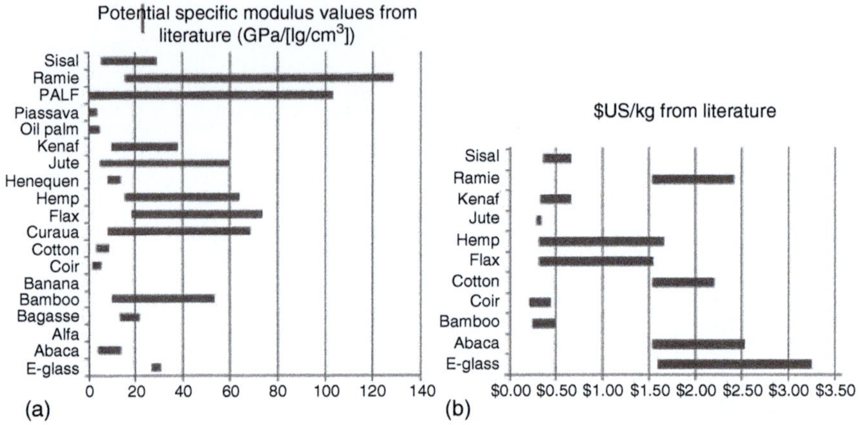


**Figure 4.4** (a) Specific modulus of commonly used plant fibers and E-glass fibers. (b) The cost of some natural fibers and E-glass E-Glass: Synthetic fiber (glass fibers, type E). Source: Campilho [2].

**Table 4.1** Physical and mechanical properties of some commercial natural fibers.

| Fiber | Density (g/cm$^3$) | Diameter (μm) | Length (mm) | Tensile strength (MPa) | Young's modulus (GPa) | Elongation at break (%) | Moisture content (%) |
|---|---|---|---|---|---|---|---|
| Abaca | 1.5 | 10–30 | 4.6–5.2 | 430–813 | 31.1–33.6 | 2.9 | 14 |
| Bamboo | 0.6–1.1 | 25–88 | 1.5–4 | 270–862 | 17–89 | 1.3–8 | 11–17 |
| Banana | 1.35 | 12–30 | 0.4–0.9 | 529–914 | 27–32 | 5–6 | 10–11 |
| Coir | 1.20 | 7–30 | 0.3–3 | 175 | 6 | 15–25 | 10 |
| Cotton | 1.21 | 12–35 | 15–56 | 287–597 | 6–10 | 2–10 | 33–34 |
| Flax | 1.38 | 5–38 | 10–65 | 343–1035 | 50–70 | 1.2–3 | 7 |
| Hemp | 1.47 | 10–51 | 5–55 | 580–1110 | 30–60 | 1.6–4.5 | 8 |
| Jute | 1.23 | 5–25 | 0.8–6 | 187–773 | 20–55 | 1.5–3.1 | 12 |
| Kenaf | 1.20 | 12–36 | 1.4–11 | 295–930 | 22–60 | 2.7–6.9 | 6.2–12 |
| Pineapple | 1.50 | 8–41 | 3–8 | 170–1627 | 60–82 | 1–3 | 14 |
| Ramie | 1.44 | 18–80 | 40–250 | 400–938 | 61.4–128 | 2–4 | 12–17 |
| Sisal | 1.20 | 7–47 | 0.8–8 | 507–855 | 9–22 | 1.9–3 | 11 |

Source: Peças et al. [7]. Licensed under CC BY 4.0.

5. Good processing and less abrasive damage to processing equipment such as extruder or injection molding machines compared to inorganic fillers; does not lead to skin irritation.
6. Outstanding mechanical properties (especially tensile strength and modulus) relative to their density.
7. Unlike with synthetic fibers during production, the negative impact on humans is much less.

8. Good thermal and acoustic insulation (good sound and heat insulation).
9. Recycling and problem-free disposal, low-emission of toxic fumes when fired or exposed to high temperatures.
10. Low density, high modulus, high specific strength, and stiffness.
11. Better energy retrieval.
12. High electrical resistance.

The disadvantages of plant fibers are as follows:

1. Lower strength properties, particularly lower impact strength than that of synthetic fibers.
2. Poor compatibility with hydrophobic polymers.
3. Varying plant quality geographic conditions such as weather, plant growing, and soil properties.
4. Inferior dimensional stability and moisture absorption caused by hydroxyl groups. They negatively affect the mechanical properties and interfacial bonding between the polymer matrix and the fiber.
5. Limited maximum processing temperature. Higher processing temperatures above 200–220 °C cause thermal degradation, limiting the polymer matrix options.
6. Lower biological durability; modification of fibers increases the durability.
7. Low fire resistance.
8. The production cost fluctuates due to agricultural politics and harvesting conditions.

Softwood and hardwood fibers have different structural characteristics. For example, softwood has lower density, longer fibers, low acidity, low extractives, higher lignin content, as well as lower hemicellulose content. The processing of the softwood fibers in the extruder is generally better than that of hardwoods due to lower abrasion effect for extruder screw and faster production due to higher lignin content. In particular, pinewood is one of the most used softwood fibers for wood plastic composite production. The characteristics of natural fibers significantly affect the quality of the final product [18–22] (Figure 4.5). Typical raw materials and compounds are presented in Figure 4.5.

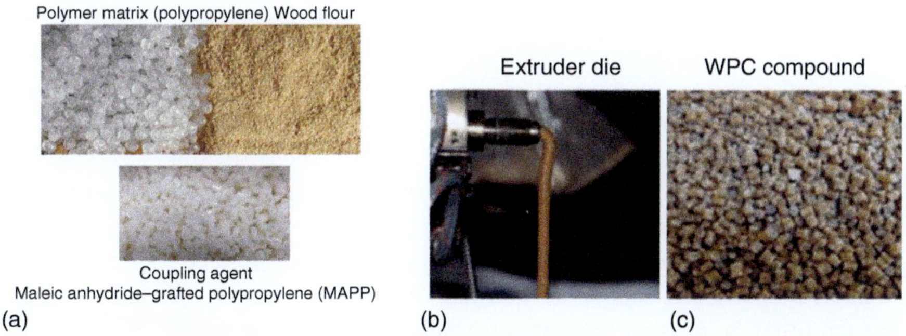


**Figure 4.5** (a) Raw materials for WPC. (b) WPC compounding in the twin screw extruder. (c) WPC compound.

## 4.2 Major Factors Influencing Quality of Biocomposites

The main factors influencing the quality of biocomposites are as follows [13]:

- Fiber type and its properties
- Polymer matrix selection
- Polymer/natural fiber compatibility and interfacial bond strength
- Fiber dispersion
- Fiber orientation
- Manufacturing process of biocomposite
- Presence of porosity and microvoids

### 4.2.1 Selection of Natural Fibers

The mechanical properties of the natural fibers depend mainly on their chemical composition and structure. The structure and chemical properties of the plant fibers are influenced by many factors such as the source of fibers, soil properties, extraction method, maturity, growing conditions, and harvesting time. Thus, these parameters greatly affect the physicomechanical properties of the plant fibers.

The fibers are obtained from different locations of plants such as seed, leaf, fruit, bast, and stalk. Bast fibers have higher tensile modulus and strength than the fibers obtained from other locations of plant fibers mentioned above. The bast fibers most used in the polymer composite industry are flax, hemp, ramie, jute, and kenaf [2]. Hybrid fibers/thermoplastic composites can be produced with a combination of two or more different plant fibers. The hybrid use of plant fibers aims to improve the physical, mechanical, and thermal properties of biocomposites. The hygroscopicity of natural fibers also affects the physical and mechanical properties of wood plastic composites (WPCs). The natural fibers must be dried in a dryer to 0–1% moisture content before blending with the polymer matrix in the extruder. Similarly, the compounds must be dried before injection or hot-press molding to prevent microgaps, voids, and gaps.

The thermal stability of hemicelluloses and lignin is lower than that of cellulose. Physical or chemical modification of natural fibers is one way of reducing the percentage content of hemicelluloses and lignin or even eliminating them completely. In particular, alkali treatment removes hemicelluloses, cleans the impurities on the surface, and increases the crystallinity of cellulose, which improves the tensile strength of the fiber. Edeerozey et al. [23] modified kenaf fibers with different concentrations of NaOH (3%, 6%, and 9% NaOH) and determined the mechanical properties of the fibers. They reported that 6% and 9% NaOH treatment gave better results. The concentration of NaOH and the treatment time and temperature are important factors for this treatment.

The thermal stability of natural fiber is one important reason for choosing to produce natural fiber composites. The thermal stabilities of natural fibers may differ from each other. In general, thermal degradation of the most natural fibers starts above 200 °C. Thus, the high temperature limits the use of natural fibers in polymer composites [24].

Wood is still the most used raw material in the production of the filament. The optimum properties of wood material are given below:

1. Wood should not be decayed by fungi.
2. Moisture content of wood during production should be 1% or lower.
3. The extractive content in wood should be low.
4. Wood having a density between 0.35 and 0.65 g/cm$^3$ is suitable.
5. Coniferous tree species are preferred because of the faster travel in the extruder and faster composite profile drawing.
6. Wood without bark should be used.
7. The particle size of the wood should be appropriate.

In WPC production, wood flour is used as a by-product (planar and shavings) in the manufacturing of timber and workshop, furniture, doors, and windows (Figure 4.6). Even paper residues and sanding trimming waste of medium density fibreboard (MDF) and particle board can be used in the production of WPC. Production residues from the manufacture of WPC as well as WPC at the end of its useful life can be recycled again, for which the products are first ground, subsequently melted again, and finally processed into new products. Furthermore, agricultural wastes such as rice husk, wheat straw, and sugar bagasse are also used in the production of thermoplastic composites. The properties of WPCs made with wood fibers, up to a certain fiber content, are considerably better than the WPCs with wood flour because

**Figure 4.6** Wood flour used in WPC production (for injection molding, wood flour size should be above 80 mesh. As for the extrusion process, generally 40 mesh size is suitable).

**Figure 4.7** Wood fibers used in WPC production.

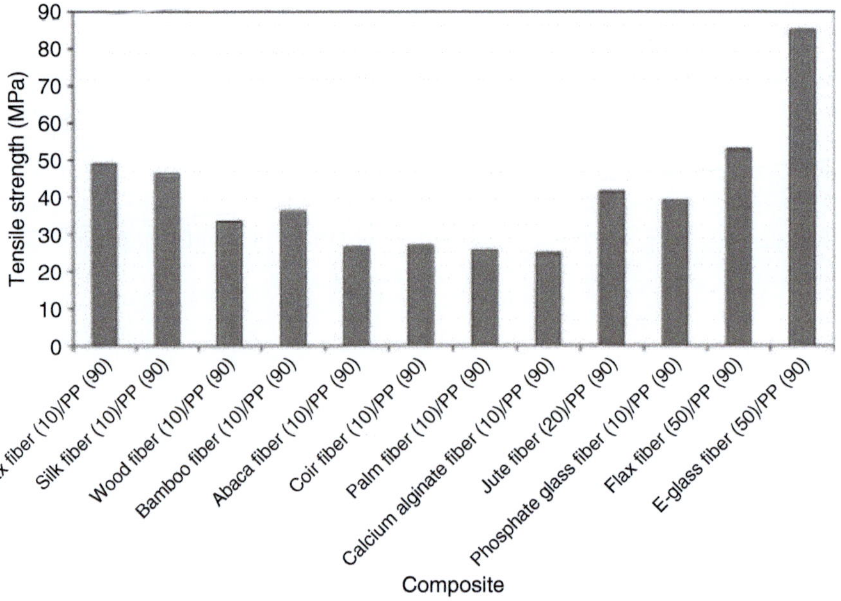


**Figure 4.8** Tensile strength of different fiber-reinforced polypropylene composites. Source: Shubhra et al. [21].

of the aspect ratio (Figure 4.7). However, higher amount of fibers and increasing fiber length have a problem that can increase their tendency to agglomerate, hence resulting in nonhomogeneous dispersion of fibers in the matrix [13].

The tensile strength of biocomposites produced with wood and various plant fibers determined by Shubhra et al. [21] is given in Figure 4.8. Among the natural fibers, the highest tensile strength was found in flax fiber/polymer composites, followed by silk fiber, jute-fiber-reinforced polymer composites, and so on. The tensile strength of the E-glass fiber-reinforced polymer composites was significantly higher than that of the natural fiber polymer composites, but the density of the composites was higher and it is not an environmentally friendly composite.

### 4.2.2 Effect of Fiber/Particle Size on the Physical and Mechanical Properties of Biocomposites

Physical and mechanical properties of biocomposites are considerably affected by the aspect ratio, length, and diameter of the fibers used, their distribution in the matrix, the volumetric percentage of the fibers, and their orientation within the matrix [24, 25]. As the external load is applied to the biocomposite, stress is transferred from fiber to fiber by the polymer matrix. Thus, aspect ratio, the fiber orientation, and distribution of fibers affect the degree of load transferred from the polymer matrix to the natural fibers. Randomly aligned fibers in the polymer matrix tend to transfer external loads better. Therefore, hand-laid natural fiber composites have better mechanical performance [24–26]. The best tensile strength is obtained as the natural fiber is aligned parallel to the direction of the load applied. Some alignment of the short fibers can be carried out during injection molding and the alignment can be partly controlled by polymer matrix viscosity and mold design.

The fibers are positioned in the polymer matrix in three different ways: (i) aligned-continuous, (ii) aligned-discontinuous, (iii) randomly oriented fiber-reinforced composites (Figure 4.9). The longitudinally aligned natural fibers in the biocomposites have lower compression strength because of the fact that fibers buckle while the transversely aligned natural fibers in the biocomposites have lower tensile strength. However, the technological properties of the randomly oriented biocomposites are not uniform as in aligned fibers because it is not easy to distribute the fibers uniformly in the polymer matrix [24]. Consequently, the parameters mentioned

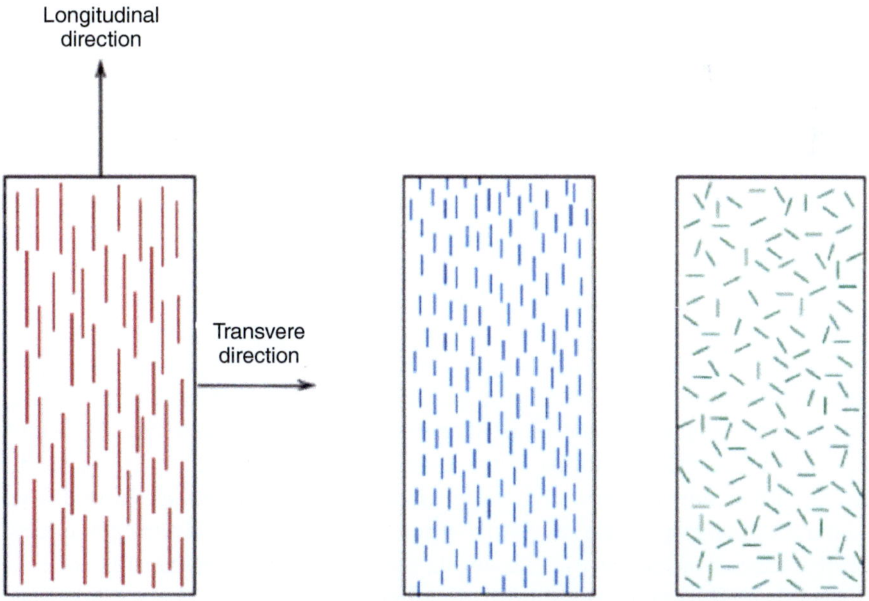


**Figure 4.9** Orientation of fibers in the composites of (a) aligned-continuous, (b) aligned-discontinuous, and (c) polymer composites reinforced with randomly oriented natural fibers.

**Table 4.2** Technological properties of polypropylene/wood composites depending on the particle size and coupling agent.

| Particle size | | PP-g-MA (%) | Aspect ratio[a)] (l/d) | Tensile modulus[b)] (GPa) | Tensile strength (MPa) | Elongation at break (%) | Notched izod (J/m) |
|---|---|---|---|---|---|---|---|
| Mesh | mm | | | | | | |
| 40 | 0.42 | 0 | 3.5 ± 0.3 | 1.3 ± 0.2 | 18.1 ± 1.4 | 4.15 ± 0.6 | 11.7 ± 1.1 |
| 40 | 0.42 | 2 | 3.5 ± 0.3 | 1.4 ± 0.3 | 21.3 ± 1.2 | 4.33 ± 0.5 | 15.3 ± 0.8 |
| 50 | 0.30 | 0 | 4.2 ± 0.3 | 1.4 ± 0.3 | 20.6 ± 2.1 | 4.17 ± 1.2 | 10.4 ± 1.3 |
| 50 | 0.30 | 2 | 4.2 ± 0.3 | 1.6 ± 0.4 | 25.2 ± 2.4 | 4.23 ± 0.7 | 13.5 ± 2.1 |
| 60 | 0.25 | 0 | 4.4 ± 0.3 | 1.4 ± 0.2 | 22.8 ± 1.7 | 4.12 ± 0.5 | 9.7 ± 1.5 |
| 60 | 0.25 | 2 | 4.4 ± 0.3 | 1.6 ± 0.1 | 27.1 ± 2.3 | 4.32 ± 1.4 | 12.1 ± 1.7 |

a) The ratio of fiber length to its diameter.
b) All values are reported as mean (N≥6).
Source: Nourbakhsh et al. [27]. © 2010, SAGE Publications.

should be monitored during the preparation of raw materials and production of the biocomposites to obtain good physical and mechanical properties.

Nourbakhsh et al. [27] reported that the best technological properties of wood flour/polypropylene (PP) composites was found in the specimens having smaller wood particle with an average diameter of 0.25 mm (Table 4.2). Furthermore, the tensile modulus improved with decreasing wood particle diameter, which enhanced the interfacial bonding between the wood particle and the polymer matrix. They determined that the strength properties of the wood/polypropylene moderately improved with the addition of 2 wt.% PP-g-MA. The tensile modulus of the specimens produced with smaller particles increased more as compared to the specimens with larger particles. In addition, increasing aspect ratio of the wood particles positively affected the technological properties of the composites. Their results showed that the elongation at break of the specimens was not changed by increasing the particle size. The polypropylene composites with wood particles displayed lower impact strength compared to the neat polypropylene.

The slenderness ratio of the filler has a significant effect on the tensile properties of the biocomposites, in particular at higher filler content. Increasing slenderness ratio increases the tensile strength and modulus. Especially at lower filler content, the effect of fiber length is more pronounced on the tensile properties of the composites [25]. Gu and Kokta [28] investigated the effect of wood fiber size on the mechanical properties of polypropylene composites (Figure 4.10). Their results showed that for different fiber proportions, aspect ratio had an important effect throughout the WPC manufacturing processes. The tensile modulus of the specimens decreased with fiber content above 40 wt.% due to nonuniform distribution of fibers. Fibers that are too long may cause decreases in the mechanical properties because the increase in fiber length results in agglomeration, namely nonuniform dispersion in the polymer matrix. On the other hand, better interfacial bond can be obtained by short fibers

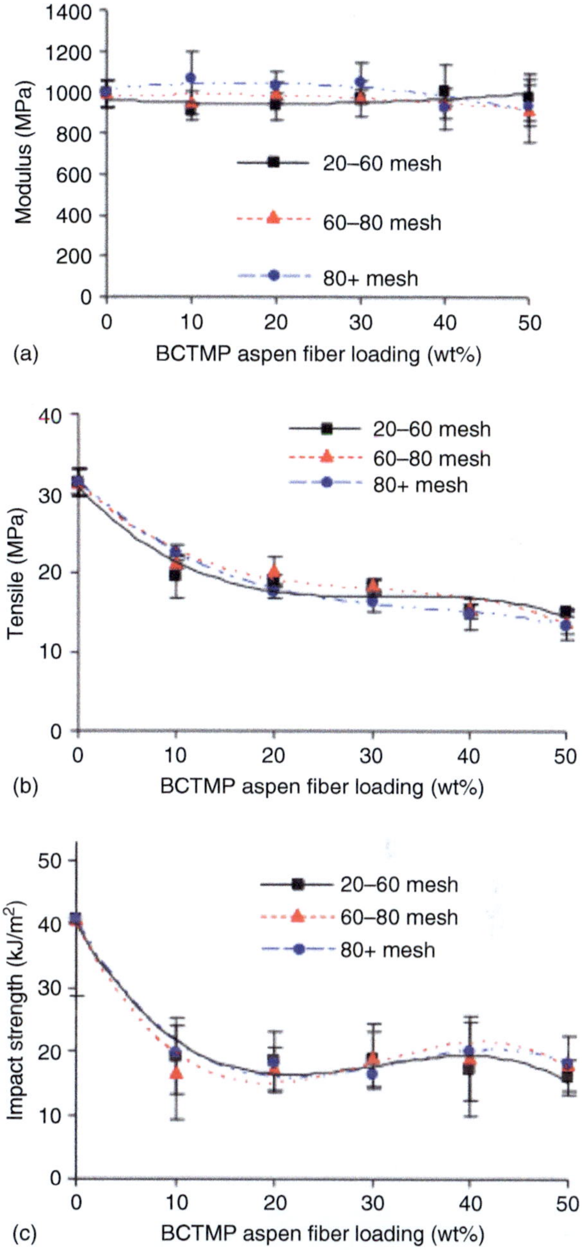


**Figure 4.10** The effect of fiber on the tensile strength and impact strength of wood-filled polypropylene composites (BCTMP, bleached chemi-thermomechanical pulp of aspen). (a) Tensile modulus. (b) Tensile strength. (c) Impact strength. Source: Gu and Kokta [28].

because the polymer matrix penetrates more into the vessels of the wood fibers and improves interfacial bonding as compared to the longer fibers. A similar result was also reported by Stark and Rowlands [25]. Gu and Kokta [28] reported that wood flour mesh size had a significant effect on the impact energy. In general, the notched impact energy increases with increasing particle size as a function of increase in the fracture of surface area. On the other hand, the increase in the particle size results in lower unnotched impact energy due to increase in the stress concentrations in the composite [25].

The effect of sisal fiber length (from 1 to 30 mm) on the hot-press-molded sisal/polypropylene composites loaded with 25 wt.% content of the fibers was determined by Jayaraman [29]. He reported that the notched impact strength of composites progressively increased with increasing fiber length. In another study, Mohanty et al. [30] investigated the effect of fiber length (3, 6, and 10 mm) on the compression molded jute/polypropylene composites with 30 wt.% fiber content and 0.5 wt.% maleic anhydride grafted polypropylene (MAPP). The polymer composites reinforced with 6 mm length fibers gave the highest notched impact strength. Bledzki et al. [31] determined the tensile properties of petroleum-based polymer PP and bio-based polymer polylactic acid (PLA) and Polyamide 6.10 with natural abaca fibers. The bio-based polymer composites showed higher tensile strength and modulus than the petroleum-based polypropylene/abaca composites. It was observed that PLA exhibited higher strength and modulus compared to polyamide/abaca composites.

### 4.2.3 Effect of Filler Content on the Mechanical Properties of Biocomposites

Generally, the bending strength of WPCs can be improved by increasing the content of wood flour up to 50 wt.% while the bending modulus can be improved by increasing the content of wood flour up to 70 wt.% when the coupling agent is used [32]. Basiji et al. [33] investigated the influence of fiber length and fiber loading on the mechanical properties of the WPCs. They reported that the flexural properties of the WPCs significantly improved with increasing fiber length and content. From a previous study, the effect of the processing method and flake content on the flexural modulus of the high density polyethylene (HDPE) composites is presented in Figure 4.11.

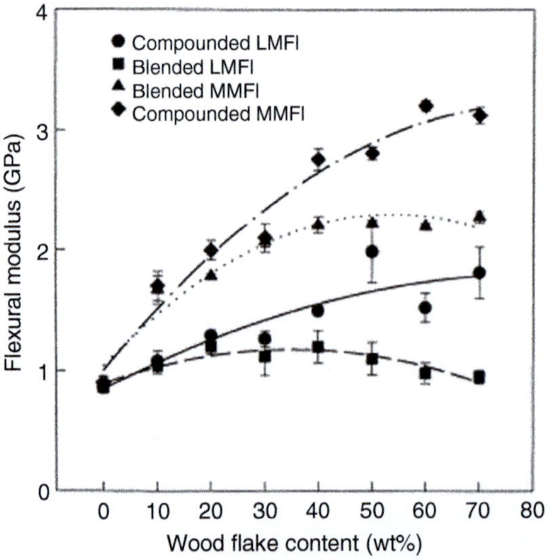


**Figure 4.11** Effect of processing method and filler loading on the flexural modulus of the HDPE composites. Source: Balasuriya and Ye Mai [34].

**Table 4.3** Mechanical properties of WPCs at 40% wood fiber or wood flour content.

| Material | Tensile properties | | | Flexural properties | | Izod impact energy | |
|---|---|---|---|---|---|---|---|
| | Strength (MPa) | Modulus (MPa) | Elongation (MPa) | Strength (MPa) | Modulus (MPa) | Notched (J/m) | Unnotched (J/m) |
| Wood flour | 25.4 | 3.87 | 1.9 | 44.2 | 3.03 | 22.2 | 73 |
| Wood flour + MAPP | 32.3 | 4.10 | 1.9 | 53.1 | 3.08 | 21.2 | 78 |
| Wood fiber | 28.2 | 4.20 | 2.0 | 47.9 | 3.25 | 23.2 | 91 |
| Wood fiber + MAPP | 52.3 | 4.23 | 3.2 | 12.5 | 3.22 | 21.6 | 162 |
| Polypropylene | 28.5 | 1.53 | 5.9 | 38.3 | 1.19 | 20.9 | 656 |

MAPP, 3% maleated polypropylene.
Source: Stark and Rowlands [25].

Stark and Rowlands [25] studied the effect of particle aspect ratio on the properties of WPCs. They produced wood/polypropylene composites using wood flour and wood fibers with and without MAPP. The mechanical properties of the WPCs produced with wood fibers were significantly higher than those of the WPCs produced with wood flour at 40 wt.% reinforcing filler content. The results clearly revealed that the slenderness ratio positively affected the mechanical properties of the composites because the wood fibers gave better stress transfer than wood flour. The incorporation of the maleated polypropylene significantly improved the interfacial bond between wood fibers and the polymer matrix (Table 4.3).

Nourbakhsh et al. [35] investigated impact bending strength of the aspen fiber/polypropylene composites with different amounts of the aspen fibers. They reported that the impact strength of the untreated aspen fiber/polypropylene composites steadily decreased when the fiber content was increased in the polymer matrix (Figure 4.12). This was due to the ductile mobility of polymer molecules and increasing rigidity of the composites as a function of increasing fiber content.

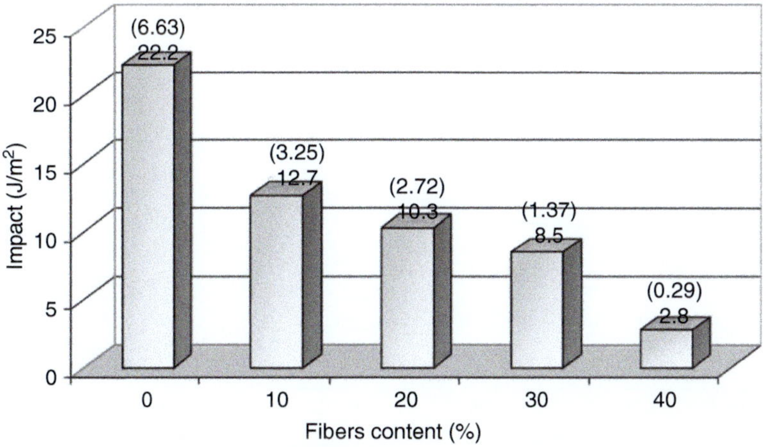


**Figure 4.12** Impact bending strength of the WPC depending on the fiber content (the values in parentheses are standard deviations). Source: Nourbakhsh et al. [35].

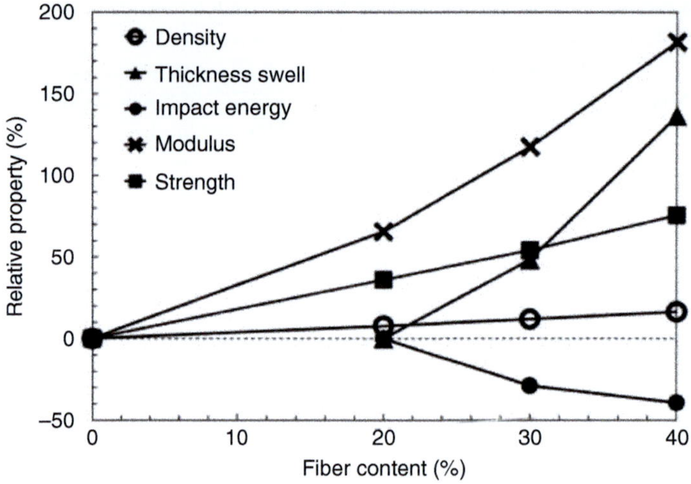


**Figure 4.13** The changes in the relative properties of WPCs at different fiber contents. Source: Migneault et al. [22].

The increasing fiber content decreases the energy absorption of the composite and makes it brittle. The neat plastic has higher energy absorption and therefore it has the highest impact bending strength (Figure 4.12). Mostly, such regions can occur at the end of fibers or in the regions having microvoids or gaps between polymer matrix and fiber due to lack of compatibility.

The variations in the technological properties of the WPCs are presented in Figure 4.13 [22]. The modulus of the WPCs is considerably increased after the 20 wt.% addition of wood flour into the WPC while thickness swelling is significantly increased as compared to the neat HDPE matrix. The impact energy is

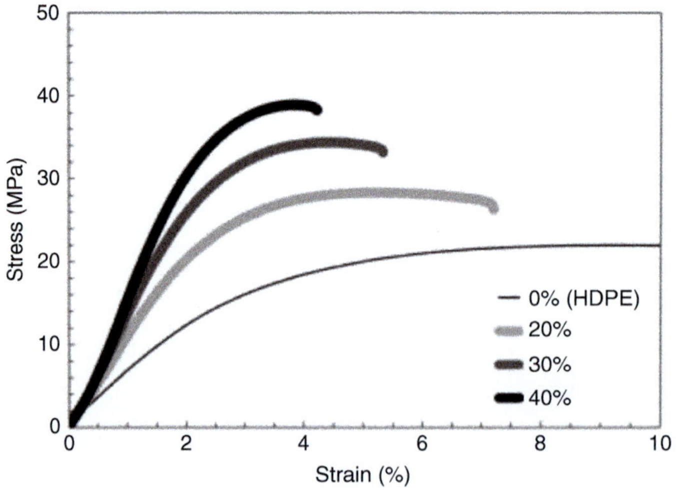


**Figure 4.14** Tensile stress versus strain curves for WPCs having different amounts of aspen wood fibers. Source: Migneault et al. [22].

decreased with increasing wood particles. Tensile stress versus strain for the WPC having different amounts of the natural fiber is illustrated in Figure 4.14 [22].

Migneault et al. [22] investigated the potential utilization of fillers derived from different sourced residues such as sawmill sawdust, bark, recycled biocomposite, paper and paper sludge, and underused wood species in the production of WPC. The WPCs produced with residues had lower mechanical and physical properties than those produced with clean wood, with some exceptions. WPCs with the kraft sludge showed the best mechanical properties in terms of tensile strength and impact bending strength. The polymer composites filled with deinking paper sludge gave the toughest property and best dimensional stability. The polymer composites with the waste of wood-based panels showed similar characteristics to the WPCs filled with clean wood. The lowest mechanical properties were found in the bark-filled polymer composites.

### 4.2.4 Compatibility Between Natural Fiber/Polymer Matrix and Surface Modification

The surface of natural fillers is polar (hydrophilic) while the surface of polymer matrix is apolar (hydrophobic). The physical and mechanical properties of the natural fiber composites are significantly affected by the interfacial bond between the polymer matrix and natural fiber [36–51]. A good interfacial bond contributes to the stress transfer from the matrix to the filler when the external load is applied. If the surface of the natural fiber is not modified with a coupling agent, the interaction between the fibers and the hydrophobic polymer matrix will be weak. For this reason, coupling agents are used to improve compatibility between the surfaces of the hydrophilic plant fiber and hydrophobic polymer surfaces [52]. There are some modification techniques that considerably improve the interfacial compatibility. These modifications are physical modification – electric discharge (corona or cold plasma); chemical modification – coupling agents (MAPP and MAPE), alkaline treatment, acetylation; thermal treatment modification; epoxide, silanes, and other chemical treatments (i.e. stearic acid $CH_{32}(CH)_{16}COOH$ in ethyl alcohol solution and permanganate treatment) [14]. Sodium hydroxide (alkali) treatment contributes to increase in the crystalline region of cellulose chain, cleaning the impurities on the surface of the fiber and removing the free hydrophilic hydroxyl groups, which enhances the properties of the biocomposites.

The water absorption and dimensional stability of biocomposites are mainly influenced by a lot of factors such as fiber size, fiber source, the chemical properties of the fiber, loading level of fibers, polymer type, production method, processing parameters, in particular temperature and pressure, moisture in the fiber, microvoids, cracks, and gaps between polymer matrix and fiber, use of coupling agent, and interfacial surface [15, 53–56].

The ambient relative humidity and temperature affect the moisture absorption of biocomposites because water molecules diffuse into the biocomposite and attach onto hydrophilic OH groups of cellulose and hemicellulose of the fiber cell wall, which weakens the interfacial bond between the natural fiber and polymer matrix.

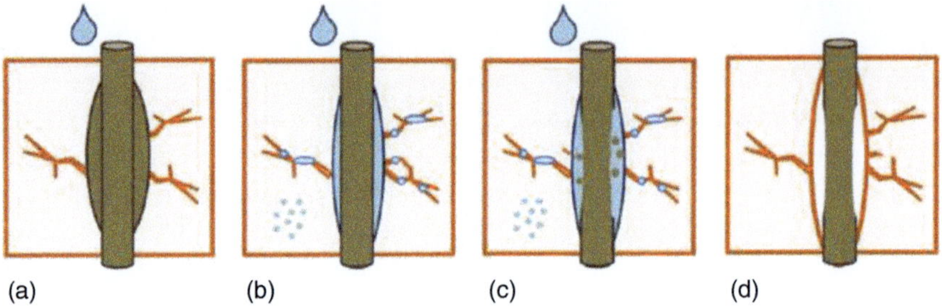


**Figure 4.15** The development of mechanical cracks caused by water or moisture penetration into the biocomposites: (a) development of microcracks due to expansion of swollen natural fiber. (b) Water molecules penetrate into microcracks between the polymer matrix and the natural fiber. (c) Water-soluble components leach from the fiber. (d) The maximum debonding occurs between the matrix and the fiber. Source: Azwa et al. [57].

In the first phase, the water absorbed into the biocomposites is classified into bound and free water. At first, the water is bound to OH groups of wood cell, and is known as bound water. After the wood cell is fully saturated with water molecules, the water moves to microcracks and voids. The dimensional stability of the biocomposites is caused by the bound water because the natural fibers swell. Hence, the swollen wood fibers create stress between the polymer matrix and fiber surface and cause debonding at the interfacial surface. The penetration mechanism of water molecules into the biocomposite is presented in Figure 4.15.

Decreasing polymer content as a function of increasing wood flour content in the biocomposites is mainly responsible for the lower water resistance of biocomposites [14, 15]. This is because of the hydrophobic character of the thermoplastic and thus chemical inactivity. When the thermoplastics melt during processing, they cover the surface of the wood fiber and penetrate the vessels of the fibers. The fibers wrapped with the polymer matrix will be less exposed to the water. Melt flow index (MFI) of the polymer matrix affects its viscosity and the increase in the MFI value decreases the viscosity of the polymer [45]. As the amount of the filler is increased in the WPC, the complete wrapping of the filler will decrease due to decreasing amounts of the plastic content. Typical mass uptake and thickness swelling of the WPC depending on filler content and the immersion time are presented in Figure 4.16.

Polypropylene and polyethylene are the most used polyolefin family thermoplastics. Maleic anhydride-grafted-polyethylene (MAPE) and MAPP coupling agents are commonly used in biocomposites. These coupling agents are chemically bonded to free hydroxyl groups of the natural fiber and blended by wetting in the polymer chain (Figure 4.17).

The coupling agents enhance the compatibility between the natural fiber and the polymer matrix by ester linkages. The MFI values of the coupling agents are quite high because they melt fast and spread on the natural fibers and the polymer matrix. Since the coupling agent improves the interfacial bond between the natural fiber and the polymer matrix by ester bonds, it decreases the possibility of microcracks mainly caused by moisture or water molecules. As a result, it increases the mechanical properties and dimensional stability of the biocomposites considerably. Usually,

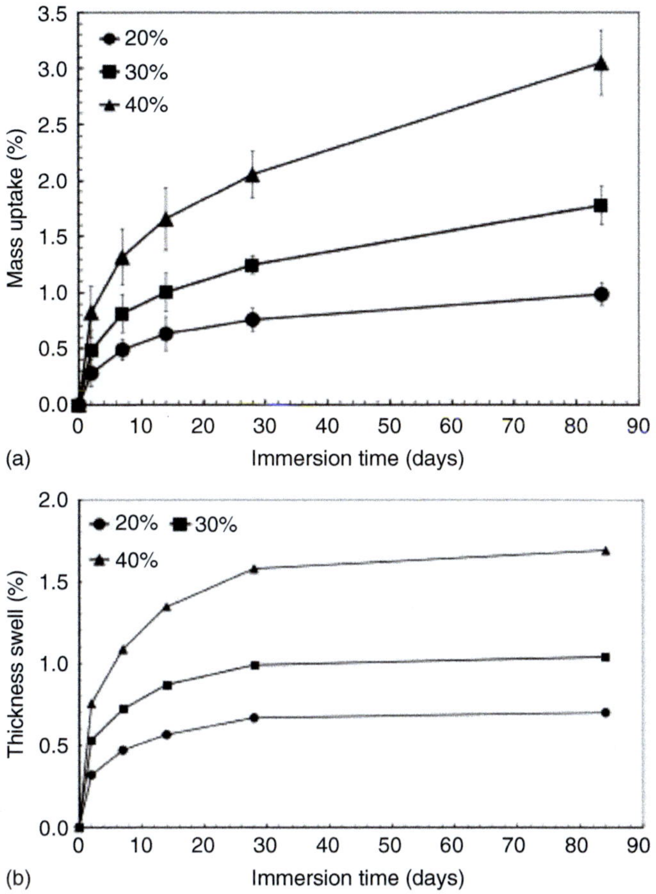


**Figure 4.16** Water absorption (a) and thickness swelling (b) for the water-immersed WPCs made with three contents of the fibers. Source: Migneault et al. [22].

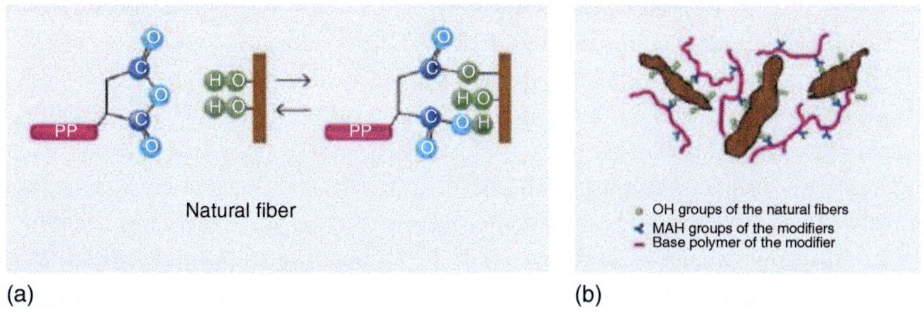


**Figure 4.17** (a) Reaction of coupling agent with natural fibres (PP: polypropylene). (b) Incorporation of the natural fibres into the polymer matrix.

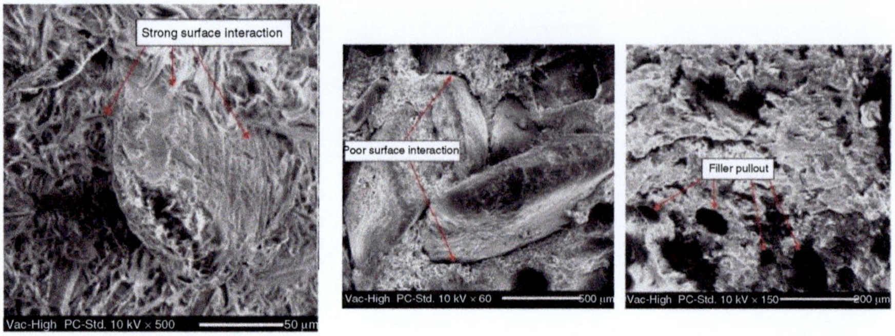


(a) (b)

**Figure 4.18** Effect of coupling agent on the interfacial bond of wood and polymer matrix. (a) Strong bonding between wood and polypropylene matrix by the coupling agent (MAPP). (b) Poor interfacial bonding without the coupling agent MAPP. Source: Ayrilmis et al. [19].

the amount of the coupling agents (MAPP or MAPP) is between 1 and 4 wt.%. Of course, the amount of the coupling agent to be used in the biocomposite depends on the addition of the filler content in the matrix. As the amount of the filler is increased in the polymer matrix, the addition of the coupling agent is increased. The maleic anhydride (MAH) content is also one of the important factors determining its effectiveness and its content in the composite. Mostly, 2–3 wt.% of coupling agent is used in the biocomposites (10–50 wt.% filler content). The effect of the coupling agent on the interfacial bond is presented in Figure 4.18 [19].

Mijiyawa et al. [58] investigated the effect of coupling agent and loading level of birch fibers on the tensile properties of the WPCs (Figure 4.19). The tensile modulus of the WPCs improved with increasing fiber content, especially for 30 and 40 wt.% filler contents. Good correlation ($R^2 = 0.98$) was found between fiber content and modulus of elasticity of the composites with MAPP. The incorporation of the MAPP into the composite increased the tensile strength but there was no significant improvement in the tensile modulus.

A comparison of the mechanical properties of wood flour and wood-fiber-reinforced polypropylene composites with and without coupling agent was carried out by Stark and Rowlands [25]. The properties of the composites reinforced with the wood fibers were found to be considerably higher than those of the composites produced with wood flour (Table 4.4).

Generally, the impact strength of the natural fiber/thermoplastic composites is negatively affected by increasing fiber content. However, the addition of the coupling agent into the polymer matrix improves the impact bending strength. Nourbakhsh et al. [35] investigated the effect of coupling agent on the impact strength of bio-based composites. However, the addition of MAPP and polybutadiene isocyanate (PBNCO) into the composite improved the impact strength but it was still lower than that of the unfilled PP (Figure 4.20). The results showed that interfacial bond strength enhanced when the PBNCO was added into the composition. Their results clearly showed that the modification of wood fibers helped energy absorption.

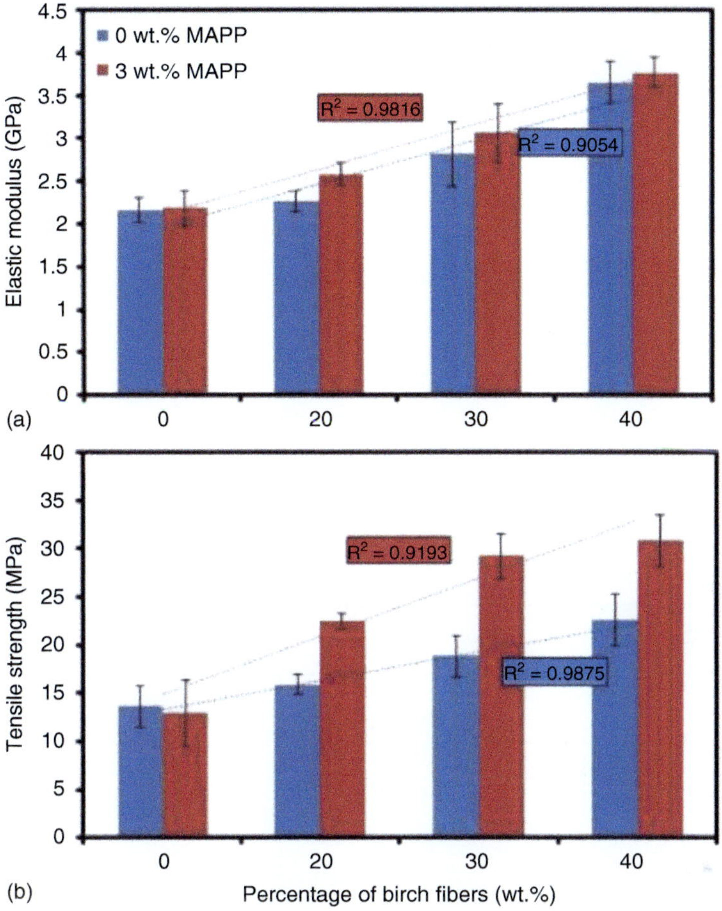


**Figure 4.19** The effect of birch fiber content and coupling agent (MAPP) on the tensile modulus (a) and tensile strength (b) (MAPP, maleate polypropylene; PP, polypropylene). Source: Mijiyawa et al. [58]. © 2014, SAGE Publications.

## 4.2.5 Type of Polymer Matrix

Polymer matrix holds the natural fibers together and protects them against weathering effects and biological effects in the biocomposite. It acts as a stress distributor in the composite and transfers the load to the reinforcements. Thermoplastics or thermosets can be used in the production of biocomposites. The physical, mechanical, and thermal performances of the biocomposites are greatly affected by the MFI of the thermoplastics. The classification of polymer matrices is presented in Figure 4.21.

The processing method and viscosity of the polymer matrix have an important role in the physical and mechanical properties of biocomposites. As the viscosity of the matrix is decreased by increasing the MFI value, the matrix flows well around the fibers and may penetrate into the vessels of the natural fibers. A polymer matrix having lower MFI has comparatively lower modulus than a polymer matrix

**Table 4.4** Main mechanical properties of polypropylene/wood composites with or without coupling agent.

| Property | Unit | PP | PP + 40 wt.% wood flour | PP + 40% wood flour + 3% coupling agent | PP + 40% wood fiber | PP + 40% wood fiber + 3% coupling agent |
|---|---|---|---|---|---|---|
| Specific gravity | — | 0.9 | 1.05 | 1.05 | 1.03 | 1.03 |
| Tensile strength | MPa | 28.5 | 25.4 | 32.3 | 28.2 | 52.3 |
| Tensile modulus | GPa | 1.53 | 3.87 | 4.10 | 4.20 | 4.23 |
| Bending strength | MPa | 38.3 | 44.20 | 53.10 | 47.90 | 72.40 |
| Bending modulus | GPa | 1.19 | 3.03 | 3.08 | 3.25 | 3.22 |

PP, polypropylene. Coupling agent: maleic anhydride grafted PP (MAPP).
Source: Stark and Rowlands [25].

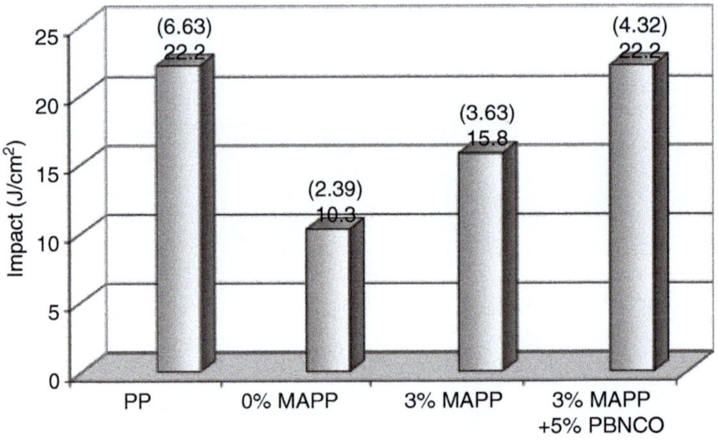


**Figure 4.20** The effect of coupling agents on the impact strength of composites (values in parentheses are standard deviations). PBNCO, polybutadiene isocyanate. Source: Nourbakhsh et al. [35]. © 2008, SAGE Publications.

having higher MFI. Hence, a low-viscosity matrix causes lower modulus for the biocomposites [34]. To improve the properties of biocomposites and decrease the cost of the biocomposite, polymer blends can be used in the production of biocomposites such as PP/PE blends/natural fiber, PLA/PBAT/natural fiber. Polymer blends may create synergic effect on the biocomposite properties as compared to the individual effect of singular polymers. It may have synergic effect on the performance of the composites, which is better than the performance of the singular polymers [24].

### 4.2.6 Processing Conditions in the Manufacture of Biocomposite

The physical and mechanical properties of biocomposites are significantly affected by processing methods and processing conditions. Four techniques are the most used

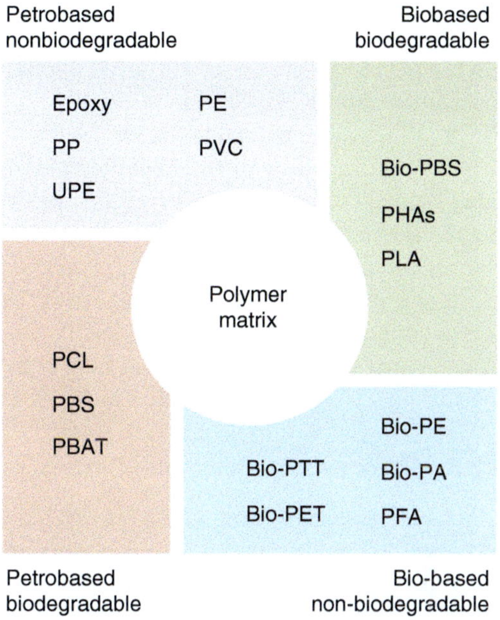


**Figure 4.21** Classification of polymer matrix. Source: Mohanty et al. [59].

in the production of biocomposites – injection molding, extrusion profile, compression molding, and resin transfer molding. Generally, the compression applied in injection molding is higher than the extrusion pressure. Hence, the density of the natural fibers may increase up to 1.30 g/cm$^3$ (the density of wood cell is 1.5 g/cm$^3$), depending on the injection pressure. As a result, the density of the biocomposite may be higher than that of the biocomposites produced with extrusion pressure. Similarly, the pressure in the hot-press molding determines the density of the biocomposites.

Injection molding is generally limited to composites of less than 40 wt.% fiber content due to the viscosity requirements. The length of the fiber is reduced during the extrusion process. Compression molding is useful for the biocomposites with loose chopped fiber or short/long natural fibers either randomly oriented or aligned or mats. Hot-press molding is mostly used in the production of natural fiber/thermoset composites. The natural fiber/thermoset prepreg sheets are stacked in together and then hot-pressed. Especially for thick composites, during the pressing and heating process the viscosity of the matrix should be carefully controlled because the polymer matrix should completely melt and penetrate the space within the fiber composite [13].

The properties of biocomposites can be determined by changing the process parameters, which are processing pressure, viscosity, holding time, and process temperature. Natural fibers are sensitive to thermal degradation above 200 °C due to their natural components, cellulose, lignin, and hemicelluloses. Above 200 °C, the thermal degradation of natural fibers considerably increases; thus, the technological properties of the biocomposites decrease with increasing processing temperature, as well as color change (brownish color) of the natural fibers.

### 4.2.7 Presence of Voids and Porosity

Voids and porosity are two of the most important factors that affect the physical and mechanical properties of biocomposites. The voids occur in the biocomposite during the processing. The main reasons for the microvoids between the natural fiber and polymer matrix are the moisture of the fibers, poor interfacial bonding (incompatibility between polymer and matrix), the rate of cooling during the processing, gases (bubbles) emitted from natural fibers (especially volatile organic compounds above 200 °C) and polymer matrix, and uncontrolled process parameters such as processing temperature and speed. Degassing should be carried out during the processing to prevent microvoids [60]. Porosity significantly affects the mechanical properties of biocomposites. This is due to the presence of air and gas bubbles during processing, thermal degradation of fibers, and low wettability of fibers.

### 4.2.8 Nanocellulose-Reinforced Biocomposites

Utilization of nanocellulose as reinforcement in the biocomposites has considerably increased in the last two decades due to its unique properties. The main property of nanocellulose is its very high surface area. Therefore, use of a very small quantity of nanocellulose such as 1–2 wt.% in the composites is enough. There are many studies on the use of nanocellulose in the polymer matrix. The use of nanocellulose in these studies varied from 1 to 5 wt.%, mostly between 1 and 3 wt.%. A higher amount of nanocellulose negatively affects the properties of the composites because of the agglomeration of the nanocellulose in the matrix. The uniform distribution of nanocellulose in the matrix using different techniques such as ultrasonic process and high shear mixer is noteworthy.

Cellulose is the most abundant natural polymer on earth, and its utilization has many environmental benefits. Cellulose microfibrils are bundles of stretched cellulose chain molecules and are the smallest structural units of lignocellulosic bioresources [61]. The word "nanocellulose" generally refers to cellulosic materials with one dimension in the nanometer range. They exist in nature as nanostructured materials, the so-called cellulose nanofibers (CNFs) [61]. The significance of forest biomaterials has been sharply increasing due to their abundance and substitutability and as a potential alternative to petroleum resources. Thus, high-quality utilization of forest biomaterial and their developments have intensively been requested for the next generation to ensure environmentally friendly biomaterials in future. The research field of forest biomaterials has been dealing with many aspects of the utilization, significance, understanding, and promotion of forest resources and industries [62]. The structure of wood cell and nanocellulose is illustrated in Figure 4.22.

**Advantages of nanocellulose for biocomposites [61–64]**

- Abundantly available
- Low weight
- Biodegradable
- Nontoxic and optical transparency
- Cheaper
- Renewable

- Unique mechanical properties
- Good thermal stability
- High specific surface area
- Modest abrasivity during processing
- Highly absorbent
- very high tensile strength – 8 to 10 times that of steel
- Stiffer than Kevlar®
- Electrically conductive

**Drawbacks of nanocellulose for biocomposites [61–64]**

- Substantial hydrophilicity
- Poor compatibility

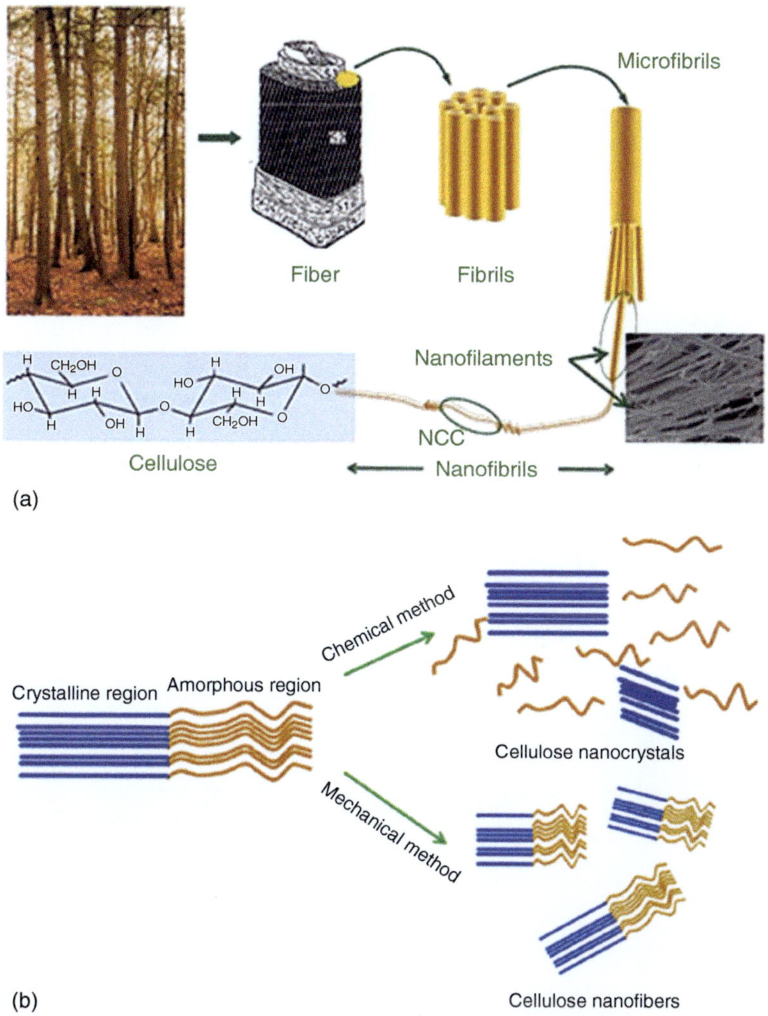


**Figure 4.22** (a) Structure of wood cell. (b) Nanocellulose. Source: Perumal et al. [63].

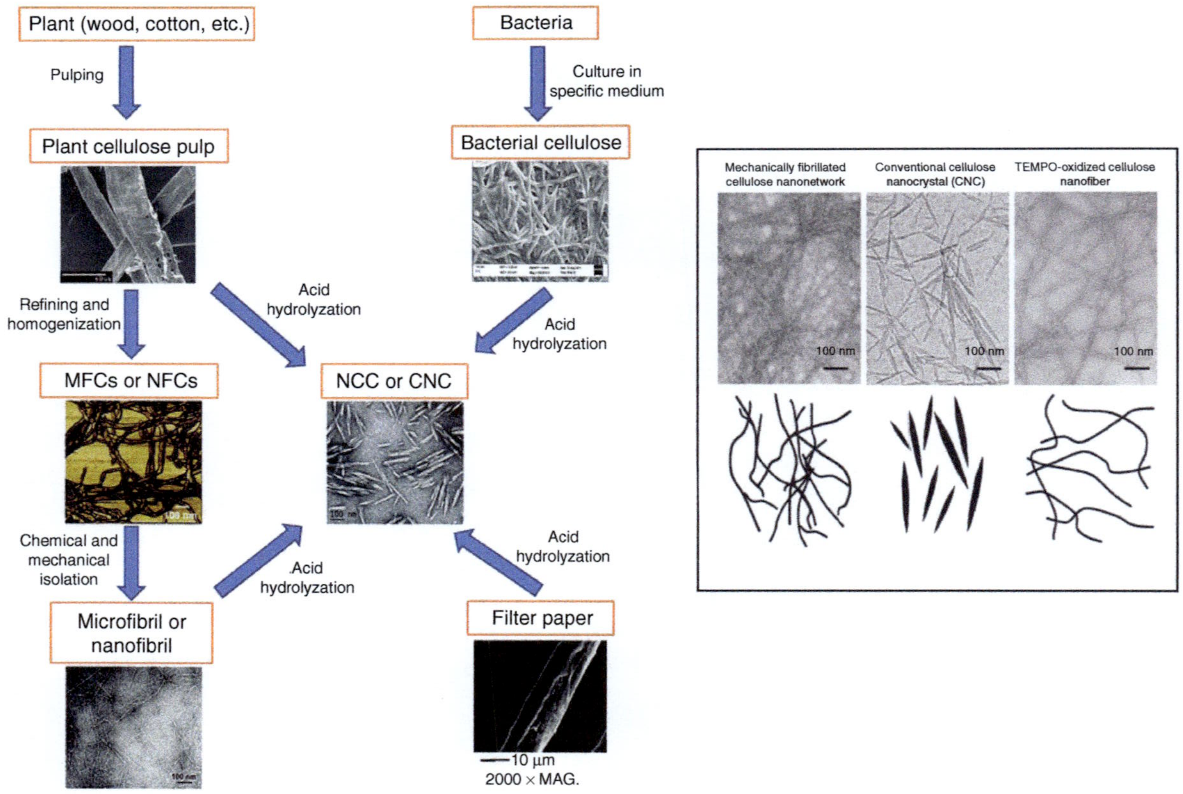


**Figure 4.23** Relationship between different kinds of nanocelluloses. Source: Wei et al. [64].

#### 4.2.8.1 Preparation and Properties of Cellulose Nanofibers

Generally, there are two representative CNFs, nanocrystalline cellulose (NCC) and microfibrillated cellulose (MFC). NCC is generally prepared by acid hydrolysis under strict conditions and has high crystallinity and low aspect ratio. The MFC shows gel-like characteristics while the NCC suspensions show liquid crystalline characteristics. Another nanocellulose known as bacterial nanocellulose (BNC) is produced using acetic acid bacteria of the genus Gluconacetobacter [61]. The relationship between different kinds of nanocelluloses is presented in Figure 4.23.

Typical NCC solutions at 5 and 12 wt.% and NCC powder (by spray-dried) are presented in Figure 4.24.

The NCCs have very attractive fundamental properties, i.e. high strength and stiffness, light density, biodegradability, and extremely low thermal expansion property. It has a very high Young's modulus (130–140 Gpa) [61] (Table 4.5).

#### 4.2.8.2 Industrial Applications of Cellulose Nanofibers

Recently, researchers the world over have focused on the inherent strength and performance of CNFs and on the applications for a new class material (Figure 4.25). Among those applications, lightweight and high-performance CNF-reinforced nanocomposites have attracted great attention [68].

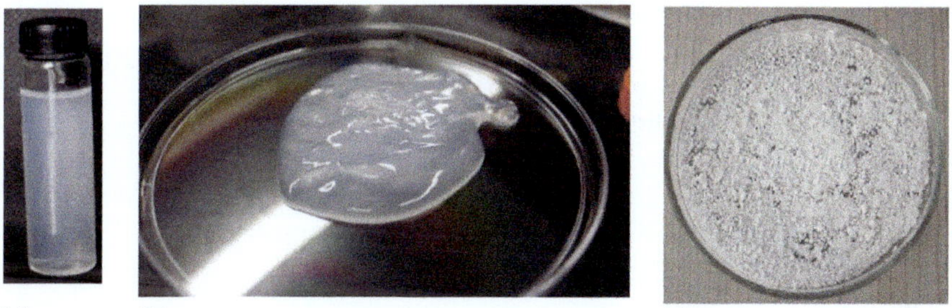

(a) (b) (c)

**Figure 4.24** NCC at 5 wt.% solution; at 12% wt solution; and NCC powder (spray-dried). Source: Abe et al. [65].

**Table 4.5** Comparison of average tensile properties and cost of nanocrystalline cellulose (NCC) and other synthetic fibers.

| Material | Density (g/cm$^3$) | Tensile strength (MPa) | Tensile modulus (GPa) | Cost ($/kg) |
|---|---|---|---|---|
| High-strength steel | 7.9 | 600 | 210 | ~1 |
| Aluminum 6061-TL | 2.7 | 275 | 70 | ~2 |
| Glass fiber (E-type) | 2.5 | 3500 | 80 | ~2 |
| Carbon fiber | 1.8 | 4000 | 230 | ~20 |
| Nanocrystalline cellulose | 1.5 | 7500 | 135 | 4-10 |

Source: Nelson et al. [66].

**Figure 4.25** Nanocrystalline cellulose (NCC). Source: Anonymous [67].

Some industrial usage areas of nanocellulose in the market are given below:

- Biocomposites
- Improved paper and packaging products
- Reinforced polymers
- Thermoplastic and thermosetting resins
- Advanced composite materials
- Body armor applications
- Lithium batteries, screen/barrier films, packaging films
- Hydrofracking and oil drilling fluids
- High-performance spun fibers and textiles
- Paints, adhesives, varnish, lacquers, coatings, paints
- Optical devices
- Pharmaceutical and drug delivery
- Enhanced building products
- Additives for foods, cosmetics
- Aerospace and transportation

Lightweight and high-performance CNF-reinforced nanocomposites have attracted great attention. Sehaqui et al. [69] developed a fast preparation technique for cellulose and cellulose/inorganic nanopaper structures for industrial applications (Figure 4.26a). Chun et al. [68] developed environmentally friendly cellulosic nanofiber paper-derived separator membranes (CNP separators) for use in lithium ion batteries (Figure 4.26b). The CNP separator enhances the ionic conductivity, electrolyte wettability, and thermal shrinkage of the battery.

The defense industry in many countries has focused on NCC-reinforced polymer composites due to their great properties such as very low weight and durability against impact effects as an alternative to aramid fibers, especially for body armor and ballistic glass [14]. Owing to the crystal structure, the nanocellulose is incredibly tough and its strength is nearly eight times higher than that of stainless steel (Figure 4.26c). Thus, NCC will likely be the building material with low weight and high performance for the future body armor material [70].

(a) (b) (c)

**Figure 4.26** (a) Transparent film from nanocellulose. (b) Nanofiber paper-derived separator membranes for use in lithium ion batteries. (c) Body armor applications of NCC. Source: Anonymous [67, 70].

## 4.3 Conclusions

The market for natural-fiber-reinforced thermoplastic composites has rapidly increased in the last decade. Owing to the less environmental impact of synthetic fibers, the proportion of the natural fibers used in the thermoplastic and thermoset composites will increase in the near future. Modification with natural fibers and some property enhancers improves the thermal and technological properties of biocomposites. In particular, nanocellulose applications in the thermoplastic composites have gained a great deal of attention in scientific research and industry. In the near future, the use of nanocellulose in the high-value-added thermoplastic composites will increase as a function of decreasing price and increasing production capacity of nanocellulose. Furthermore, bioplastics will be partly replaced with petroleum plastics in the production of natural-fiber-reinforced polymer composites used in indoor applications because it is an environmentally friendly biomaterial.

## References

1 Saxena, M., Pappu, A., Sharma, A. et al. (2011). Composite materials from natural resources: recent trends and future potentials. In: *Advances in Composite Materials: Analysis of Natural and Man-Made Materials*, Chapter 6 (ed. P. Tesinova), 121–162. Crotia: Intech Publisher.

2 Campilho, R.D.S.G. (2015). Introduction to natural fiber composites. In: *Natural Fiber Composites* (ed. R.D.S.G. Campilho), 65–88. Boca Raton, FL: CRC Press Taylor.

3 Ansell, M.P. (2015). *Wood Composites*. Kidlington: Woodhead Publishing.

4 AL-Oqla, F.M., Salit, M.S., Ishak, M.R., and Aziz, N.A. (2015). Selecting natural fibers for bio-based materials with conflicting criteria. *Am. J. Appl. Sci.* 12: 64–71.

5 Akil, H.M., Mohd, O., Mazuki, M. et al. (2011). Kenaf fiber reinforced composites: a review. *Mater. Des.* 32: 4107–4121.

6 Sripaiboonkij, P., Sripaiboonkij, N., Phanprasit, W., and Jaakkola, M.S. (2009). Respiratory and skin health among glass microfiber production workers: a cross-sectional study. *Environ. Health* 8: 36.

**7** Peças, P., Carvalho, H., Salman, H., and Leite, M. (2018). Natural fiber composites and their applications: a review. *J. Compos. Sci.* 2: 66.
**8** Bledzki, A., Sperber, V., and Faruk, O. (2002). *Natural and Wood Fibre Reinforcement in Polymers*. Smithers Rapra Press. ISBN: ISBN-10: 1859573592.
**9** Ayrilmis, N., Kwon, J.H., Han, T.H., and Durmus, A. (2015). Effect of wood-derived charcoal content on properties of wood plastic composites. *Mater. Res.* 18: 654–659.
**10** Ayrilmis, N., Tasdemir, M., and Akbulut, T. (2017). Water absorption and mechanical properties of PP/HIPS hybrid composites filled with wood flour. *Polym. Compos.* 38: 863–869.
**11** Ayrilmis, N. and Ashori, A. (2014). Lignocellulosic fibers and nanocellulose as reinforcing filler in thermoplastic composites. *Euroasian J. Forest Sci.* 2 (2): 1–6.
**12** Ayrilmis, N. (2014). Waste chestnut shell as a source of reinforcing fillers for polypropylene composites. *J. Thermoplast. Compos. Mater.* 27: 1054–1064.
**13** Pickering, K.L., Efendy, M.G.A., and Le, T.M. (2016). A review of recent developments in natural fiber composites and their mechanical performance. *Composites Part A* 83: 98–112.
**14** Ayrilmis, A. and Ashori, A.R. (2015). Alternative solutions for reinforcement of thermoplastic composites. In: *Natural Fiber Composites*, Chapter 3, 1e (ed. R.D.S.G. Campilho), 65–92. CRC Press.
**15** Oksman Niska, K. and Sain, M. (2007). *Wood-Polymer Composites*. Cambridge: Woodhead Publishing Limited.
**16** Kaymakci, A. and Ayrilmis, N. (2017). Preparation and characterization of high performance wood polymer nanocomposites using multi walled carbon nanotubes. *J. Compos. Mater.* 51: 1187–1195.
**17** Özdemir, F., Ayrilmis, N., and Mengeloğlu, F. (2017). Effect of dolomite powder on combustion and technological properties of WPC and neat polypropylene. *J. Chilean Chem. Soc.* 64: 3716–3720.
**18** Kaymakci, A., Ayrilmis, N., Ozdemir, F., and Gulec, T. (2013). Utilization of sunflower stalk in manufacture of thermoplastic composite. *J. Polym. Environ.* 21: 1135–1142.
**19** Ayrilmis, N., Kaymakci, A., and Ozdemir, F. (2013). Physical, mechanical, and thermal properties of polypropylene composites filled with walnut shell flour. *J. Ind. Eng. Chem.* 19: 908–914.
**20** Ayrilmis, N. and Kaymakci, A. (2013). Fast growing biomass as reinforcing filler in thermoplastic composites: *Paulownia elongata* wood. *Ind. Crops Prod.* 43: 457–464.
**21** Shubhra, Q.T.H., Alam, A.K.M.M., and Quaiyyum, M.A. (2013). Mechanical properties of polypropylene composites: a review. *J. Thermoplast. Compos. Mater.* 26: 362–391.
**22** Migneault, S., Koubaa, A., and Perre, P. (2014). Effect of fiber origin, proportion, and chemical composition on the mechanical and physical properties of wood-plastic composites. *J. Wood Chem. Technol.* 34: 241–261.

**23** Edeerozey, A.M.M., Akil, H.M., Azhar, A.B., and Ariffin, M.I.Z. Chemical modification of kenaf fibers. *Mater. Lett.* 61 (10): 2023–2202.

**24** Shesan, O.J., Stephen, A.C., Chioma, A.G. et al. (2019). Improving the mechanical properties of natural fiber composites for structural and biomedical applications. In: *Composites from Renewable and Sustainable Materials* (eds. A.B. Pereira and F.A.O. Fernandes). IntechOpen. https://doi.org/10.5772/intechopen.85252.

**25** Stark, N. and Rowlands, R. (2003). Effects of wood fiber characteristics on mechanical properties of wood/polypropylene composites. *Wood Fiber Sci.* 35: 167–174.

**26** Kabir, M.M., Wang, H., Aravinthan, T. et al. (2011). Effects of natural fibre surface on composite properties: a review. *1st International Postgraduate Conference on Engineering, Designing and Developing the Built Environment for Sustainable Wellbeing (eddBE2011)* (27–29 Apr 2011), Brisbane, Australia.

**27** Nourbakhsh, A., Karegarfard, A., and Ashori, A. (2010). Effects of particle size and coupling agent concentration on mechanical properties of particulate-filled polymer composites. *J. Thermoplast. Compos. Mater.* 23: 169–174.

**28** Gu, R. and Kokta, B.V. (2010). Mechanical properties of PP composites reinforced with BCTMP aspen fiber. *J. Thermoplast. Compos. Mater.* 23: 513–542.

**29** Jayaraman, K. (2003). Manufacturing sisal–polypropylene composites with minimum fiber degradation. *Compos. Sci. Technol.* 63: 367–374.

**30** Mohanty, S., Nayak, S.K., Verma, S.K., and Tripathy, S.S. (2004). Effect of MAPP as a coupling agent on the performance of jute–PP composites. *J. Reinf. Plast. Compos.* 23: 625–637.

**31** Jaszkiewicz, A., Bledzki, A.K., and Feldmann, M. (2010). High-strength bio-based composites -mechanical performance and capability study - PP vs. PLA vs. PA 6.10. In: *8th Global WPC and Natural Fiber Composites Congress and Exhibition (22–23 June 2010), Stuttgart, Germany* (eds. A. Jaszkiewicz, A.K. Bledzki and M. Feldmann), B14.

**32** Li, Y. (2012). Effect of coupling agent concentration, fiber content, and size on mechanical properties of wood/HDPE composites. *Int. J. Polym. Mater. Polym. Biomater.* 61 (11): 882–890.

**33** Basiji, F., Safdari, V., Nourbakhsh, A., and Pilla, S. (2010). The effects of fiber length and fiber loading on the mechanical properties of wood-plastic (polypropylene) composites. *J. Turk. Agric. For.* 34: 191–196.

**34** Balasuriya, P.W. and Ye Mai, Y.W. (2001). Mechanical properties of wood flake–polyethylene composites. Part I: effects of processing methods and matrix melt flow behavior. *Composites Part A* 32: 619–662.

**35** Nourbakhsh, A., Kokta, B., Ashori, A., and Jahan-Latibari, A. (2008). Effect of a novel coupling agent, polybutadiene ısocyanate, on mechanical properties of wood-fiber polypropylene composites. *J. Reinf. Plast. Compos.* 27: 1679–1687.

**36** Ayrılmış, N., Ozdemir, F., and Kaymakçı, A. Sunflower seed cake as reinforcing filler in thermoplastic composites. *J. Appl. Polym. Sci.* 129 (3): 1170–1178.

**37** Kaymakci, A., Ayrilmis, N., and Gulec, T. (2012). Surface properties and hardness of polypropylene composites filled with sunflower stalk flour. *Bioresources* 8: 592–602.

**38** Buyuksari, U., Ayrilmis, N., and Akbulut, T. (2012). Compression wood as a source of reinforcing filler for thermoplastic composites. *J. Appl. Polym. Sci.* 123: 1740–1745.

**39** Hosseinihashemi, S.K., Arwinfar, F., Najafi, A. et al. (2016). Long-term water absorption behavior of thermoplastic composites produced with thermally treated wood. *Measurement* 86: 202–208.

**40** Hosseinihashemi, S.K., Eshghi, A., Kord, B. et al. (2016). Hardness, decay and water resistance of polypropylene/montmorillonite/almond shell flour composites. *Mater. Res.* 19: 440–445.

**41** Arwinfar, F., Hosseinihashemi, S.K., Latibari, A.J. et al. (2016). Mechanical properties and morphology of wood plastic composites produced with thermally treated beech wood. *Bioresources* 11: 1494–1504.

**42** Ayrilmis, N., Kaymakci, A., and Güleç, T. (2015). Potential use of decayed wood in production of wood plastic composite. *Ind. Crops Prod.* 74: 279–284.

**43** Ayrilmis, N. (2015). Combined effect of acetylation and wax emulsion on physical and mechanical properties of particleboard. *Holz als Roh- und Werkstoff* 73: 845–847.

**44** Ayrilmis, N., Dundar, T., Kaymakci, A. et al. (2014). Mechanical and thermal properties of wood-plastic composites reinforced with hexagonal boron nitride. *Polym. Compos.* 35: 194–200.

**45** Ayrilmis, N., Benthien, J.T., Thoemen, H., and White, R.H. (2011). Properties of flat-pressed wood plastic composites containing fire retardants. *J. Appl. Polym. Sci.* 122 (5): 3201–3210.

**46** Ayrilmis, N. (2013). Combined effects of boron and compatibilizer on dimensional stability and mechanical properties of wood/HDPE composites. *Composites Part B* 44 (1): 745–749.

**47** Ayrilmis, N., Jarusombuti, S., Fueangvivat, V. et al. (2012). Coir fiber reinforced polypropylene composite panel for automotive interior applications. *Fibers Polym.* 12: 919–926.

**48** Ayrilmis, N. and Jarusombuti, S. (2011). Flat-pressed wood plastic composite as an alternative to conventional wood-based panels. *J. Compos. Mater.* 45: 103–112.

**49** Ayrilmis, N., Buyuksari, U., and Dundar, T. (2010). Waste pine cones as a source of reinforcing fillers for thermoplastic composites. *J. Appl. Polym. Sci.* 117: 2324–2330.

**50** Ayrilmis, N. and Buyuksari, U. (2010). Utilization of olive mill sludge in the manufacture of fiberboard. *Bioresources* 5: 1859–1867.

**51** Ayrilmis, N. and Buyuksari, U. (2010). Utilization of olive mill sludge in manufacture of lignocellulosic/polypropylene composite. *J. Mater. Sci.* 45: 1336–1342.

**52** Ayrilmis, N., Jarusombuti, S., Fueangvivat, F., and Bauchongkol, P. (2011). Effect of thermal-treatment of wood fibers on properties of flat-pressed wood plastic composites. *Polym. Degrad. Stab.* 96: 818–822.

**53** Tufan, M. and Ayrilmis, N. (2016). Potential use of hazelnut husk in recycled high-density polyethylene composites. *Bioresources* 11: 7476–7489.

**54** Hosseinihashemi, S.K., Karimi, S., Latibari, A.J. et al. (2016). Use of black locust/poplar wood as filler in thermoplastic composites. *Turk. J. Agric. For.* 40: 327–334.

**55** Tufan, M., Güleç, T., Peşman, E., and Ayrilmis, N. (2016). Technological and thermal properties of thermoplastic composites filled with heat-treated alder wood. *Bioresources* 11: 3153–3164.

**56** Al-Maharma, A.Y. and Al-Huniti, N. (2019). Critical review of the parameters affecting the effectiveness of moisture absorption treatments used for natural composites. *J. Compos. Sci.* 3: 27.

**57** Azwa, Z., Yousif, B., Manalo, A., and Karunasena, W. (2013). A review on the degradability of polymeric composites based on natural fibers. *Mater. Des.* 47: 424–442.

**58** Mijiyawa, F., Koffi, D., Kokta, B.V., and Erchiqui, F. (2014). Formulation and tensile characterization of wood-plastic composites. *J. Thermoplast. Compos. Mater.* 28: 1675–1692.

**59** Mohanty, A.K., Vivekanandhan, S., Pin, J.-M., and Misra, M. (2018). Composites from renewable and sustainable resources: challenges and innovations. *Science* 362: 536–542.

**60** Mehdikhani, M., Gorbatikh, L., Verpoest, I., and Lomov, S. (2018). Voids in fiber-reinforced polymer composites: a review on their formation, characteristics, and effects on mechanical performance. *J. Compos. Mater.* 53: 1579–1669.

**61** Jang, J.H. (2015). Nanoscopic morphology and physicochemical properties of lignocellulose nanofibers and their utilization. PhD thesis. Graduate School, Kangwon National University.

**62** Dufrence, A. (2013). Nanocellulose: a new ageless bionanomaterial. *Mater. Today* 16: 220–227.

**63** Perumal, A.B., Sellamuthu, P.S., Nambiar, R.B. et al. (2019). Biocomposite reinforced with nanocellulose for packaging applications. In: *Green Biopolymers and Their Nanocomposites*, Materials Horizons: From Nature to Nanomaterials (ed. D. Gnanasekaran), 83–123. Singapore: Springer.

**64** Wei, H., Rodriguez, K., Renneckar, S., and Vikesland, P.J. (2014). Environmental science and engineering applications of nanocellulose-based nanocomposite. *Environ. Sci.: Nano* 1: 302–316.

**65** Abe, K., Iwamoto, S., and Yano, H. (2007). Obtaining cellulose nanofibers with a uniform width of 15 nm from wood. *Biomacromolecules* 8: 3276–3278.

**66** Nelson, K. et al. (2016). American process: production of low cost nanocellulose for renewable, advanced materials applications. In: *Materials Research for Manufacturing*, Springer Series in Materials Science, vol. 224 (eds. L. Madsen and E. Svedberg), 267–302. Cham: Springer.

**67** Anonymous. https://forcetoknow.com/science/japanese-created-clear-transparent-paper.html (trans paper) (accessed on 1 March 2020).

**68** Chun, S.J., Choi, E.S., Lee, E.H. et al. (2012). Eco-friendly cellulose nanofiber paper-derived separator membranes featuring tunable nanoporous network channels for lithium-ion batteries. *J. Mater. Chem.* 22: 16618–16626.

**69** Sehaqui, H., Liu, A., Zhou, Q., and Berglund, L.A. (2010). Fast preparation procedure for large, flat cellulose and cellulose/inorganic nanopaper structures. *Biomacromolecules* 11 (9): 2195–2198.

**70** Anonyomus. https://nanografi.com/nanoparticles/cellulose-nanofiber-cellulose-nanofibril-nanofibrillated-cellulose-cnfs (accessed on 5 March 2020).

# 5

# Machinability Analysis on Biowaste Bagasse-Fiber-Reinforced Vinyl Ester Composite Using *S/N* Ratio and ANOVA Method

*Balasubramaniam Stalin[1], Ayyanar Athijayamani[2], and Rajini Nagarajan[3]*

[1]*Anna University, Department of Mechanical Engineering, Regional Campus Madurai, Madurai 625019, Tamil Nadu, India*
[2]*Government College of Engineering, Department of Mechanical Engineering, Bodinayakanur 625582, Tamil Nadu, India*
[3]*Kalasalingam University, Department of Mechanical Engineering, Anand Nagar, Krishnankoil, Virudhunagar 626126, Tamil Nadu, India*

## 5.1 Introduction

In recent years, cellulosic natural fibers and their wastes have been used as fillers in polymers to achieve cost savings and to impart some applicable properties [1–3]. Owing to their high functional strength and strength-to-weight ratio, polymer composites strengthened by synthetics have been replacing many of the traditional structural materials, such as wood and steel. Despite the various benefits, synthetic fibers such as carbon, glass, and aramid are health hazards and pollute the environment. Materials researchers and engineers have recently recognized plant-based natural fibers as suitable substitute materials for synthetic fibers. Natural fibers based on plants have the following series of benefits: low cost, lightweight, high specific strength, environmentally safe and biodegradable, nontoxic, non-abrasive for processing equipment, and nonhazardous to humans. As a suitable reinforcement for polymer matrix composites, they have recently become attractive among the material researchers and scientists because of these noted merits of the natural fibers [4–6]. Among the various natural fibers, BFs are increasingly used as reinforcement in polymer composites and are getting popular with many material scientists and designers. Bagasse fiber are the sugarcane residue after the juice has been extracted and have been utilized as raw resources in many sectors such as pulp, sugarcane, heat power generation system, and fiber panel. Several works have been reported on BF-reinforced composites of polymers, such as bagasse-aliphatic polyester [7], bagasse-unsaturated polyester [8], bagasse-polyethylene [9], bagasse-polypropylene [10], bagasse-poly(vinyl) chloride [11], bagasse-poly(hydroxybutyrate) [12], bagasse-epoxy [13], and bagasse-polystyrene [14]. Such fibers are usually thrown out as solid wastes after use. They can pollute water and soil. In view of this, an attempt is made to use these biowaste fibers as reinforcement in polymer matrix

*Mechanical and Dynamic Properties of Biocomposites,* First Edition. Edited by Senthilkumar Krishnasamy, Rajini Nagarajan, Senthil Muthu Kumar Thiagamani, and Suchart Siengchin.

composites preparation. In addition, no machinability analysis studies have been performed of VE composites reinforced with biowaste BFs. The current research was designed to create a composite of the vinyl ester matrix using the biowaste BFs as reinforcement. Different proportions of biowaste hybrid fiber reinforcements (bagasse/coir) with VE matrix composite have been reported [15]. To achieve product realization, the modeling and optimization of the machining parameters in the production processes is used. This enhances their function in the manufacturing and industrial engineering processes more quickly, more reliably, and more effectively [16]. The torque and thrust force forecast and evaluation were studied in the drilling of hybrid fiber (roselle/sisal) composites using optimization models, and the response variables were examined [17]. Drilling is one of the manufacturing methods commonly used to fit fasteners for the assembly of composites laminates. The parameters of drilling and specimen were examined based on specimen feed rate, speed, thickness, and drill size. The findings of the delamination investigation established thickness, feed rate, and cutting speed as the most significant factors in the drilling process. In the drilling of glass fiber reinforced plastics (GFRP) composite, signal-to-noise ($S/N$) ratio is engaged to examine the effect of various parameters on delamination [18]. The Taguchi methodology optimizations were utilized in drilling at the entry and exit of the holes, concurrently reducing the delamination factor [19]. The Taguchi design and analysis of variance (ANOVA) are utilized to evaluate the experimental results and indicated that the most responsive parameters for thrust force/torque are feed and speed [20]. It is also used for predicting and estimating the thrust force in composite material drilling [21]. The extensive use of these composites and the urge to integrate them has indicated that there may be an increasing demand for machining of composites. Some of the most common machining approaches are drilling, turning, and milling. Among all machining operations, drilling is one of the fundamental machining operations that might be currently completed on fiber-reinforced composite materials [22]. An important concern is the thrust force and torque evolved during the drilling operation. For the composite industry, control of thrust force and torque is needed in drilling. The thrust force and torque evolved in the drilling operation are proportional to the quantity of damage that takes place in composite materials around the hole. When the device thrust determined in machining is more, it results in the damage occurring within the composite materials. The quantity of torque and thrust force is to be reduced when drilling composite materials. For decreasing the torque and thrust force, modeling and optimization of procedure parameters are required. The effects of machining parameters on thrust pressure and torque is an essential concern [23].

Drilling of composite materials is analyzed by many researchers [24–26]. Most have been targeted closer to examine the torque and thrust force. Although much research has been done on the effects of machining parameters on the responses during drilling of diverse herbal fiber bolstered polymer composites, not much work has been done on the drilling of bagasse/vinyl ester composite experimentally. In this study, an experimental attempt is made to examine the effects of the machining parameters, i.e. feed rate ($fr$), cutting speed ($s$), and drill diameter ($d$) on torque and induced thrust force in bagasse/vinyl ester composite drilling using Taguchi and ANOVA techniques.

## 5.2 Experimental Methodology

### 5.2.1 Materials

After extraction of the sugar-bearing juice, bagasse fibers remain as biowaste materials. These fibers were collected from the fiber industry at Solavanthan, Madurai, Tamil Nadu, India, for this investigation. The matrix material used was vinyl ester resin provided by GVR Traders, Madurai, Tamilnadu, India. Methyl ethyl ketone, cobalt naphthenate, and *N*,*N*-dimethylaniline, respectively, were used as accelerator, catalyst, and promoter.

### 5.2.2 Specimen Preparation

The weight percentage and length of the fibers were 35 wt.% and 5 mm, respectively. The releasing agent was applied on the inner surface of the mold box (150 mm × 150 mm × 3 mm) prior to processing to ensure effective removal of the cured composite rod. The fibers in the bagasse were distributed randomly into the mold. Then, to prepare the resin mixture, the vinyl ester resin was combined with the accelerator and catalyst. Prior to the process, the releasing agent was applied on the inner side of the mold box (150 mm × 150 mm × 3 mm) to ensure the easy removal of the cured composite rod. The bagasse fibers were randomly distributed into the mold. Then, the vinyl ester resin was mixed with the accelerator and catalyst to prepare the resin mixture. After pouring, the mold box was closed and kept under pressure for 24 hours.

### 5.2.3 Machining of the Composite Specimen

In this study, the drilling operations were performed using regular HSS twist drills on bagasse/vinyl ester composites utilizing the MAXMILL CNC machining center. During the drilling processes, strain gauge drill tool dynamometer was utilized to measure the torque and thrust force, respectively. The specifications of the strain gauge drill tool dynamometer are given in Table 5.1.

Figure 5.1 displays the schematics layout for the experimental setup. Table 5.2 illustrates the constant geometry of HSS twist drills utilized in this analysis.

### 5.2.4 Selection of Orthogonal Array

According to the Taguchi quality loss function for the present investigation, an $L_9$ orthogonal array was selected. Taguchi optimization technique began with orthogonal array selection, with a distinct number of levels specified for each of the factors. $N_{min} = (n - 1)\,k + 1$ has given the minimum number of tests in the sequence, where

**Table 5.1** Specifications of strain gauge drill tool dynamometer (PolyLab-Make).

| **Drill tool dynamometer-POLYLAB make digital display (2 channel)** |
|---|
| • Sensing unit consisting of cylinder with strain gauge cemented<br>• Accurate sensing of axial and radial thrusts independently<br>• Suitable to drill holes up to 25 mm in mild steel<br>• Power supply unit-stabilized DC power to the two bridge circuits<br>• Separate bridge balancing the power unit<br>• Suitable to measure two independent force components during drilling process<br>• Digital strain gauge indicator to measure the balance voltage of the bridge<br>• Calibrated using proving ring and standard weights<br>• Stabilized single-phase 230 V AC power supply, and suitable stabilizer |

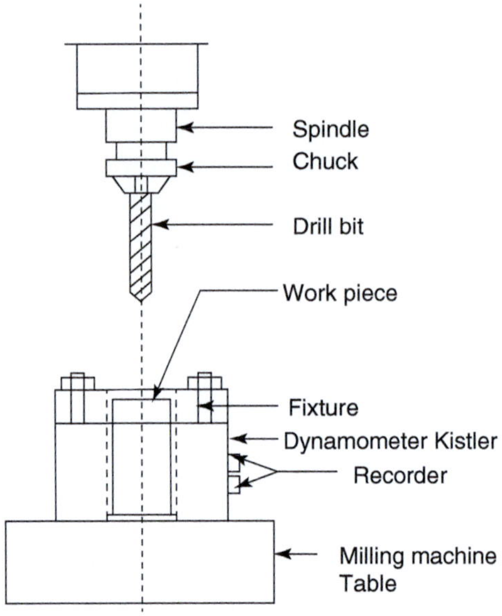


**Figure 5.1** Schematic layout of the drilling setup.

**Table 5.2** Constant tool geometry of HSS twist drill bits utilized in this study.

| **Drill diameter (DD) (mm)** | **Point angle (PA) (°)** | **Helix angle (HA) (°)** | **Rake angle (RA) (°)** | **Clearance angle (CA) (°)** |
|---|---|---|---|---|
| 6, 8, 10 | 118 | 30 | 30 | 12 |

**Table 5.3** Levels and factors of machining parameters.

| Code | Factor | Levels | | |
|---|---|---|---|---|
| | | I | II | III |
| A | Feed rate (rev/min) | 0.1 | 0.2 | 0.3 |
| B | Cutting speed (rpm) | 600 | 1200 | 1800 |
| C | Drill diameter (mm) | 6 | 8 | 10 |

**Table 5.4** $L_9$ orthogonal array and experimental values of torque and thrust force.

| Trial | Levels | | | Thrust force (N) | Torque (N-m) |
|---|---|---|---|---|---|
| | 1 | 2 | 3 | | |
| 1 | 0.1 | 600 | 6 | 14.71 | 0.98 |
| 2 | 0.1 | 1200 | 8 | 19.62 | 2.94 |
| 3 | 0.1 | 1800 | 10 | 39.24 | 2.94 |
| 4 | 0.2 | 600 | 8 | 49.05 | 3.92 |
| 5 | 0.2 | 1200 | 10 | 73.58 | 2.94 |
| 6 | 0.2 | 1800 | 6 | 53.96 | 2.45 |
| 7 | 0.3 | 600 | 10 | 98.1 | 6.38 |
| 8 | 0.3 | 1200 | 6 | 83.39 | 3.92 |
| 9 | 0.3 | 1800 | 8 | 132.44 | 5.89 |

$k$ is the number of factors and $n$ is the number of levels. The identified machining parameters affecting the drilling process of the composite specimen and their levels are presented in Table 5.3. Table 5.4 shows the $L_9$ orthogonal array and experimental values of torque and thrust force.

### 5.2.5 Development of Multivariable Nonlinear Regression Model

The data obtained from the experiments were utilized to construct a mathematical model utilizing regression. The proposed relation between the variables of the response and the parameters of machining can be expressed in the following form:

$$Y = f\,(fr, v, d) \tag{5.1}$$

where $v$, $fr$, and $d$ are the cutting speed, feed rate, and drill diameter respectively of the machining process. $Y$ is the observed response.

The correlation between responses and machining parameters is modeled as follows:

$$F = k \times fr^x \times v^y \times d^z \tag{5.2}$$

where $k$, $x$, $y$, and $z$ are constant and $f$, $v$, $and$ $d$ are the machining parameters.

## 5.3 Results and Discussion

### 5.3.1 Influence of Machining Parameters on Thrust Force and Torque

Drilling on the composite cylindrical rod was done in the MAXMILL CNC for three different drill diameters (6, 8, and 10 mm). Figure 5.2a–c shows the effects on torque and thrust force of the machining parameters – diameter of the drill bits, cutting speed, and feed rate. The thrust force in the drilling of bagasse/vinyl ester composites is increased not only by the feed rate but also by the cutting speed and drill diameter.

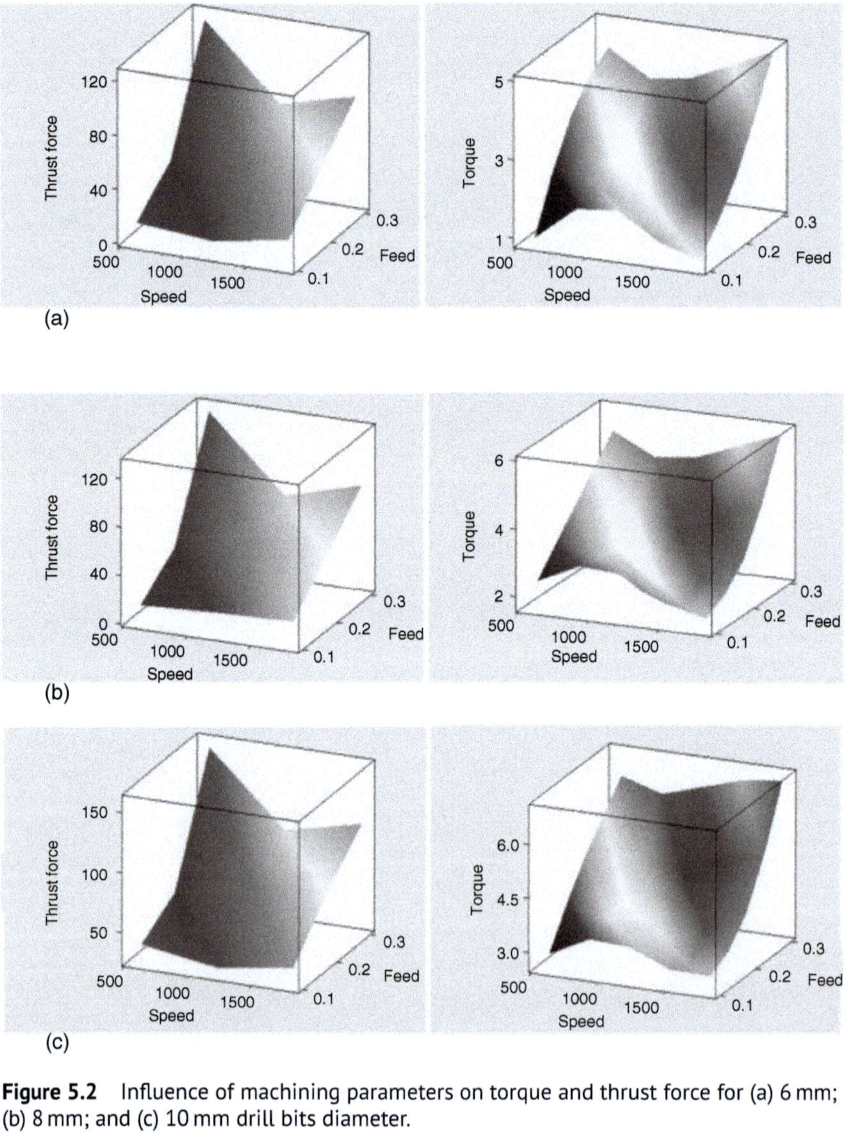


**Figure 5.2** Influence of machining parameters on torque and thrust force for (a) 6 mm; (b) 8 mm; and (c) 10 mm drill bits diameter.

The maximum thrust force was obtained at 0.3 mm/rev feed rate for 6, 8, and 10 mm drill diameters. But the maximum torque was obtained in the maximum level of machining parameters for 6, 8, and 10 mm drill diameter. The torque for 0.3 mm/rev feed rate and 1800 rpm cutting speed were 4.91, 5.89, and 6.87 N-m for 6, 8, and 10 mm drill diameter. A scatter in the values of thrust force for 0.1 and 0.3 mm/rev feed rate was observed with increasing cutting speed for 6, 8, and 10 mm drill bits diameters. The thrust forces increased with the cutting speed for 0.2 mm/rev feed rate for all drill diameters. In all cases, a scatter in the torque values was observed for all sets of machining parameters.

### 5.3.2 *S/N* Ratio

Experimental torque and thrust force values have been converted into the corresponding *S/N* ratios using Eq. (5.1). For torque and thrust force, a smaller-the-better criterion has been selected and expressed as

$$\frac{S}{N} = -10 \log \frac{1}{n} \left( \sum y^2 \right) \tag{5.3}$$

The *S/N* ratios for torque and thrust force are given in Table 5.5. Based on the *S/N* ratio, the optimum thrust force parameters are the level 3 feed rate (0.3 mm/rev), level 3 cutting speed (1800 rpm), and level 2 drill diameter (8 mm), respectively. Similarly, the optimal torque parameters are the level 3 feed rate (0.3 mm/rev), the level 1 cutting speed (600 rpm), and level 3 drill diameter (10 mm).

### 5.3.3 ANOVA

Analysis of variance is a widely used technique. It enables us to divide the entire variation into a group of parts that are ascribable to different factors and a residual

**Table 5.5** *S/N* ratio for torque/thrust force.

| S. No. | Levels | | | S/N – thrust force | S/N – torque |
|---|---|---|---|---|---|
| | 1 | 2 | 3 | | |
| 1 | 0.1 | 600 | 6 | −23.352 | 0.175 |
| 2 | 0.1 | 1200 | 8 | −25.854 | −9.367 |
| 3 | 0.1 | 1800 | 10 | −31.875 | −9.367 |
| 4 | 0.2 | 600 | 8 | −33.813 | −11.866 |
| 5 | 0.2 | 1200 | 10 | −37.335 | −9.367 |
| 6 | 0.2 | 1800 | 6 | −34.641 | −7.783 |
| 7 | 0.3 | 600 | 10 | −39.833 | −16.096 |
| 8 | 0.3 | 1200 | 6 | −38.422 | −11.866 |
| 9 | 0.3 | 1800 | 8 | −42.440 | −15.402 |

**Table 5.6** ANOVA responses for thrust force.

| Factors | Degree of freedom (DOF) | Sum of squares (SS) | Mean squared (MS) | *F*-test | % Contribution (C) |
|---|---|---|---|---|---|
| A | 2 | 266.9 | 133.4 | 104.0 | 83.3 |
| B | 2 | 24.2 | 12.1 | 9.450 | 7.6 |
| C | 2 | 26.7 | 13.3 | 10.390 | 8.3 |
| Error | 2 | 2.6 | 1.3 | | 0.8 |
| Total | | 320.4 | 160.2 | | 100.0 |

**Table 5.7** ANOVA responses for torque.

| Factors | Degree of freedom (DOF) | Sum of squares (SS) | Mean squared (MS) | *F*-test | % Contribution (C) |
|---|---|---|---|---|---|
| A | 2 | 103.4 | 51.69 | 6.3 | 56.54 |
| B | 2 | 3.8 | 1.91 | 0.2 | 2.09 |
| C | 2 | 59.3 | 29.64 | 3.6 | 32.42 |
| Error | 2 | 16.4 | 8.17 | | 8.94 |
| Total | | 182.9 | 91.43 | | 100.00 |

variation that could not be accounted for by any of these factors. The aim of using ANOVA is to investigate the design parameters that have a significant impact on the product or process's best characteristics. A complete data analysis was carried out to determine the effect on thrust force and torque of different machining parameters (feed rate, drill diameter, and cutting speed). It gives the proportional contribution of each machining parameter on torque and thrust force. The ANOVA results show that the feed rate appears to make a larger contribution to the thrust pressure than reducing pace and drill diameter, as shown in Table 5.6. Similarly, the feed rate was also seen as having the greatest contribution to the torque compared to the speed and drill diameter as seen in Table 5.7.

### 5.3.4 Correlation of Machining Parameters with Responses

Using multivariable nonlinear regression analysis, the correlation between the machining parameters and responses (thrust force and torque) was obtained:

$$\text{Thrust force (N)} = 7.36 \times f^{1.35} \times v^{0.42} \times d^{0.65} \tag{5.4}$$

$$\text{Torque (N} - \text{m)} = 3.23 \times f^{0.87} \times v^{-0.05} \times d^{0.91} \tag{5.5}$$

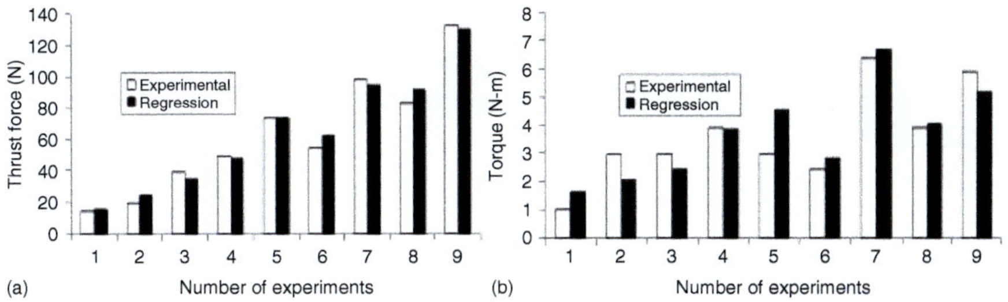


**Figure 5.3** Comparison graph of (a) thrust force, and (b) torque between prediction and experimental values.

The squared residual values ($R^2$) for regression equations of thrust force and torque were 0.96 and 0.81 respectively. Figure 5.3a,b show comparisons of prediction and experimental torque and thrust force, respectively. The regression method was found to be a feasible and effective way to evaluate and predict drilling-induced thrust force and torque while considering the $R^2$.

### 5.3.5 Confirmation Test

Additionally, there were 10 more experiments conducted to validate the regression model results. Comparison of experimental confirmation with the expected torque and thrust force using the regression model is shown in Table 5.8.

The mean absolute percentage error between the experimental confirmation and predicted torque and thrust force values using regression model were 5.81%

**Table 5.8** Experimental confirmation and predicted torque and thrust force.

| Trial | Thrust force (N) | | Error (%) | Torque (N-m) | | Error (%) |
|---|---|---|---|---|---|---|
| | EX | RM | | EX | RM | |
| 1 | 36.10 | 41.39 | −14.7 | 1.98 | 2.36 | −19.3 |
| 2 | 41.89 | 48.00 | −14.6 | 3.25 | 3.41 | −4.95 |
| 3 | 48.50 | 54.34 | 7.1 | 2.69 | 3.51 | 4.89 |
| 4 | 57.62 | 63.03 | −9.4 | 3.87 | 4.39 | −5.43 |
| 5 | 95.10 | 102.45 | −7.7 | 4.52 | 4.79 | −5.93 |
| 6 | 66.98 | 71.88 | −7.3 | 3.45 | 4.40 | −27.6 |
| 7 | 95.88 | 100.19 | 5.3 | 4.55 | 5.20 | −14.3 |
| 8 | 115.50 | 122.57 | 2.3 | 5.46 | 6.49 | −18.9 |
| 9 | 67.50 | 71.72 | −6.2 | 3.98 | 3.75 | 5.85 |
| 10 | 26.89 | 30.38 | −12.9 | 2.89 | 2.80 | 3.04 |
| Mean absolute percentage error | | | 5.81 | | | 8.26 |

and 8.26%, respectively. However, the observed thrust force and torque of both experimental confirmation and regression model showed deviation within 9%. It was observed clearly that the regression model was a better and effective method for the prediction of torque and thrust force in the drilling of bagasse fiber-reinforced vinyl ester composites.

## 5.4 Conclusions

This study reports the outcome of machining parameters on the torque and thrust force produced by HSS twist drill bits during the fiber-reinforced short bagasse vinyl ester composite drilling process. Prediction and comparison of the torque and thrust force are also provided. Thrust force and torque in bagasse fiber-reinforced vinyl ester composite drilling increased not only with feed rate but also with cutting speed and diameter of drill bits.

Torque and thrust force in the drilling of bagasse fiber-reinforced vinyl ester composite increased not only with feed rate but also with cutting speed and the diameter of the drill bits. A greater contribution to the torque and thrust force was seen from the feed rate than from the drill diameter and the cutting speed. The optimal thrust force parameters are the level 3 feed rate (0.3 mm/rev), level 3 cutting speed (1800 rpm), and level 2 drill diameter (8 mm) per drill. The optimal parameters for torque are the level 3 feed rate (0.3 mm/rev), level 1 cutting speed (600 rpm), and level 3 (10 mm) drill diameter. Confirmation testing showed that the regression model is a better and efficient method for predicting torque and thrust force in drilling bagasse fiber-reinforced vinyl ester composite.

## References


1 Athijayamani, A., Thiruchitrambalam, M., Winowlin, J.J.T., and Alavudeen, A. (2008). Analysis of chopped roselle and sisal fiber hybrid polyester composite. *Int. J. Plast. Technol.* 12: 1031–1038.

2 Athijayamani, A., Thiruchitrambalam, M., Natarajan, U., and Pazhanivel, B. (2009). Effect of moisture absorption on the mechanical properties of randomly oriented natural fibers/polyester hybrid composites. *Mater. Sci. Eng., A* 517: 344–353.

3 Athijayamani, A., Thiruchitrambalam, M., Natarajan, U., and Pazhanivel, B. (2010). Influence of alkali-treated fibers on the mechanical properties and machinability of roselle and sisal fiber hybrid polyester composite. *J. Polym. Compos.* 31: 723–731.

4 Thiruchitrambalam, M., Athijayamani, A., Sathiyamurthy, S., and Thaheer, A.S.A. (2010). A review on the natural fiber-reinforced polymer composites for the development of roselle fiber-reinforced polyester composite. *J. Nat. Fibers* 7 (4): 307–323.

5 Adekunle, K., Cho, S.W., Ketzscher, R., and Skrifvars, M. (2012). Mechanical properties of natural fiber hybrid composites based on renewable thermoset resins derived from soybean oil, for use in technical applications. *J. Appl. Polym. Sci.* 124 (6): 4530–4541.

6 Manickam, C., Kumar, J., Athijayamani, A., and Samuel, J.E. Effect of various water immersions on mechanical properties of roselle fiber–vinyl ester composites. *Polym. Compos.* 36 (9): 1638–1646. https://doi.org/10.1002/pc.23072.

7 Cao, Y., Shibata, S., and Fukumoto, I. (2006). Mechanical properties of biodegradable composites reinforced with bagasse fibre before and after alkali treatments. *Composites Part A* 37 (3): 423–429. https://doi.org/10.1016/j.compositesa.2005.05.045.

8 Oladele, I.O. (2014). Effect of bagasse fibre reinforcement on the mechanical properties of polyester composites. *J. Assoc. Prof. Eng. Trinidad and Tobago* 42 (1): 12–15.

9 Agunsoye, J.O. and Aigbodion, V.S. (2013). Bagasse filled recycled polyethylene bio-composites: morphological and mechanical properties study. *Results Phys.* 3: 187–197. https://doi.org/10.1016/j.rinp.2013.09.003.

10 Cerqueira, E.F., Baptista, C.A.R.P., and Mulinari, D.R. (2011). Mechanical behaviour of polypropylene reinforced sugarcane bagasse fiber composite. *Procedia Eng.* 10: 2046–2055. https://doi.org/10.1016/j.proeng.2011.04.339.

11 Huang, Z., Wang, N., Zhang, Y. et al. (2012). Effect of mechanical activation pretreatment on the properties of sugarcane bagasse/poly(vinyl chloride) composites. *Composites Part A* 43 (1): 114–120.

12 Silva Pinto, C.E.S., Arizaga, G.G.C., Fernando, W. et al. (2009). Studies on the effect of pressure and incorporation of sugarcane bagasse fibers on the structure and properties of poly(hydroxybutyrate). *Composites Part A* 40: 573–582.

13 Tewari, M., Singh, V.K., Gope, P.C., and Chaudhary, A.K. (2012). Evaluation of mechanical properties of bagasse-glass fiber reinforced composite. *J. Mater. Environ. Sci.* 3 (1): 171–184.

14 Al Bakri, A.M.M., Liyana, J., Norazian, M.N. et al. (2013). Mechanical properties of polymer composites with sugarcane bagasse filler. *Adv. Mater. Res.* 740: 739–744.

15 Stalin, B. and Athijayamani, A. (2016). The performance of bio waste fibres reinforced polymer hybrid composite. *Int. J. Mater. Eng. Innov.* 7 (1): 15–25. https://doi.org/10.1504/IJMATEI.2016.077312.

16 Rao, R.V. (2011). *Advanced Modeling and Optimization of Manufacturing Processes*. London: Springer-Verlag London Limited. https://doi.org/10.1007/978-0-85729-015-1.

17 Athijayamani, A., Natarajan, U., and Thiruchitrambalam, M. (2010). Prediction and comparison of thrust force and torque in drilling of natural fibre hybrid composite using regression and artificial neural network modelling. *Int. J. Mach. Mach. Mater.* 8 (1/2): 131–145. https://doi.org/10.1504/IJMMM.2010.034492.

18 Mohan, N.S., Kulkarni, S.M., and Ramachandra, A. (2007). Delamination analysis in drilling process of glass fiber reinforced plastic (GFRP) composite

materials. *J. Mater. Process. Technol.* 186 (1): 265–271. https://doi.org/10.1016/j.jmatprotec.2006.12.043.

**19** Gaitonde, V.N., Karnik, S.R., and Davim, J.P. (2008). Taguchi multiple-performance characteristics optimization in drilling of medium density fibreboard (MDF) to minimize delamination using utility concept. *J. Mater. Process. Tech.* 196 (1–3): 73–78.

**20** Chavan, V.R., Dinesh, K.R., Veeresh, K. et al. (2017). Taguchi's orthogonal array approach to evaluate drilling of GFRP particulate composites. *Mater. Today: Proc.* 4 (10): 11245–11250. https://doi.org/10.1016/j.matpr.2017.09.046.

**21** Tsao, C.C. and Hocheng, H. (2008). Evaluation of thrust force and surface roughness in drilling composite material using Taguchi analysis and neural network. *J. Mater. Process. Technol.* 203 (1–3): 342–348. https://doi.org/10.1016/j.jmatprotec.2006.04.126.

**22** Subramanian, K. and Cook, N.H. (1997). Sensing of drill wear and prediction of drill life. *Trans. ASME, J. Eng. Ind.* 99: 295–301.

**23** Singh, R.V.S., Latha, B., and Senthilkumar, V.S. (2009). Modeling and analysis of thrust force and torque in drilling GFRP composites by multi-facet drill using fuzzy logic. *Int. J. Recent Trends Eng.* 1 (5): 66–70.

**24** Ho-Cheng, H. and Puw, H.Y. (1992). On drilling characteristics of fiber reinforced thermoset and thermoplastics. *Int. J. Mach. Tools Manuf.* 32 (4): 583–592.

**25** Khashaba, U.A. (2004). Delamination in drilling GFR-thermoset composites. *Compos. Struct.* 63: 313–327.

**26** Khashaba, U.A., Seif, M.A., and Elhamid, M.A. (2007). Drilling analysis of chopped composites. *Composites Part A* 38: 61–70.

# 6

# Mechanical and Dynamic Properties of Kenaf-Fiber-Reinforced Composites

*Brijesh Gangil[1], Lalit Ranakoti[2], and Pawan K. Rakesh[2]*

*[1]H.N.B. Garhwal University, Mechanical Engineering Department, Garhwal, 246174, Srinagar, India*
*[2]NIT UK, Department of Mechanical Engineering, Srinagar 246174, Uttarakhand, India*

## 6.1 Introduction

Research with regard to composite material has always been associated with the properties of reinforcement, matrix, or both. The final properties of the composite material depend on these [1]. Fabricating a composite without prior knowledge of the reinforcement and the matrix may lead to wastage. However, composites are still fabricated and investigated in very large numbers in the research community in the hope of encouragement, motivation, and optimistic outputs [2]. The reinforcing material used in the composite may be natural or synthetic depending upon the requirement of the investigation. Synthetic fiber has very high strength and shows better adhesion with the matrix material than natural fiber because of which high-strength polymer composites are generally based on synthetic fibers [3]. Nevertheless, various findings reveal that the chemical treatment of natural fiber has been a vital aspect for the enhancement of natural fiber–polymer adhesion. The current situation demands increasing exploration of natural fibers in the polymer composite material due to its eco-friendly characteristics, low-cost, and ease of availability [4]. Basically, it has been seen that composite materials based on natural fiber are explicitly manufactured for their mechanical and tribological performance. Owing to this prerequisite, varieties of natural fibers have been subjected to composite fabrication [5]. Kenaf fiber (KF) is one of the natural fibers that have been examined for their potential in composite materials. Results reported in the past encompass mechanical, tribological, dynamic, and biodegradability characterization.

One of the oldest natural fibers used as reinforcement in the composite is kenaf fiber [6]. It has been used for ropes and sackcloth for over 5 millennia. The beginning of the domestication of Kenaf fiber is credited to North Africa, but later its production started all over the world [7]. Kenaf is a lightweight and the strongest natural fiber known to humans. Composite materials based on kenaf fiber have been recognized as a most likely alternative for synthetic fiber composites. Literature suggests that a wide range of polymers have been used for composite fabrication

*Mechanical and Dynamic Properties of Biocomposites,* First Edition. Edited by Senthilkumar Krishnasamy, Rajini Nagarajan, Senthil Muthu Kumar Thiagamani, and Suchart Siengchin.

with kenaf fiber as reinforcement [8]. Composites manufactured in the future with kenaf fiber will have to be based on the results reported in the past. In order to yield the maximum possible information from the work done in the past in relevance to kenaf-fiber-reinforced polymer composite, a review of the mechanical and dynamic properties of the kenaf-fiber-reinforced polymer composite would be helpful in the development of future composite fabrications.

## 6.2 Mechanical Properties of Kenaf-Fiber-Reinforced Polymer Composite

A material is always characterized by its mechanical properties, such as tensile, compression, flexural, impact, and wear characteristics. The ability of a material to perform under critical or extreme situations is characterized by its properties. To make the material viable for industrial applications, the mechanical properties of the material need to be thoroughly investigated.

At present, plenty of work has been conveyed till date by several scholars in the field of kenaf fiber–polymer composite. For instance, kenaf-fiber-based polylactic acid (PLA) composite exhibits better mechanical properties as compared to the other natural fiber–PLA composites. Kenaf fiber in the PLA matrix enhances the tensile and flexural strength of the matrix material till 50% of KF but as the percentage of kenaf fiber in the PLA increases beyond 50%, the said properties start decreasing. The Young's modulus of kenaf fiber–PLA composite also shows significant enhancement up to 70% of fiber content but decreases as the fiber content exceeds 70% (Figure 6.1) [9].

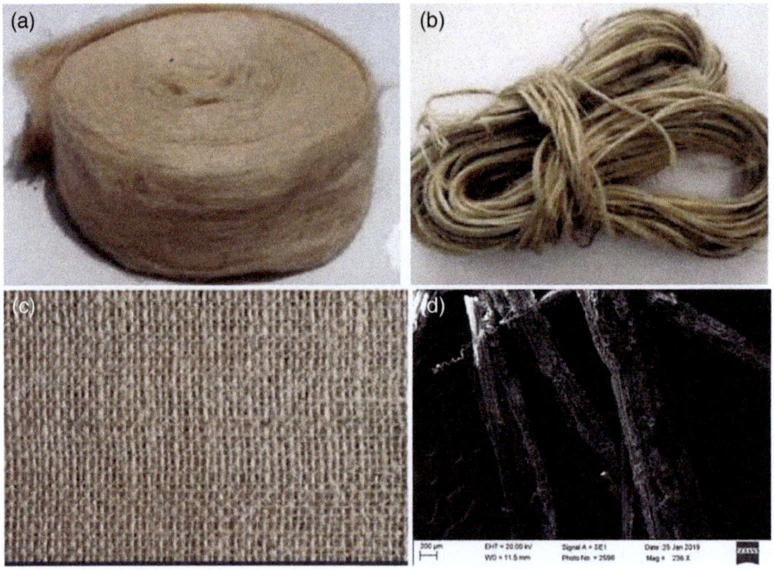


**Figure 6.1** Different forms of kenaf fiber (a) non-woven, (b) fiber strands, (c) bi-directional mat and (d) Scanning electron image of kenaf fiber.

The reason behind the reduction in the property may be attributed to the deficiency in matrix material, which further leads to increase in the void content in the composite. The ductility of the matrix material helps in efficient transfer of stress from the matrix to the fiber, leading to improved mechanical properties of the composite. Kenaf fiber–polypropylene (PP) composites can be prepared by incorporating short and long fibers. Short kenaf fibers (20 mm) can be mixed with PP in a mixer to achieve homogeneity while long KF fibers (130 mm) can be placed in between layers of PP [10]. In spite of achieving better mechanical strength, the optimized fabrication technique was recognized as a layered shifting technique in which PP in the form of powder is placed uniformly in the kenaf fiber to make alternate layers of fiber and matrix. The properties of the kenaf-fiber-based polymer composite are incredibly influenced by the coupling agents due to the enhanced interfacial adhesion between the matrix and the fiber [11]. In such instances, on treating the kenaf fiber with maleic anhydride polypropylene (MAPP), the flexural, tensile, and impact strength were found to be superior to those of the untreated kenaf fiber composite [12]. Fiber loading also plays a vital role in the final strength of the composite material. While increasing the fiber content alone cannot always be advantageous optimizing the fiber loading with respect to the matrix and fabrication technique is. For example, by using hot-press technique for long kenaf fiber at 50% fiber loading, the composite shows improved strength but short kenaf fiber composite fabricated by similar technique at equal fiber loading does not. This happens due to the difference in the fiber length, inhomogeneity, and poor bonding of fiber and matrix [13]. Sometimes, changing the process of fabrication may result in a better and durable composite. For instance, kenaf fiber–PP composite, fabricated in the form of sandwich structure, exhibits better wetting between the matrix and the fiber. This leads to enhancement in flexural modulus as confirmed by scanning electron microscopy (SEM) [14]. Apart from the technique used in the fabrication, heating time and curing also play very crucial roles in governing the mechanical properties of the kenaf fiber–polymer composite. Longer curing time gives higher ductility while a shorter one gives rigidness. While preparing a composite by induction heating, the heating parameter should be such that it does not go beyond the melting point of the matrix. One such study focused on the bonding performance of the kenaf fiber and matrix heated the polymer at elevated temperature, and reported enhanced flexural strength of the composite [15, 16]. The issue of nonhomogeneous fiber–polymer composite is very common in natural fiber composites. In this regard, a composite of kenaf fiber–PP was fabricated with the help of compression molding technique [17]. It was found that compaction pressure and cooling process are the essential parameters to decide the strength of the composite. For the composites fabricated with kenaf fiber and PLA numerous observations have been reported with respect to the processing parameters of the fabrication. However, optimized processing parameters to obtain the maximum strength were reported as heating at 185 °C for 15 minutes [18]. It should be noted that heating above 185 °C for a longer period of time may result in thermal degradation of PLA.

When kenaf fiber is used in mat form, the strength of the composites could be as good as the material used in automobile applications. For this, excellent experimental mechanical results such as tensile strength, flexural strength, and elongation were reported for kenaf fiber–PLA composites in which the kenaf fiber was used as a unidirectional mat. Mechanical strength of the kenaf fiber–polymer composite can be improved by using coupling agents [19]. Almost 60% improvement was reported for the tensile strength for the kenaf fiber–PLA composite with the addition of ammonium polyphosphate (APP) [20]. However, compatibility between the kenaf fiber and PLA decreases, due to which hardness and flexural strength reduce. With the increasing growth in the area of nanocomposites, kenaf fiber has been reinforced in the nanofiber form for the development of nanocomposite. Nanocomposites composed of nanofibers of kenaf fiber with PLA matrix were investigated at fiber loading of 3%, 5%, 8%, and 10%. The mechanical properties of the nanocomposite get significantly improved at nanofiber loading above 3% as reported in the study [21].

Studies regarding the orientation of kenaf fiber have been examined by several researchers. The results show that both parallel and perpendicular directions of kenaf fiber reinforcement have their own specific advantages. For instance, strength and modulus are enhanced for parallel and perpendicular direction of reinforcement respectively. The fiber brings anisotropy to the polymer matrix, which is further reflected in the mechanical properties of the composite [22]. The compatibility of fiber matrix depends on the chemical constituents present in the fiber. Since kenaf fiber is a natural fiber, its affinity toward water is high due to which a large amount of water is generally found in kenaf fiber. The mechanical properties of kenaf fiber–polymer composite are generally degraded due to the presence of water; thus various compatibilizers and chemical processes are available to extract the water from the fiber, making it compatible with the polymers [23]. The results related to the improvement in the mechanical properties for treated kenaf fiber as compared to untreated kenaf fiber polymer composites are available in the past literature. Alkaline treatment at 6% concentration subsequently makes the fiber rough, and improves the interlocking efficiency of the fiber and the matrix. Polymer composites comprising of kenaf fiber, nanofillers, and polymer matrix are now extensively under investigation. Nano-silica filler, organically modified montmorillonite (ommt), single- and multiwall carbon nanotubes, etc. are some of the commonly known nanofillers that have been used with kenaf fiber for the development of hybrid nanocomposites [24]. Fracture toughness, impact strength, and modulus of rupture increase significantly with the combination of kenaf fiber and nanofillers [25].

## 6.3 Dynamic Mechanical Analysis

Dynamic mechanical analysis (DMA) is a nondestructive material testing, considered to be the most effective technique for the analysis of viscoelastic behavior of polymer composites [26]. Parameters such as temperature, time, stress, and frequency are used to define the storage modulus ($E'$), loss modulus ($E''$), and tan $\delta$. $E'$ is defined as the ability of a material to absorb energy in the elastic limit while $E''$ is the dissipation of energy in the form of heat. Tan $\delta$ is the ratio of loss modulus and storage modulus, which defines the damping parameter of the material. These properties are measured under varying load and temperature, making them the main parameters of DMA. To have deep insights of DMA, the glass transition temperature $T_g$ should be clearly understood because most of the critical values are directly linked to the $T_g$. A polymer has higher damping factor than the fiber due to its high viscosity and low elasticity but the final damping factor of the composite depends upon the composition of the fiber and the polymer [27]. The DMA of a kenaf fiber composite is discussed below.

## 6.4 Storage Modulus (*E*′) of Kenaf Fiber–Polymer Composite

One of the most important parameters for the DMA analysis, storage modulus ($E'$) for kenaf fiber polymer composite, has been reported in several studies. For instance, the kenaf fiber–epoxy based composite was examined for $E'$. The $E'$ decreases as the temperature increases but becomes maximum at the $T_g$. The $E'$ further decreases gradually as the temperature $> T_g$ because the random mobility of the molecule results in loosened packing arrangement [28]. It should be kept in mind that large variations were observed for $E'$ in glass region as compared to the rubbery region. Further, addition of nanofillers in kenaf fiber–polymer composite leads to enhancement in the value of $E'$. The nanofiller enhances the interfacial bonding of Kenaf fiber and epoxy matrix due to the modulus mismatch of the filler and the fiber [29]. $E'$ can be enhanced by the hybrid kenaf fiber composite. The oil palm empty fruit bunch (OPEFB)-based kenaf fiber–epoxy hybrid composite shows higher value of $E'$ as compared to kenaf fiber–epoxy and almost comparable to ommt-filled Kenaf fiber–epoxy nanocomposite. For both the regions, glassy and rubbery, ommt-based kenaf fiber–epoxy composite shows the highest value of $E'$ due to the insertion of nanoclay of ommt in between the epoxy matrix, leading to efficient stress transfer. Increasing the length of the kenaf fiber also leads to increase in the value of $E'$ as long kenaf fiber (30 mm)–epoxy composite exhibits higher $E'$ than short kenaf fiber (5 mm)–epoxy composite. Another reason for the increased storage modulus is the restriction of the mobility of chain and the improvement in the mechanical interlocking as observed in the report when the kenaf is reinforced in the copolyester (CP) matrix [30]. It has also been seen that $E'$ varies inversely with temperature. $E'$ decreases with increase in the temperature and vice versa. For instance, a kenaf fiber–polymer composite treated with NaOH shows 13 MPa of $E'$ at 145 °C and 40 MPa of $E'$ at 60 °C.

## 6.5 Loss Modulus (*E*″) of Kenaf Fiber–Polymer Composite

The dissipation of energy in the form of heat is known as loss modulus ($E''$). The variation of $E''$ with temperature for kenaf fiber composite, hybrid kenaf fiber, and

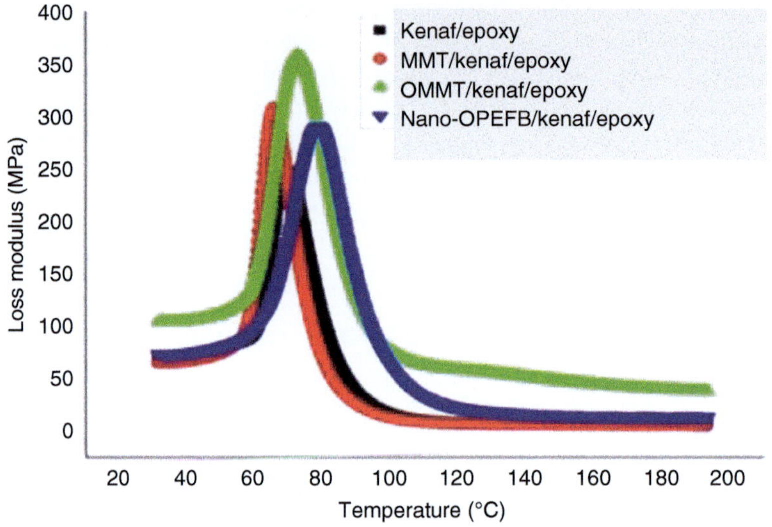


**Figure 6.2** Effect of nanofiller loading in kenaf/epoxy composites on loss modulus. Source: Based on Saba et al. [31].

kenaf fiber nanocomposite, as shown in Figure 6.2, follows a certain pattern of first increasing, reaching a maximum, and then decreasing, almost similar to the $E'$. The plot clearly shows that the nanofiller-filled kenaf fiber composite exhibit the maximum $E''$ [31]. The addition of nanofillers increases the overall surface area and internal friction, which results in improved energy dissipation. Among the nanofillers, the maximum peak of $E''$ is demonstrated for ommt followed by montmorillonite (MMT) and nano-OPEFB.

The improved interfacial attraction among the reactants due to the presence of polar and nonpolar groups leads to enhancement in $E''$ for the kenaf fiber nanocomposites [32]. Incorporation of stiffer and phyllosilicate minerals in the form of layer (MMT, ommt) into the kenaf/epoxy composites improves the complex modulus ($E$) of the kenaf/epoxy hybrid nanocomposites as compared to kenaf/epoxy composites. Treatment of kenaf fiber such as mercerization has a significant effect on the $E''$ of composite but only above the $T_g$ [33]. There is no obvious reason but $E''$ increases in the plastic region and decreases in the rubbery region as the temperature increases further.

## 6.6 Damping Factor (Tan $\delta$)

Tan δ of kenaf–epoxy composites and kenaf–epoxy nanohybrid nanocomposites has been shown in Figure 6.3. The kenaf/epoxy composites exhibit the maximum tan $\delta$ among all the nanohybrid composites. Tan $\delta$ increases with increment in temperature, reaches a peak in the transition area, and decreases in the rubbery region for all kenaf fiber composites. The value of tan $\delta$ is found to be low below $T_g$ because of the frozen chain segments and it is high due to the high mobility of the molecule.

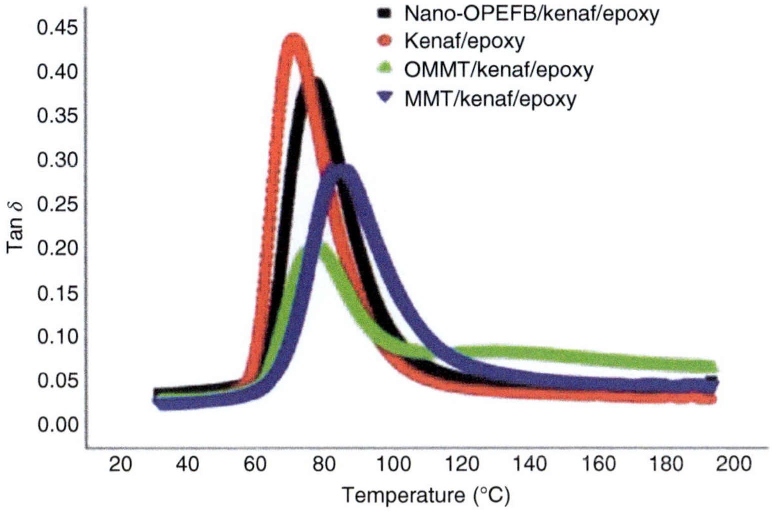


**Figure 6.3** Effect of nanofiller loading in kenaf/epoxy composites on damping factor. Source: Based on Shinoj et al. [34].

As in case of $E''$, the tan $\delta$ of kenaf fiber nanocomposites increases with the addition of nanofillers. However, in the case of OPEFB nanofillers, tan $\delta$ decreases for both glassy and rubbery regions [34] as the addition of OPEFB causes larger increment in the value of $E'$ as compared to $E''$ due to increased stiffness of the kenaf fiber nanocomposite by the inclusion of hard particles. Tan $\delta$ is a function of $T_g$, which can be understood by the fact that increasing the temperature beyond $T_g$ causes changes in the tan $\delta$ pattern due to the reduction of friction in between the molecular chains.

The peak of tan $\delta$ as can be seen from Figure 6.2 is broader for the nanohybrid kenaf fiber composite due to the obstruction imposed by the nanofiller to the polymer matrix by making a cross-linking bond. The kenaf fiber hybrid composite comprising NaOH-treated fiber exhibits a higher value of tan $\delta$ than the untreated kenaf fiber hybrid composite. Moreover, early degradation of the composite takes place for the untreated kenaf fiber composite than the treated kenaf fiber hybrid composite [35]. Interestingly, silane-treated kenaf fiber composite exhibits two relaxation peaks, which is neither found in the case of NaOH-treated fiber nor in the pure copolyester [36]. Tan $\delta$ as a function of temperature for CP and its kenaf fiber composites was also studied. It was observed that the damping peak occurs between −20 and −30 °C with the β- or glass transition of CP. The treated kenaf fiber composite shows an α-transition between 70 and 90 °C associated with the molecular movement in crystalline phase [37]. There is no relationship between this peak's position or intensity and the type of composite or composite treatment.

## 6.7 Glass Transition Temperatures ($T_g$)

Glass transition is the method by which a polymeric material property is characterized. It is the temperature range where a substance softens from a

**Table 6.1** Peak height of tan $\delta$ curve and $T_g$ obtained from $E''$ and tan $\delta$ curve peak for kenaf/epoxy composites and all kenaf/epoxy composites [25, 28, 31, 38–40].

| Sample | Peak height of tan $\delta$ curve | $T_g$ from tan $\delta$ curve (°C) | $T_g$ from $E''$ curve (°C) |
|---|---|---|---|
| Kenaf/epoxy 05/25 | – | 64.49 | 53.27 |
| Kenaf/epoxy 10/25 | – | 64.68 | 58.57 |
| Kenaf/epoxy 30/25 | – | 67.17 | 61.03 |
| Kenaf/epoxy 50/25 | – | 68.43 | 62.47 |
| Neat resin | 0.734 | 69.1 | 60.5 |
| Untreated kenaf-UD | 0.824 | 73.8 | 63.1 |
| Treated kenaf–UD | 0.320 | 87.0 | 78.3 |
| Untreated kenaf-MAT | 0.810 | 71.5 | 62.0 |
| Treated kenaf-MAT | 0.305 | 84.3 | 72.5 |
| Kenaf/epoxy | 0.43 | 73.01 | 70.01 |
| Nano OPEFB/kenaf/epoxy | 0.36 | 81.00 | 80.6 |
| MMT/kenaf/epoxy | 0.27 | 90.00 | 70.0 |
| ommt/kenaf/epoxy | 0.20 | 80.02 | 80.0 |
| Kenaf/epoxy | 0.49 | 77.42 | 73.08 |

Sources: Saba et al. [25, 28, 31], Kumar et al. [38], Fiore et al. [39], Chee et al. [40].

rigid, glass-like state to a rubber-like state. During temperature scan heating the onset decreases in the storage modulus, accompanied by a peak in tan δ. Table 6.1 shows the values of peak height of the tan $\delta$ curve, $T_g$ values from the tan $\delta$, and $T_g$ from $E''$ values for various composites. Values of $T_g$ obtained from $E''$ are comparatively lower than the values of $T_g$ obtained for tan $\delta$.

Various natural fiber (American wood fibers natural fillers and Kenaf fibers)-filled Polypropylene matrix composites were developed and the basic information on the effects of the contents on the thermal, mechanical, and viscoelastic behavior of such composites collected through DMA. As evident from Table 6.2, Kenaf fiber epoxy composite exhibits lower value of $T_g$ as compared to kenaf fiber hybrid composite. Kenaf-fiber-reinforced polyester-pultruded composites have a large dependence on the fiber loading [28]. As the fiber loading increases, the dynamic modulus decreases till the temperature remains below $T_g$ but increases as the temperature goes beyond $T_g$. However, as the frequency shifts, the $T_g$ shifts to higher temperatures, leading to good fiber/matrix interaction. Enhancement in the $T_g$ value and reduction in tan $\delta$ peak height for kenaf/epoxy hybrid nanocomposites indicate that molecular mobility of the epoxy matrix decreases in the composite.

$T_g$ is governed by several factors; some tend to increase it while others decrease it. It can be established that the shearing of materials dominates the studied systems during the mixing of fiber and matrix. Variable processing conditions for the

**Table 6.2** Storage modulus, loss modulus, and tan δ values for different composites [36].

| Temperature | Composites formulation | Storage modulus | Loss modulus | Tan δ |
|---|---|---|---|---|
| −60 | | 5.530 | 0.164 | 0.030 |
| −20 | | 5.118 | 0.156 | 0.030 |
| 20 | PP-KF-25 | 3.126 | 0.189 | 0.060 |
| 60 | | 1.835 | 0.117 | 0.064 |
| 100 | | 1.047 | 0.096 | 0.092 |
| −60 | | 7.349 | 0.219 | 0.030 |
| −20 | | 6.701 | 0.198 | 0.030 |
| 20 | PP-KF-50 | 4.733 | 0.225 | 0.048 |
| 60 | | 3.132 | 0.182 | 0.058 |
| 100 | | 2.001 | 0.166 | 0.083 |
| −60 | | 5.668 | 0.155 | 0.027 |
| −20 | | 5.060 | 0.148 | 0.029 |
| 20 | PP-NP-25 | 3.002 | 0.173 | 0.058 |
| 60 | | 1.658 | 0.110 | 0.066 |
| 100 | | 0.906 | 0.088 | 0.097 |
| −60 | | 6.536 | 0.183 | 0.028 |
| −20 | | 6.252 | 0.180 | 0.029 |
| 20 | PP-NP-50 | 4.100 | 0.210 | 0.051 |
| 60 | | 2.646 | 0.156 | 0.059 |
| 100 | | 1.658 | 0.143 | 0.086 |
| −60 | | 5.019 | 0.130 | 0.026 |
| −20 | | 4.486 | 0.125 | 0.028 |
| 20 | PP-RH-25 | 2.595 | 0.166 | 0.064 |
| 60 | | 1.306 | 0.096 | 0.074 |
| 100 | | 1.181 | 0.107 | 0.090 |
| −60 | | 5.364 | 0.150 | 0.028 |
| −20 | | 4.776 | 0.152 | 0.032 |
| 20 | PP-RH-50 | 3.052 | 0.158 | 0.051 |
| 60 | | 1.918 | 0.120 | 0.063 |
| 100 | | 1.181 | 0.107 | 0.090 |
| −60 | | 5.687 | 0.135 | 0.027 |
| −20 | | 5.065 | 0.146 | 0.029 |
| 20 | PP-wood flour (WF)-25 | 3.046 | 0.180 | 0.060 |
| 60 | | 1.659 | 0.114 | 0.069 |

*(continued)*

**Table 6.2** (Continued)

| Temperature | Composites formulation | Storage modulus | Loss modulus | Tan $\delta$ |
|---|---|---|---|---|
| 100 | | 1.181 | 0.107 | 0.090 |
| −60 | | 6.852 | 0.237 | 0.035 |
| −20 | | 6.270 | 0.218 | 0.035 |
| 20 | PP-WF-50 | 4.234 | 0.227 | 0.054 |
| 60 | | 2.850 | 0.162 | 0.057 |
| 100 | | 1.816 | 0.150 | 0.083 |

Source: Based on Mazuki et al. [36].

preparation of test specimen can be another factor influencing the $T_g$. Additional results are required to authenticate the reasons. In this regard, Natural-fiber-filled Polypropylene matrix composites were manufactured and examined for thermal, mechanical, and viscoelastic behavior through DMA. From Table 6.2 it is found that natural-fiber-filled polypropylenes act more elastically than their pure counterparts. Hybrid composite of kenaf fibers and rice hulls displayed the maximum and the minimum storage modulus values, respectively, representing improved reinforcement efficiency. Rice-hull-filled composites exhibit maximum loss factor while the minimum is obtained for kenaf-fiber-filled composite at all ranges of temperature. Moreover, the stiffness increases and damping reduces with increase in the fiber content of the composite. The glass transition temperature in natural polypropylene fiber composites is governed by various factors. Some of these factors tend to escalate the transition temperature ($T_g$), while others lower the temperature $T_g$. It was concluded that lowering of $T_g$ might be due to the materials sheared off the fibers during the mixing process. The potential impact of variable processing conditions for preparing the test specimens can be considered as another factor influencing the glass transition temperature. Further work is required to confirm such an effect.

## 6.8 Conclusion

Kenaf fiber has been considered as a prominent plant fiber having mechanical properties comparable to those of synthetic fibers. Composite materials based on kenaf fiber have great scope for replacing synthetic fiber composites.

Mechanical results show that the properties of polymers, such as tensile strength and flexural strength, are significantly enhanced by using the kenaf fiber. Nanocomposites composed of nano-kenaf fiber with PLA matrix also show better mechanical results than neat PLA. Since kenaf fiber is a natural fiber, its affinity toward water is high due to which large amount of water is generally found in kenaf fiber, which should be taken care of in the upcoming investigations of kenaf fiber polymer composites. In the case of DMA, $E'$ increases with increase in the length of the kenaf fiber and also with the incorporation of nanofibers. The variation of $E''$

with temperature for kenaf fiber composite, hybrid kenaf fiber, and kenaf fiber nanocomposite follows a certain pattern of first increasing, reaching a maximum, and then decreasing, almost similar to $E'$. Kenaf hybrid composites exhibit high value of $\tan \delta$ than the untreated kenaf fiber hybrid composite. Moreover, early degradation of the composite takes place for the untreated kenaf fiber composite than the treated kenaf fiber hybrid composite.

## References


**1** Schultz, J.A., Lavielle, L., and Martin, C. (1987). The role of the interface in carbon fibre-epoxy composites. *J. Adhes.* 23 (1): 45–60.

**2** Amić, A., Marković, Z., Marković, J.D. et al. (2017). Radical scavenging potency of 4-hydroxyphenylpropionic acid: a theoretical approach. *25th Croatian Meeting of Chemists and Chemical Engineers with International Participation 3rd Symposium "Vladimir Prelog".*

**3** Lalit, R., Mayank, P., and Ankur, K. (2018). Natural fibers and biopolymers characterization: a future potential composite material. *Strojnícky časopis - J. Mech. Eng.* 68 (1): 33–50.

**4** Ranakoti, L., Gupta, M.K., and Rakesh, P.K. (2019). Analysis of mechanical and tribological behavior of wood flour filled glass fiber reinforced epoxy composite. *Mater. Res. Express* 6 (8): 085327.

**5** Jogur, G., Nawaz Khan, A., Das, A. et al. (2018). Impact properties of thermoplastic composites. *Text. Prog.* 50 (3): 109–183.

**6** Sanjay, M.R., Arpitha, G.R., Naik, L.L. et al. (2016). Applications of natural fibers and its composites: an overview. *Nat. Resour.* 7 (3): 108–114.

**7** Islam, M. (2019). Kenaf (*Hibiscus cannabinus* L., Malvaceae) research and development advances in Bangladesh: a review. *J. Nutr. Food Process.* 2 (1).

**8** Ahmad, F., Choi, H.S., and Park, M.K. (2015). A review: natural fiber composites selection in view of mechanical, light weight, and economic properties. *Macromol. Mater. Eng.* 300 (1): 10–24.

**9** Huda, M.S., Drzal, L.T., Misra, M., and Mohanty, A.K. (2006). Wood-fiber-reinforced poly (lactic acid) composites: evaluation of the physicomechanical and morphological properties. *J. Appl. Polym. Sci.* 102 (5): 4856–4869.

**10** Batouli, S.M., Zhu, Y., Nar, M., and D'Souza, N.A. (2014). Environmental performance of kenaf-fiber reinforced polyurethane: a life cycle assessment approach. *J. Cleaner Prod.* 66: 164–173.

**11** Cho, D., Lee, H.S., and Han, S.O. (2009). Effect of fiber surface modification on the interfacial and mechanical properties of kenaf fiber-reinforced thermoplastic and thermosetting polymer composites. *Compos. Interfaces* 16 (7–9): 711–729.

**12** Dieu, T.V., Phai, L.T., Ngoc, P.M. et al. (2004). Study on preparation of polymer composites based on polypropylene reinforced by jute fibers. *JSME Int. J. Ser. A: Solid Mech. Mater. Eng.* 47 (4): 547–550.

**13** Graupner, N., Herrmann, A.S., and Müssig, J. (2009). Natural and man-made cellulose fibre-reinforced poly (lactic acid)(PLA) composites: an overview about mechanical characteristics and application areas. *Composites Part A* 40 (6–7): 810–821.

**14** Chaudhary, V., Bajpai, P.K., and Maheshwari, S. (2018). Studies on mechanical and morphological characterization of developed jute/hemp/flax reinforced hybrid composites for structural applications. *J. Nat. Fibers* 15 (1): 80–97.

**15** Alagirusamy, R., Fangueiro, R., Ogale, V., and Padaki, N. (2006). Hybrid yarns and textile preforming for thermoplastic composites. *Text. Prog.* 38 (4): 1–71.

**16** Ranakoti, L., Gupta, M.K., and Rakesh, P.K. (2019). Silk and silk-based composites: opportunities and challenges. In: *Processing of Green Composites* (eds. P.K. Rakesh and I. Singh), 91–106. Singapore: Springer.

**17** Prakash, K.G. and Thotappa, C. (2018). A comparative study on properties of natural fiber reinforced composites: a review. *Int. J. Res.* 5 (20): 1576–1583.

**18** Ramesh, M. (2016). Kenaf (*Hibiscus cannabinus* L.) fibre based bio-materials: a review on processing and properties. *Prog. Mater Sci.* 78: 1–92.

**19** Wambua, P., Ivens, J., and Verpoest, I. (2003). Natural fibres: can they replace glass in fibre reinforced plastics? *Compos. Sci. Technol.* 63 (9): 1259–1264.

**20** Zhang, Z.X., Zhang, J., Lu, B.X. et al. (2012). Effect of flame retardants on mechanical properties, flammability and foamability of PP/wood–fiber composites. *Composites Part B* 43 (2): 150–158.

**21** Jonoobi, M., Harun, J., Mathew, A.P., and Oksman, K. (2010). Mechanical properties of cellulose nanofiber (CNF) reinforced polylactic acid (PLA) prepared by twin screw extrusion. *Compos. Sci. Technol.* 70 (12): 1742–1747.

**22** Hassan, E., Wei, Y., Jiao, H., and Huo, Y.M. (2012). Plant fibers reinforced poly(lactic acid) (PLA) as a green composites. *Int. J. Eng. Sci. Technol.* 4 (10): 4429–4439.

**23** Sudha, P., Nasreen, K., and Vinodhini, P.A. (2015). *Natural Fiber Composites and Applications*, 335–364. Oakville: Apple Academic Press.

**24** Karger-Kocsis, J., Mahmood, H., and Pegoretti, A. (2015). Recent advances in fiber/matrix interphase engineering for polymer composites. *Prog. Mater Sci.* 73: 1–43.

**25** Saba, N., Jawaid, M., Alothman, O.Y., and Paridah, M.T. (2016). A review on dynamic mechanical properties of natural fibre reinforced polymer composites. *Constr. Build. Mater.* 106: 149–159.

**26** Ardanuy, M., Claramunt, J., and Toledo Filho, R.D. (2015). Cellulosic fiber reinforced cement-based composites: a review of recent research. *Constr. Build. Mater.* 79: 115–128.

**27** Romanzini, D., Lavoratti, A., Ornaghi, H.L. et al. (2013). Influence of fiber content on the mechanical and dynamic mechanical properties of glass/ramie polymer composites. *Mater. Des.* 47: 9–15.

**28** Saba, N., Paridah, M.T., Abdan, K., and Ibrahim, N.A. (2016). Dynamic mechanical properties of oil palm nano filler/kenaf/epoxy hybrid nanocomposites. *Constr. Build. Mater.* 124: 133–138.

**29** Han, S.I., Lim, J.S., Kim, D.K. et al. (2008). In situ polymerized poly(butylene succinate)/silica nanocomposites: physical properties and biodegradation. *Polym. Degrad. Stab.* 93: 889–895.

**30** Mokhothu, T.H., Guduri, B.R., and Luyt, A.S. (2011). Kenaf fiber-reinforced copolyester biocomposites. *Polym. Compos.* 32 (12): 2001–2009.

**31** Saba, N., Paridah, M.T., Abdan, K., and Ibrahim, N.A. (2016). Effect of oil palm nano filler on mechanical and morphological properties of kenaf reinforced epoxy composites. *Constr. Build. Mater.* 123: 15–26.

**32** Pistor, V., Ornaghi, F.G., Ornaghi, H.L., and Zattera, A.J. (2012). Dynamic mechanical characterization of epoxy/epoxycyclohexyl–POSS nanocomposites. *Mater. Sci. Eng., A* 532: 339–345.

**33** Talib, R.A., Tawakkal, I.S.M.A., and Abdan, K. (2011). The influence of mercerized kenaf fiber reinforced polylactic acid composites on dynamic mechanical analysis. *Key Eng. Mater.* 471–472: 815–820.

**34** Shinoj, S., Visvanathan, R., Panigrahi, S., and Varadharaju, N. (2011). Dynamic mechanical properties of oil palm fibre (OPF)-linear low density polyethylene (LLDPE) biocomposites and study of fibre matrix interactions. *Biosyst. Eng.* 109: 99–107.

**35** Aziz, S.H. and Ansell, M.P. (2014). The effect of alkalization and fibre alignment on the mechanical and thermal properties of kenaf and hemp bast fibre composites: Part 1 – polyester resin matrix. *Compos. Sci. Technol.* 64: 1219–1230.

**36** Mazuki, A.A.M., Akil, H.M., Saiee, S. et al. (2011). Degradation of dynamic mechanical properties of pultruded kenaf iber reinforced composites after immersion in various solutions. *Composites Part B* 42 (1): 71–76.

**37** Tajvidi, M., Falk, R.H., and Hermanson, J.C. (2006). Effect of natural fibers on thermal and mechanical properties of natural fiber polypropylene composites studied by dynamic mechanical analysis. *J. Appl. Polym. Sci.* 101 (6): 4341–4349.

**38** Kumar, R., Hashmi, S.A.R., Nimanpure, S., and Naik, A. (2017). Enhanced dynamic mechanical properties of kenaf epoxy composites.

**39** Fiore, V., Di Bella, G., and Valenza, A. (2015). The effect of alkaline treatment on mechanical properties of kenaf fibers and their epoxy composites. *Composites Part B* 68: 14–21.

**40** Chee, S.S., Jawaid, M., and Sultan, M.T. (2017). Thermal stability and dynamic mechanical properties of kenaf/bamboo fibre reinforced epoxy composites. *BioResources* 12 (4): 7118–7132.

# 7

# Investigation on Mechanical Properties of Surface-Treated Natural Fibers-Reinforced Polymer Composites

*Sabarish Radoor[1], Jasila Karayil[2], Aswathy Jayakumar[3], and Suchart Siengchin[1]*

[1] *King Mongkut's University of Technology North Bangkok, Department of Mechanical and Process Engineering, The Sirindhorn International Thai-German Graduate School of Engineering (TGGS), 1518 Wongsawang Road, Bangsue 10800, Bangkok, Thailand*
[2] *Government Women's Polytechnic College, Department of Chemistry, Malaparamba P.O., Calicut, Kerala, India*
[3] *Mahatma Gandhi University, School of Biosciences, P.D. Hills, Kottayam, Kerala, 686560, India*

## 7.1 Introduction

During the past years, there has been an increasing trend of replacing conventional synthetic fiber with natural fiber to reinforce polymer matrix as it reduces environmental issues and depletion of fossil resources. Compared to synthetic fibers, natural fibers have the advantage of being abundant, cheap, nontoxic, biodegradable, recyclable, less dense, and renewable. Natural fibers are primarily derived from plants, animals, or minerals (Figure 7.1) [1–3].

Plant fibers are lignocellulosic in nature and mainly composed of cellulose, hemicellulose, and pectin. Animal fibers precede plant fibers in their availability and production and are primarily composed of proteins. They are obtained either from the hair of animals such as sheep, bison, alpaca, and angora or from the secretion of mulberries, oak, and silkworms (silks). Mineral fibers, on the other hand, are obtained from naturally occurring minerals. They are used as fillers in thermal insulation and fireproofing materials. Asbestos is the only naturally occurring mineral [4–6].

## 7.2 Mechanical Properties of Natural Fibers

The mechanical strength of fibers depends on its composition, structure, and cellular arrangement [7–10]. The mechanical properties of some natural fibers and conventional fiber are listed in Table 7.1. It is evident from the table that the natural fiber is mechanically inferior to synthetic fiber.

*Mechanical and Dynamic Properties of Biocomposites*, First Edition. Edited by Senthilkumar Krishnasamy, Rajini Nagarajan, Senthil Muthu Kumar Thiagamani, and Suchart Siengchin.

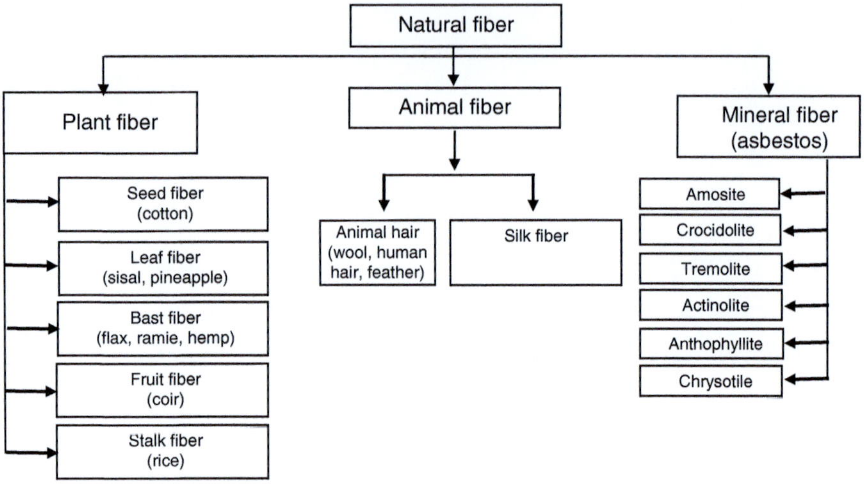


**Figure 7.1** Flow chart representation of classification of natural fibers.

**Table 7.1** Mechanical properties of natural fibers [7].

| Fiber type | Density (g/cm$^3$) | Young's modulus (GPa) | Tensile strength (MPa) | Elongation at break (%) | Moisture Constant (%) | Failure strain (%) | Microfibrillar spiral angle |
|---|---|---|---|---|---|---|---|
| Flax | 1.54 | 27.5–85 | 345–2000 | 1–4 | 10 | 1.3–3.3 | 10 |
| Ramie | 1.5–1.56 | 27–128 | 400–1000 | 1.2–3.8 | 8.5 | 3.7 | 7.5 |
| Hemp | 1.48 | 70 | 550–900 | 1.6 | 10.8 | 2–4 | 6.2 |
| Jute | 1.3–1.46 | 10–30 | 393–800 | 1.5–1.8 | 12.6 | 2–3 | 8 |
| Sisal | 1.45–1.5 | 9–22 | 350–700 | 2–7 | 11 | 2–14 | 20 |
| Coir | 1.2 | 4–6 | 175–220 | 15–30 | 8.0 | 15–40 | 41–45 |
| Cotton | 1.5–1.6 | 5.5–12.6 | 287–597 | 7–8 | 8.5 | 6–8 | — |
| Curaca | 1.4 | 11.8 | 500–1150 | 3.7–4.3 | — | — | — |
| Kenaf | 1.5 | 53 | 930 | 1.6 | 17 | — | — |
| Pineapple | 1.32 | 60–82 | 413–1627 | 2.4 | 11.8 | 0–1.6 | 14 |
| Bamboo | 0.6–1.1 | 11–17 | 140–230 | — | — | — | — |
| Banana | 1.35 | 27–32 | 529–914 | 5.3 | — | 1–3 | — |
| Kevlar | 1.44 | 60 | 3000 | 2.5–3.7 | — | — | — |
| Carbon | 1.78 | 240–425 | 3400–4800 | 1.4–1.8 | — | — | — |

Source: From Radoor et al. [7]. © 2020, BME-PT.

## 7.3 Drawbacks of Natural Fibers

The major drawbacks of natural fibers are their poor mechanical strength, high moisture adsorption, low thermal stability, incompatibility with hydrophobic polymer matrix, poor dimension stability, hygroscopic nature, high degree of swelling, and low thermal resistance [11, 12]. Full-fledged research is on to improve the surface properties of natural fibers by different chemical and physical treatments [13–15].

## 7.4 Surface Modification of Natural Fibers

### 7.4.1 Chemical Treatment

Alkali treatment (mercerization), silane treatment, maleated coupling, acetylation, permanganate peroxide treatment, benzoylation, isocyanate treatment, etc. are the frequently used chemical methods that could enhance the interfacial fiber–matrix adhesion.

### 7.4.2 Alkaline Treatment

This is one of the widely used methods for modifying the surface property of natural fibers. In this method, the natural fiber is immersed in sodium hydroxide solution for a specific period of time at a given temperature. The introduction of alkali into the fiber reduces its surface hydroxyl group, decomposes the non-cellulose constituent (wax, oil, lignin, etc.), and disrupts its hydrogen bonds. Consequently, the toughness and hydrophobicity of the fiber increase, and thus its compatibility with the hydrophobic polymer matrix improves (Figure 7.2) [16–18]. The chemical reaction of sodium hydroxide with cellulosic fiber is displayed in Eq. (7.1).

$$\text{Fiber-cell-OH} + \text{NaOH} \longrightarrow \text{Fiber-cell-O}^{-}\text{Na}^{+} + \text{H}_2\text{O} + \text{impurities} \qquad (7.1)$$

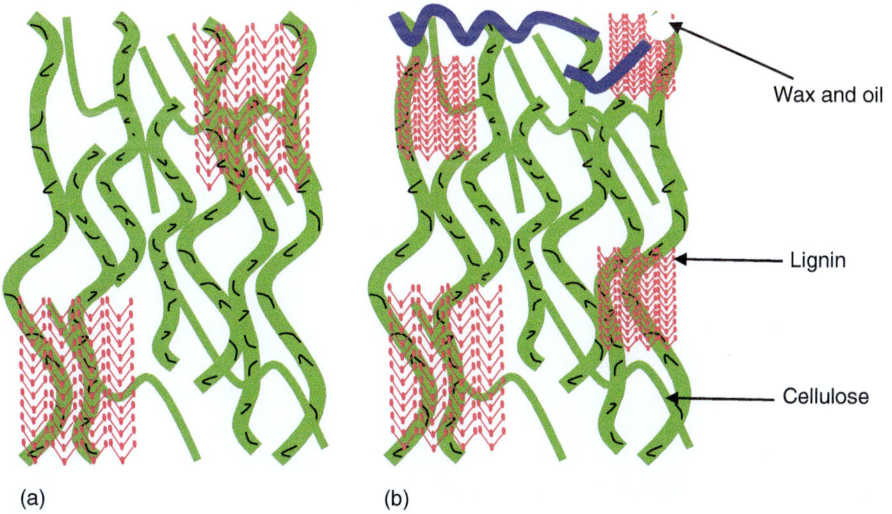


**Figure 7.2** Alkaline treatment of natural fibers: (a) untreated and (b) alkali treated. Sources: Sgriccia et al. [16], Shukor et al. [17], Kumar Sinha et al. [18].

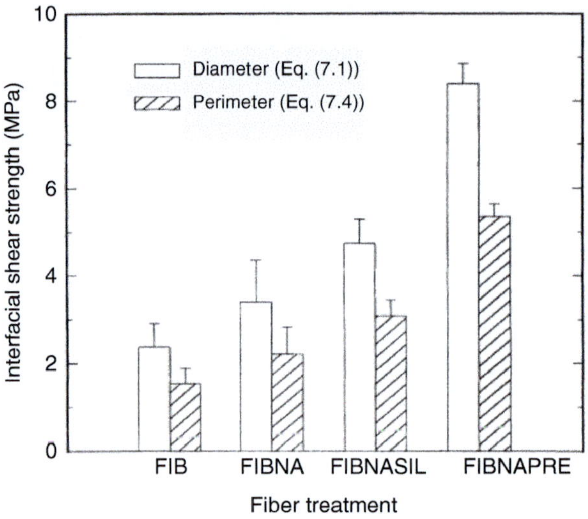


**Figure 7.3** Effect of surface treatment of henequen fiber on interfacial shear strength (FIB, no treatment; FIBNA, treated with NaOH; FIBNASIL, first treated with NaOH and then with silane coupling agent; FIBNAPRE, treated with alkali and then impregnated with matrix solution). Source: From Valadez-Gonzalez et al. [19]. © 1999, Elsevier.

Several workers have utilized alkali treatment to improve the mechanical, thermal, and water resistance of fibers. Valadez-Gonzalez et al. [19] employed the pullout test to measure the interfacial shear strength (IFSS) of alkali-treated henequen fiber. The result indicates that alkali-treated fiber exhibits good IFSS probably due to its good compatibility with the matrix (Figure 7.3).

Cai et al. [20] observed an improvement in the tensile strength and Young's modulus of abaca fiber treated with alkali. The effect of alkali concentration on the mechanical properties was also evaluated and the result indicates that alkali concentration has a significant influence on the Young's modulus of the membrane. A nonlinear stress–strain curve was noted at high concentration of sodium hydroxide and is ascribed to fiber twisting. Bachtiar et al. [21] reports that the strength of alkali plays an important role in the alkaline treatment process. In their work, they found an enhancement in the tensile strength with alkali concentration. Alkali treatment reduces lignin, hemicelluloses, and the waxy substance on the surface of the fiber, consequently enhancing the roughness and orientation of the fiber. The resultant fiber thus adheres strongly with the polymer matrix and enhances its mechanical strength. However, high concentration of alkali had a detrimental effect on the mechanical property of the fiber and is attributed to the fiber damage caused by strong alkali. This was in agreement with the report of Edeerozey et al. [22]. The authors reported that high percentage of sodium hydroxide (>9%) diminishes the tensile property of kenaf fiber. Goda et al. [23] compared the tensile strength, plastic deformation, and elongation at break of alkali-treated and untreated ramie fiber. The tensile strength of the treated fiber is found to be two to three times greater than that of the untreated fiber (Figure 7.4). High plastic deformation and elongation at break observed for the treated fiber were discussed on the basis of alkali-induced structural changes in the fiber. Akali treatment destroys the microfibril alignment of the fiber, removes the hemicelluloses, reduces the extensive hydrogen bond network of the fiber, and decreases the crystallinity of the fiber. As a result, the treated fiber becomes more plastic and exhibits high elongation at break.

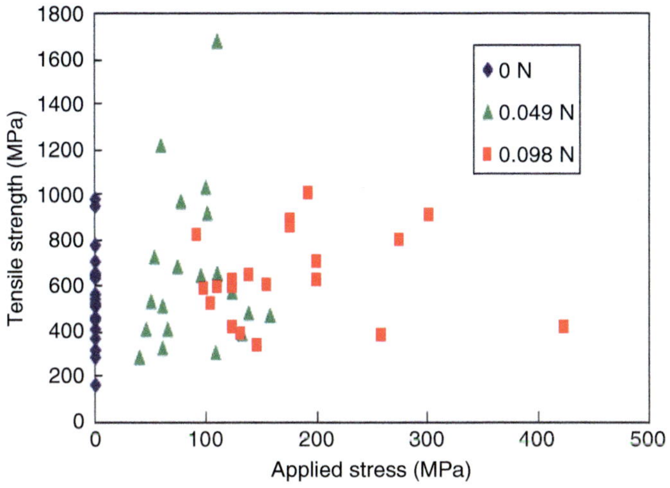


**Figure 7.4** Effect of alkaline treatment of ramie fiber on tensile strength. Source: From Goda et al. [23]. © 2006, Elsevier.

**Figure 7.5** Tensile strength of different alkaline-treated kenaf fibers (T5–3, Treated with 5% NaOH for three hours of immersion; A, B, C, D with different groups). Source: From Mahjoub et al. [25].

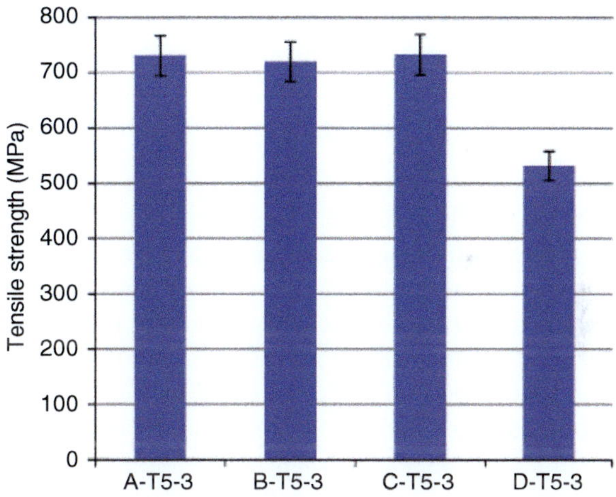


Fiore et al. [24] studied the effect of alkali immersion time on the mechanical properties of the kenaf fiber (Figure 7.5). The result indicates that the fiber treated with alkali for 48 hours had good mechanical properties, while a high immersion time (144 hours) had a negative effect on the mechanical strength of the fiber. Similar observation was reported by Mahjoub et al. [25].

Venkateshwaran et al. [26] found that 1% NaOH-treated fiber exhibits superior mechanical strength. This was due to the conversion of cellulose I to cellulose II. Meanwhile, Ibrabim et al. [27] optimized the sodium hydroxide concentration as 6% for kenaf fiber. Wong et al. [28] observed significant improvement in the mechanical properties (tensile strength, yield strength, and Young's modulus) of kenaf fiber when treated with high concentration of sodium hydroxide.

**Table 7.2** Mechanical properties of different natural fibers with alkaline treatment.

| Sample name | Tensile strength (MPa) | Tensile Modulus (GPa) | References |
|---|---|---|---|
| Sugar palm fiber/epoxy | 50 | — | [21] |
| Kenaf/epoxy | 42.5 | 4.2 | [24] |
| Untreated | 16 | 10 | [26] |
| Banana epoxy | 33 | 34 | [26] |
| Kenaf | 25.28 | — | [27] |
| Jute-coir fiber/polypropylene | 27.1 | — | [29] |

Siddika et al. [29] investigated the mechanical properties of alkali-treated jute-coir fiber/polypropylene (PP) composite. They noticed that NaOH-treated fiber displayed better mechanical properties than the raw fiber. The improvement in mechanical properties is due to the removal of hemicellulose and lignin from the fiber surface and is confirmed by FTIR analysis. We have summarized the tensile parameters of sodium hydroxide-treated natural fibers in Table 7.2.

### 7.4.3 Silane Treatment

Silane treatment employs a silane coupling agent to modify the surface of the fiber. Bifunctionality of silane enables it to interact with the fiber as well as with the polymer, thereby making it an efficient coupling agent. Hydrolysis, condensation, and bond formation are the three main steps involved in silane treatment. During silane treatment, the hydrolyzable alkoxy group of silanes gets converted into a reactive silanol. The silanol later condenses with the neighboring silanol to form an oligomer. The hydroxyl group of silanol could interact with cellulosic fiber through hydrogen bond (Eqs. (7.2) and (7.3)) while the alkyl group of silanol promotes chemical link with the polymer matrix. Therefore, the silane-treated fiber possesses high hydrophobicity (due to the reduction of hydroxyl group) and is advantageous for its interaction with hydrophobic polymer matrix [30–32].

$$CH_2CHSi\left(OC_2H_5\right)_3 \longrightarrow CH_2CHSi(OH)_3 + 3C_2H_5OH \quad (7.2)$$

$$CH_2CHSi(OH)_3 + \text{fiber} - OH \longrightarrow CH_2CHSi(OH)_2O - \text{fiber} + H_2O \quad (7.3)$$

Valadez-Gonzalez et al. [33] studied the effect of surface treatment on the mechanical properties of polymer composites. Mechanical tests revealed that henequen fiber treated with silane exhibits a higher tensile strength than untreated and sodium hydroxide-treated fiber (Figure 7.6). The surface micrograph of tensile fractured surface was complimentary to the mechanical analysis. Prominent fiber pullout and matrix tearing were observed in the SEM micrograph of untreated samples, thus indicating its poor compatibility with the polymer matrix (Figure 7.7). However,

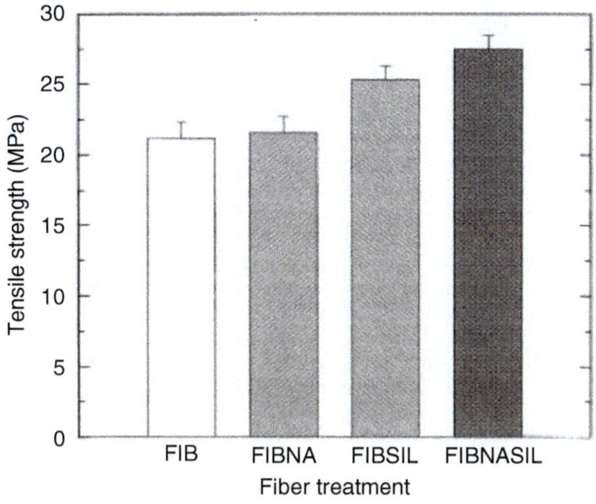


**Figure 7.6** Effect of different surface treatments on the tensile property of henequen fiber. (FIB, No treatment; FIBNA, treated with NaOH; FIBSIL, treated with silane coupling agent; FIBNASIL, first treated with alkali and then impregnated with silane coupling agent). Source: From Valadez-Gonzalez et al. [33]. © 1999, Elsevier.

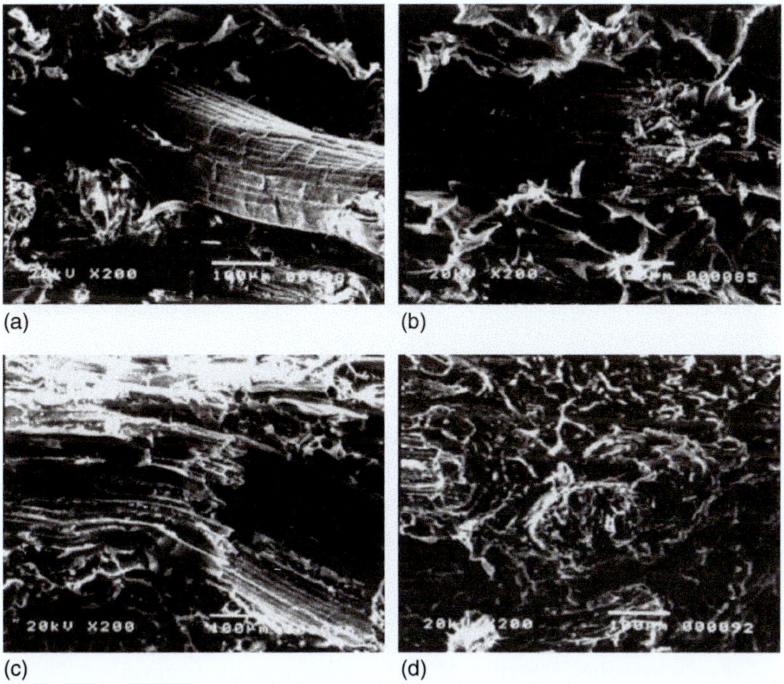


**Figure 7.7** SEM images of tensile fracture of different fibers ((a) FIB/HDPE, (b) FIBNA/HDPE, FIBSIL/HDPE, and FIBNASIL/HDPE). Source: Reproduced with permission from Elsevier, Valadez-Gonzalez et al. [33].

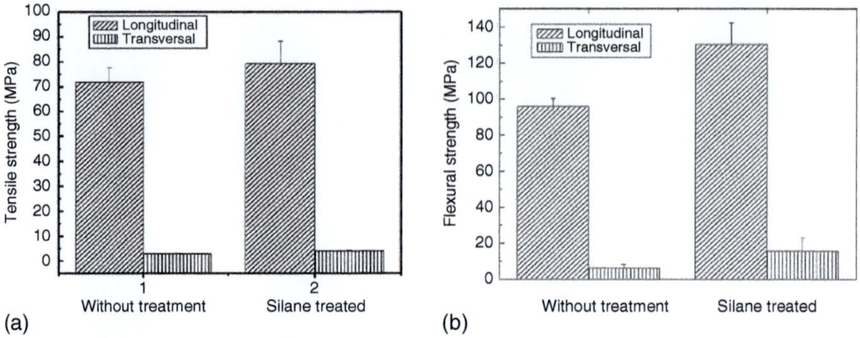


**Figure 7.8** (a) Tensile and (b) flexural strength of HDPE-reinforced henequen fiber of treated and untreated composites. Source: From Sepe et al. [34]. © 2018, Elsevier.

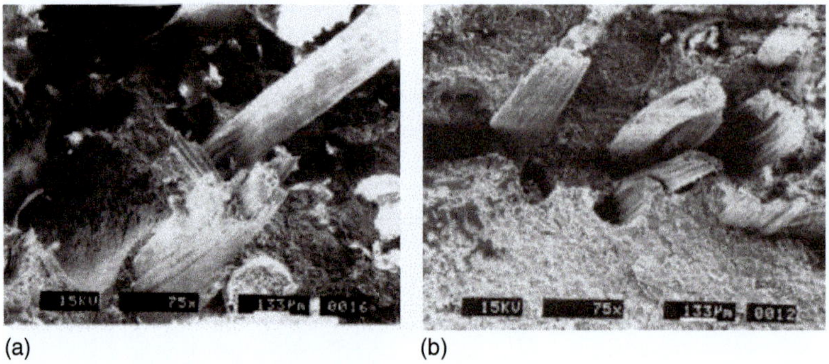


**Figure 7.9** SEM images of fracture surface of composite (a) with fibre without any surface treatment and (b) with silane treated fibre. Source: Reproduced with permission from Elsevier, Herrera-Franco et al. [34].

no evidence of fiber pullout can be observed for silane-treated fibers in the SEM micrograph, thus supporting better fiber–matrix interactions.

Herrera-Franco et al. [34] compared the mechanical properties of untreated and silane-treated high-density polyethylene (HDPE)–henequen composite. A 10% increase in longitudinal strength and a 43% increase in tensile strength were observed for silane-treated fiber. In addition to this, an improvement in mechanical properties such as flexural strength, flexural modulus, and shear strength was observed (Figure 7.8). A short fiber pullout length observed for treated fiber implies good adhesion between the fiber and the polymer matrix (Figure 7.9). The same group reported a similar mechanical behavior for silane-treated short henequen fiber [35].

Sepe et al. [36] studied the influence of silane treatment on hemp fiber. Their studies show that upon silane treatment the tensile and flexural properties of the composite increases, probably due to the improvement in fiber–matrix interaction. The optimum silane concentration for improving tensile modulus and flexural modulus is 1% and 5% respectively (Figure 7.10).

Japes and Siva [37] examine the effect of silane treatment on the mechanical properties of polyester composite reinforced with coconut sheath fiber. The result shows

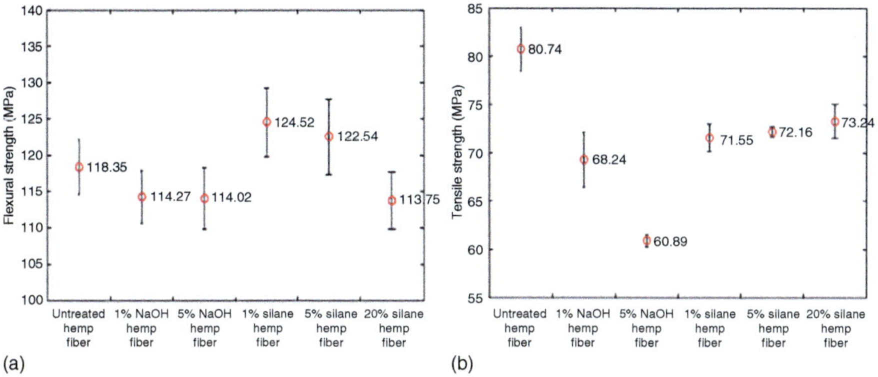


**Figure 7.10** (a) Flexural and (b) tensile strengths of hemp-reinforced composites. Source: From Sepe et al. [36]. © 2018, Elsevier.

that the mechanical strength of silane-treated composite is comparable with that of glass fiber/polymer composites and is ascribed to the surface roughness caused by silane. Liu et al. [38] developed a cellulosic fiber from corn stalk waste (CSF) and treated it with different concentrations of silane solution. The results revealed that fibers treated with optimum concentration of silane possess high tensile strength and Young's modulus (Figure 7.11). However, high silane concentration decreases the tensile strength of the fiber. This could be due to deposition of a thick layer of silane on the fiber surface, which weakens its adhesion with the polymer.

### 7.4.4 Acetylation Treatment

In acetylation treatment, an acyl group is introduced on the surface of the fiber by reacting it with acetic anhydride in the presence or absence of acid catalyst. The conversion of the hydroxyl group in the fiber into an acyl group increases its hydrophobic character and promotes its adhesion with the polymer. As a result, the mechanical and thermal properties of the acetylated fiber increases whereas the moisture absorbance diminishes [39, 40]. The reaction of acetic anhydride with natural fiber is as follows:

$$\text{Fiber}-\text{OH} + \text{CH}_3-\text{C}(=\text{O})-\text{O}-\text{C}(=\text{O})-\text{CH}_3 \longrightarrow \text{Fiber}-\text{OCOCH}_3 + \text{CH}_3\text{COOH} \tag{7.4}$$

Zaman and Khan [41] employed acetylation treatment to improve the properties of PP composite reinforced with banana empty fruit bunch fiber (BBF). When compared to the untreated composite, the acetylated composite exhibits high tensile strength, tensile modulus, flexural strength, and flexural modulus. The authors suggested that low polarity and high surface roughness of the treated fiber could be responsible for enhancing its compatibility with the polymer matrix. Good fiber–matrix interaction and homogeneous dispersion of fiber in the polymer matrix is evident from the SEM analysis. Similarly, results were reported by Bledzki et al. [42] for acetyl-treated flax fiber/polypropylene composite. However, high degree of

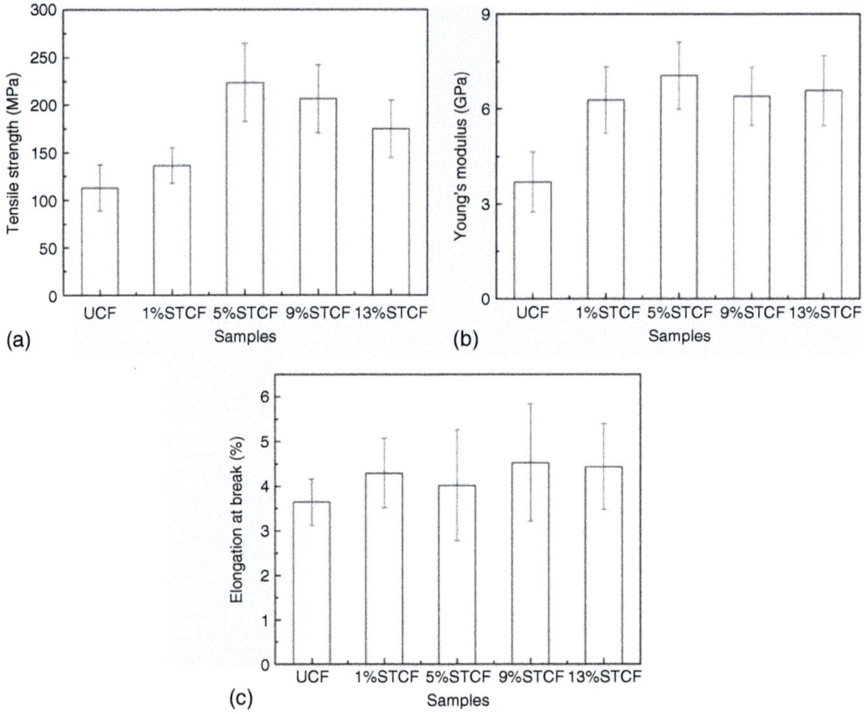


**Figure 7.11** Mechanical properties: (a) tensile Strength (b) Young's Modulus (c) elongation at break of silane-treated natural cellulose fiber: untreated CSF (UCF); 1 wt.%, 5 wt.%, 9 wt.%, and 13 wt.% silane-treated CSF. Source: From Liu et al. [38]. © 2019, Elsevier.

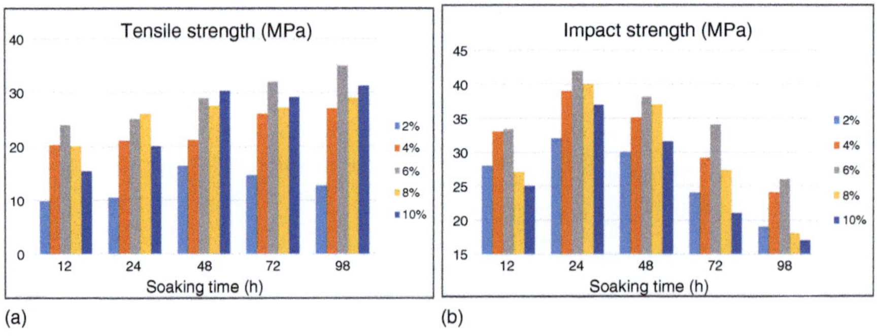


**Figure 7.12** (a) Tensile and (b) impact strength properties of palm fiber–vinyl ester composites. Source: From Senthilraja et al. [43]. © 2020, Elsevier.

acetylation was found to have an adverse effect on the mechanical properties of the fiber, probably due to the degradation of the fiber and the generation of internal cracks in the system. Senthilraja et al. [43] reported that mechanical properties of palm fiber/polymer composite could be tuned by varying the concentration of acetyl chloride and soaking time (Figure 7.12).

### 7.4.5 Benzylation Treatment

This treatment method employs benzoyl chloride to modify the fiber surface. The condensation reaction of benzoyl chloride and cellulosic fiber introduces a benzoyl group on the surface of the fiber and thus enhances its hydrophobic nature. Consequently, the benzoyl chloride-treated fiber has good adherence with the polymer [44, 45]. The reaction between the cellulosic fiber and benzoyl chloride are shown as

$$\text{Fiber-cell} - \text{OH} + \text{NaOH} \longrightarrow \text{Fiber-cell} - \text{O}^{-}\text{Na}^{+} + \text{H}_2\text{O} \tag{7.5}$$

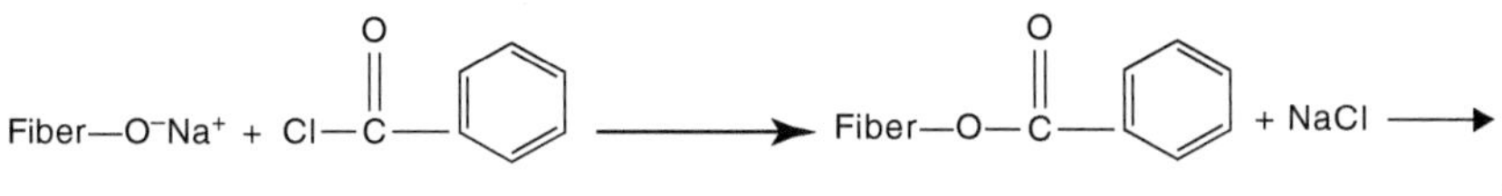


(7.6)

Nayak and Mohanty [46] studied the effect of various chemical treatments (benzoylation, acrylic acid, sodium chlorite, permanganate, and alkaline) on the properties of areca fiber. The authors reported that benzyl chloride-treated fiber exhibits better performance than fibers with other treatments, which is attributed to its improved mechanical interlocking with the polymer. Benzoylation imparts high roughness to the fiber by converting the exposed hydroxyl group of the fiber into a hydrophobic benzoyl group. The treatment also removes waxes, lignin, and other impurities from the fiber surface, which is responsible for its poor interaction with the polymer. High surface roughness of the treated fiber was confirmed by SEM analysis. This was in agreement with the results of Nair and Thomas [47]. Majid et al. [48] studied the mechanical property of benzyl chloride-modified kenaf core powder/polyvinyl chloride (PVC)/epoxidized natural rubber (ENR) composite. The benzyl chloride treatment results in a significant improvement in the mechanical properties (Young's modulus, tensile strength, and elongation at break) of the composite (Figure 7.13). Less fiber pullout observed in the SEM micrograph of benzyl chloride treated fiber is evidence of its better adhesion with

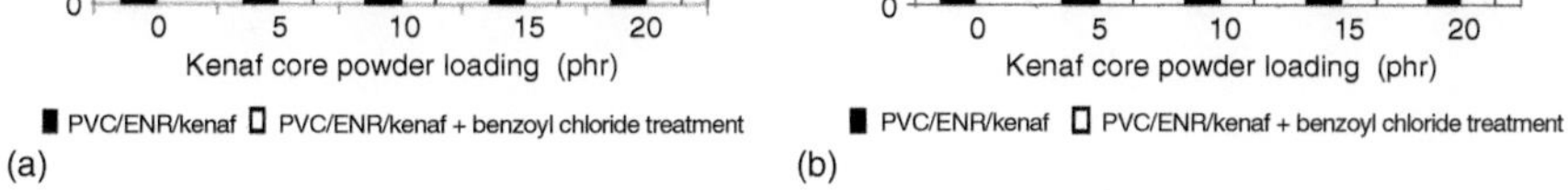


**Figure 7.13** (a) Tensile strength and (b) Young's modulus of PVC/ENR/Kenaf composite. Source: From Majid et al. [48]. © 2016, Elsevier.

the polymer. Yadav and Gupta [49] improved the adhesion of jute fiber and epoxy composite by utilizing different chemical treatments such as alkali, benzoylation, and sodium bicarbonate. The mechanical and dynamic mechanical tests suggest that the benzoyl chloride-treated fiber shows better performance than untreated fiber and fibers with other treatments.

### 7.4.6 Peroxide Treatment

Peroxide is one of the widely used chemical treatment methods, which involves the use of organic peroxides (e.g. benzoyl peroxide and dicumyl peroxide [DCP]). During this treatment, the free radical generated from organic peroxides reacts with the hydroxy group of the hydrophilic fiber and thus improves the interfacial adhesion between the fiber and the matrix (Eqs. (7.7)–(7.10)). This treatment also reduces the moisture uptake capacity and increases the thermal stability of the fiber [7, 10].

$$\mathrm{RO-OR} \longrightarrow 2\mathrm{RO}^{\cdot} \tag{7.7}$$

$$\mathrm{RO}^{\cdot} + \mathrm{PE-H} \longrightarrow \mathrm{ROH} + \mathrm{PE} \tag{7.8}$$

$$2\mathrm{RO}^{\cdot} + \mathrm{cellulose-H} \longrightarrow \mathrm{ROH} + \mathrm{cellulose} \tag{7.9}$$

$$\mathrm{PE} + \mathrm{cellulose} \longrightarrow \mathrm{PE-cellulose} \tag{7.10}$$

Sari et al. [50] employed DCP to improve the strength of pandanwangi fiber. In their study, the authors varied the concentration of DCP and observed that 4% DCP is the optimum concentration to obtain high tensile, bending, and impact strengths (Figure 7.14).

In a separate study, Joseph et al. [51] and Ahmed and Luyt [52] investigated the properties of sisal fiber-reinforced polyethylene composite. The authors concluded that peroxide-induced grafting of fiber and polymer is responsible for its superior mechanical properties. Meanwhile, Mokoena et al. [53] reported that high concentration of peroxide had a negative effect on the mechanical property of

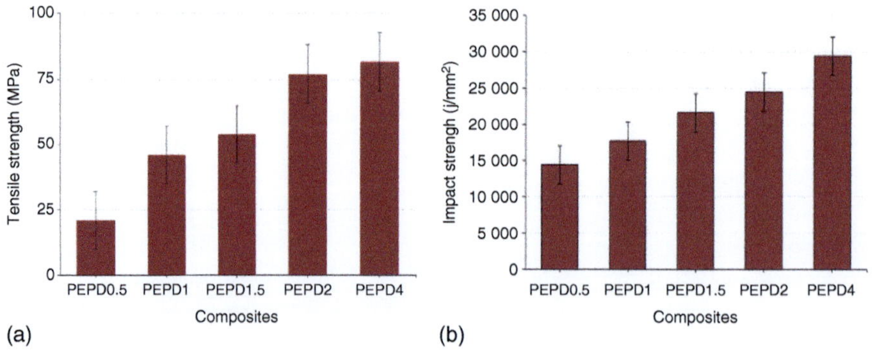


**Figure 7.14** (a) Tensile strength and (b) impact strength of pandanwangi fiber-reinforced polyethylene-PEPD. Source: From Sari et al. [50]. © 2019, Elsevier.

sisal/low-density polyethylene (LDPE) composites. Similar behavior was reported by Sapieha et al. [54] for cellulose fiber-reinforced LDPE composite. A good mechanical performance was noted for low concentration of peroxide. Madhu et al. [55] observed that peroxide-treated *Agave Americana* fiber (AAF) is a potential reinforcement for polymer composite. The tensile strength and elongation at break of peroxide treated-AAF fiber is found to be higher than that of the untreated fiber. Malunka et al. [56] utilized peroxide treatment for ethylene vinyl acetate (EVA)–sisal fiber composite. Rytlewski and group [57] studied the properties of peroxide-treated polylactic acid (PLA)/flax and PLA/hemp fiber. They observed that with increase in DCP content the tensile stress at break, elongation at break, and impact strength of the system decreased, which could be due to the development of cracks in the system.

## 7.5 Maleated Coupling Agents

In this treatment method, the fiber is treated with a maleated coupling agent such as maleic anhydride or maleic anhydride grafted polymer. Maleic anhydride reacts with the hydroxyl group present in the cellulosic fiber and makes the surface more hydrophobic. This results in the enhancement of fiber–matrix interaction [58]. The reaction of cellulose fibers with MAPP is illustrated in Figure 7.15.

Mohanty et al. [59] studied the effect of maleic anhydride-modified polypropylene (MAPP) concentration on the mechanical properties of the sisal/polypropylene composite. The results show that 1% MAPP-loaded sisal/PP composite showed better mechanical properties than the untreated fiber. When compared to the untreated composite, the tensile strength, flexural strength, and impact strength of the treated composite increase by 50%, 30%, and 60% respectively. The same group employed

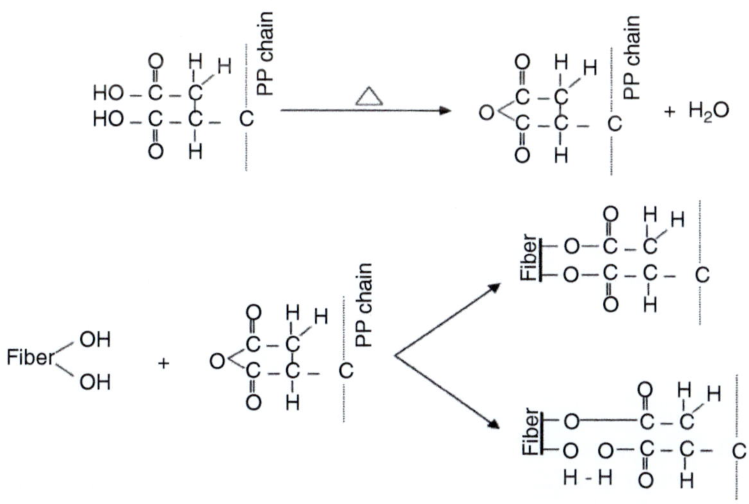


**Figure 7.15** The reaction of cellulose fibers with MAPP. Source: Based on Keener et al. [58].

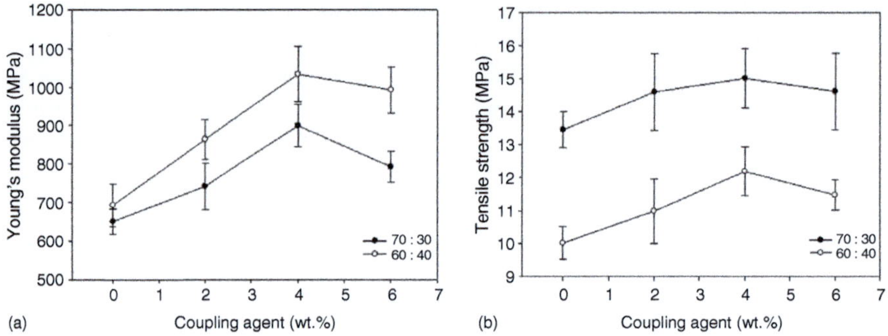


**Figure 7.16** (a) Young's modulus and (b) tensile strength of the different fiber contents with coupling agent. Source: From Kakou et al. [61]. © 2014, Elsevier.

MAPP-treated jute fiber to reinforce the PP composite [60]. Their studies revealed that composites loaded with 30% fiber and 0.5% MAPP show superior properties when compared with unmodified PP–jute fiber. Meanwhile, Kakou et al. [61] monitored the effect of MAPP concentration on the performance of polyethylene composite. Their studies revealed that the concentration of coupling agent, MAPP, plays a prominent role in determining the overall mechanical performance of the treated composite. In their study, they observed an enhancement in mechanical properties (tensile and Young's modulus) with increase in MAPP concentration and 4% MAPP is found to be the optimum concentration (Figure 7.16). This is attributed to the formation of a chemical bond (ester bond) between the fiber and the polymer, thereby increasing the compatibility between them. However, at high concentration of MAPP the authors noticed a decline in the values of tensile modulus and Young's modulus (6%), probably due to the plasticizing action of excess MAPP molecules on the polymer matrix.

Matje et al. [62] also reported 4% as the optimized MAPP concentration for improving the mechanical property of PP/hemp composite. Wong et al. [63] developed carbon fiber-reinforced polypropylene composite through injection molding. By the incorporation of maleic anhydride, a better adhesion between PP and fiber was observed, which was confirmed through SEM analysis (Figure 7.17). They report that low concentration of coupling agent shows significant improvement in tensile and impact strengths. Baykus et al. [64] report a significant improvement in the tensile strength and Young's modulus of jute-reinforced polypropylene and polyethylene composite when treated simultaneously with alkali and MAPP coupling agent.

### 7.5.1 Isocyanate

In this approach, the surface modification of natural fiber is carried out by introducing isocyanate coupling agent into the system. The reactive functional group (–NCO) of isocyanate reacts with the hydroxyl group of the fiber to form a urethane linkage (Eq. (7.11)). A secondary reaction of isocyanate with moisture reduces

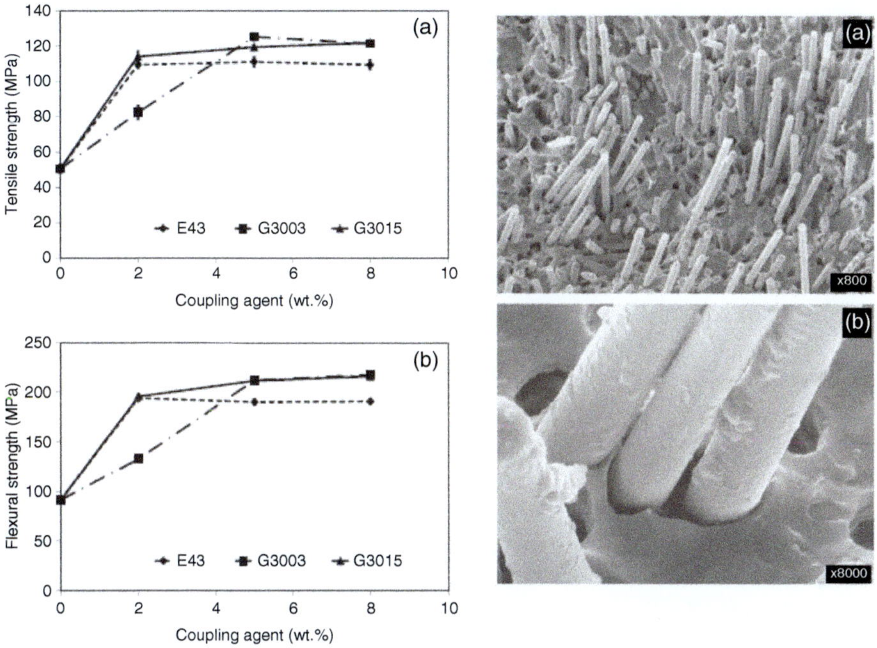


**Figure 7.17** Mechanical property and SEM images of recycled carbon fiber (RCF)-reinforced PP composite. Source: From Wong et al. [63]. © 2012, Elsevier.

the hydrophilicity of the fiber. Thus, addition of isocyanate improves the interaction between the polymer and the fiber [65].

$$\text{Fiber}-\text{OH} + \text{R}-\text{N}{=}\text{C}{=}\text{O} \longrightarrow \text{fiber}-\text{O}-\text{CO}{=}\text{NR}-\text{R} \qquad (7.11)$$

In their work, Ashaori and Nourbakhsh [66] varied the polybutadiene isocyanate (PBNCO) content (0, 2, and 4 wt.%) and examined its effect on the tensile strength and impact strength of bagasse-reinforced polypropylene composite. Their results show that composite treated with 4% PBNCO displayed better mechanical properties

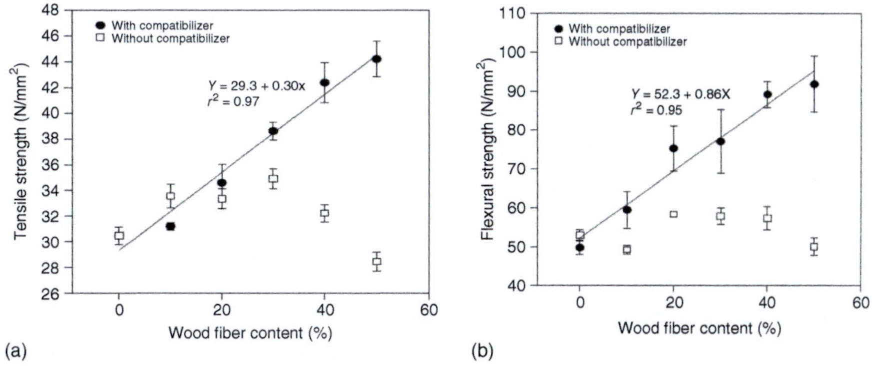


**Figure 7.18** (a) Tensile strength and (b) flexural strength of wood-fiber PP composite. Source: From Karmarkar et al. [67]. © 2007, Elsevier.

than the untreated composite. Karmarkar et al. [67] reported that incorporation of a novel compatibilizer (*m*-isopropenyl-*a*,*a*-dimethylbenzyl-isocyanate) along with isocyanate treatment significantly improves the tensile strength and flexural property of wood fiber/polypropylene composite (Figure 7.18). Raj et al. [68] improved wood fiber–polyethylene adhesion using isocyanate treatment. The isocyanate-treated composites show higher tensile strength than the untreated composite. Nourbakhsh et al. [69] employed PBNCO to improve the property of wood fiber/polypropylene composites. PBNCO treatment activates the wood fiber surface and thus leads to a better anchoring of fiber to the polymer matrix. Therefore, the treated composite exhibits high tensile strength (30 MPa) and impact strength (22 $g/m^2$). Joseph et al. [70] reported that isocyanate is one of the best treatment methods to improve the mechanical property of sisal fiber/polypropylene composite.

### 7.5.2 Permanganate Treatment

This treatment is performed by using potassium permanganate ($KMnO_4$). The $Mn^{3+}$ generated from $KMnO_4$ reacts with the hydroxyl group of fiber to form cellulose–manganate and thus initiate graft polymerization. Consequently, the fiber–matrix adhesion is increased. Permanganate treatment is also found to enhance the thermal stability of the fiber [71]. Equations (7.12) and (7.13) show the interaction of $KMnO_4$ with the hydrophilic cellulose fiber.

$$\text{Cellulose} - \text{H} + \text{KMnO}_4 \longrightarrow \text{cellulose} - \text{H} - \text{O} - \underset{\underset{\text{O}}{\|}}{\overset{\overset{\text{O}}{\|}}{\text{Mn}}} - \text{OK}^+ \tag{7.12}$$

$$\text{Cellulose} - \text{H} - \text{O} - \underset{\underset{\text{O}}{\|}}{\overset{\overset{\text{O}}{\|}}{\text{Mn}}} - \text{OK}^+ \longrightarrow \text{Cellulose} + \text{H} - \text{O} - \underset{\underset{\text{O}}{\|}}{\overset{\overset{\text{O}}{\|}}{\text{Mn}}} - \text{OK}^+ \tag{7.13}$$

Dhanalakshmi et al. [72, 73] in two separate studies observed that permanganate-treated areca fiber is a good reinforcing agent for natural rubber and epoxy composites. The treated composite exhibits higher tensile strength and flexural strength than the untreated fiber, which is due to the better mechanical interlocking of the fiber with the matrix. Mohammed and coworkers [74] studied the tensile property of sodium hydroxide pretreated sugar palm fiber at different concentrations of $KMnO_4$. A dramatic rise in tensile strength and modulus was observed for fibers treated with both $KMnO_4$ and sodium hydroxide. Addition of $KMnO_4$ improves the surface activity and roughness of the fiber, which was evident from the SEM micrograph. The surface of $KMnO_4$-treated fiber is found to be irregular due to the removal of waxy substances and lignin and thus promotes its interaction with the hydrophobic matrix. Zaman et al. [75] improved the mechanical property of jute polypropylene composite by treating it with $KMnO_4$, urea, and 2-hydroxyethyl methacrylate (HEMA). They reported that jute fabric treated with 0.03% of $KMnO_4$, 15% HMA and 1% urea possesses superior mechanical properties than untreated composites. Joseph et al. [70] utilized various chemical treatments (sodium hydroxide, isocyanates, MAPP, benzyl chloride, and permanganate) for modifying the surface property of sisal fiber. They report that concentration of $KMnO_4$ plays a major role in determining the tensile property of sisal fiber/polypropylene composites. High tensile properties were achieved even at low concentrations (0.05) of $KMnO_4$. The improvement in mechanical property of $KMnO_4$-treated composite is attributed to grafting of PP onto sisal fibers and its high surface roughness. However, high concentration of $KMnO_4$ damages the cellulose fiber and thus has a negative effect on the tensile properties of the composite. Similar results were noted by Patra and group [76] for $KMnO_4$-treated short sisal fiber/epoxy composite. Khan et al. [77] investigated the effect of $KMnO_4$ on jute fabrics in the presence of acid (oxalic acid and sulfuric acid) and alkaline media (sodium hydroxide). Excellent mechanical performance was observed for $KMnO_4$-treated composite in 5% oxalic acid.

### 7.5.3 Stearic Acid Treatment

This treatment employed stearic acid to modify the surface properties of fiber. Here, the –COOH group of stearic acid reacts with the –OH group of fiber and thus promotes the fiber–matrix interaction (Eq. (7.14)). This treatment also increases the hydrophobicity of the fiber by removing surface impurities such as pectin, wax, and lignin. [7].

$$\text{Fiber} - \text{OH} + \text{CH}_3(\text{CH}_2)_{16}\text{COOH} \longrightarrow \text{CH}_3(\text{CH}_2)_{16}\text{COO} - \text{O} - \text{fiber} + \text{H}_2\text{O} \tag{7.14}$$

Kiattipanich et al. [78] employed stearic acid for improving the tensile properties of sugarcane fiber/PP composite. The simultaneous presence of hydrophobic and hydrophilic groups in stearic acid enables it to react with the polymer as well as with the fiber and favors the fiber–matrix interaction. Thus, the stearic acid-treated composite has greater ability to transfer stress from the matrix to the fiber and possesses good tensile strength. In addition to this, stearic acid treatment also influences the shear viscosity of the composites. High concentration of stearic acid is found to lower the shear viscosity. Low value of shear viscosity implies better compatibilization. According to Nekkaa et al. [79] high tensile strength of stearic acid-treated broom fiber/polypropylene composite could be due to greater dispersion of the fiber in the polymer matrix, better wettability of the fiber by the matrix, and stearic acid-induced grafting of fiber onto the matrix. Uniform dispersion of broom fiber on the polypropylene matrix was confirmed by SEM analysis. This was in agreement with the report of Kalaprasad et al. [80]. The authors also investigated the effect of stearic acid concentration on the property of sisal/glass hybrid fiber-reinforced LDPE composite. They observed that optimum concentration of stearic acid (4%) improves the tensile strength of the composite. However, high concentration of stearic acid (>4%) adversely affects the mechanical strength, probably due to the negative effect of stearic acid on the dispersion of fiber. Enriquez et al. [81] observed good mechanical and thermal properties for stearic acid-treated HDPE/coconut fiber composite. Torres and Cubillas [82] compared the IFSS of stearic acid-treated sisal fiber/polyethylene composite with that of the untreated composite. A 23% increment in the IFSS was noted for stearic acid-treated composite.

### 7.5.4 Physical Treatment

Apart from the chemical method, physical treatment will also enhance the bonding between the fiber and the polymer and could be used to modify the surface property of the fiber. Stretching, calendaring, plasma, and electric discharge are the commonly used physical treatment methods for the surface treatment of fibers.

### 7.5.5 Plasma Treatment

Plasma treatment is a dry surface treatment and involves the use of plasma, which is comprised of free radicals, ions, and electrons (Figure 7.19). Owing to its low cost and eco-friendly nature, it is widely used to modify the surface properties of fiber. Studies show that plasma treatment alters the surface properties of fiber such as roughness (through ablation/etching), surface energy, and cross-linking power (by generating free radicals and by introducing new polar group). As a result, the fiber's roughness and hydrophobicity increase and thus the fiber could adhere more strongly to the polymer matrix. Fiber characteristics (nature and type) and plasma parameters (power, treatment time and pressure) are found to play a vital role in this treatment method [83, 84].

Bozaci et al. [85] suggested that plasma treatment is an effective surface treatment method to improve the mechanical properties of flax fiber/HDPE composite. The high IFSS value of plasma-treated composite is due to plasma-induced surface modification (roughness, chemical composition, contact area) of the flax fiber. The

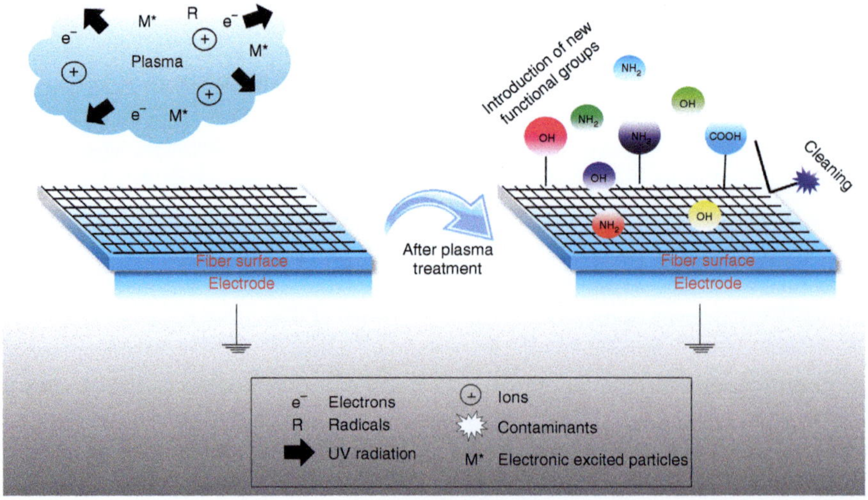


**Figure 7.19** Reaction mechanism of plasma treatment.

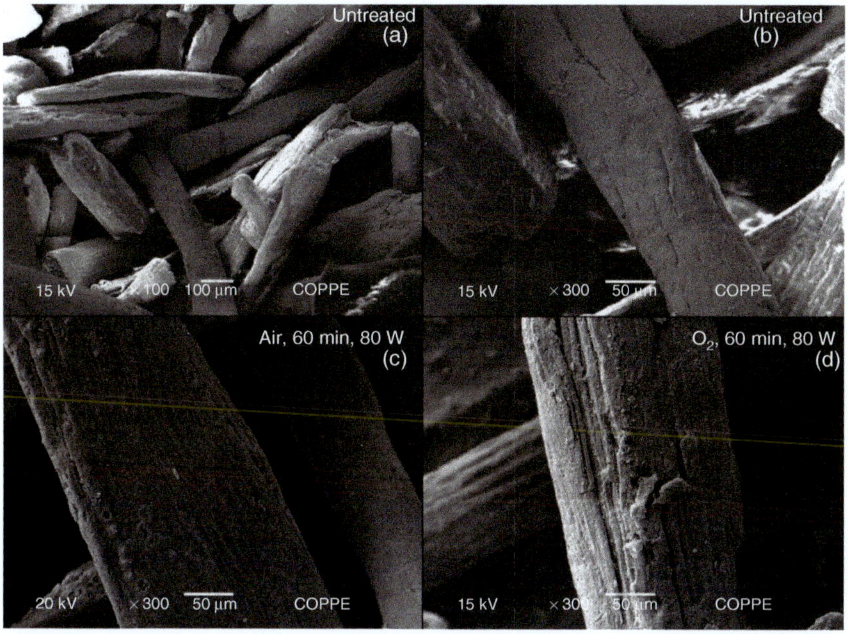


**Figure 7.20** SEM images of: (a), (b) untreated and (c), (d) plasma-treated coir fibers. Source: Reproduced with permission from Elsevier, de Farias et al. [87].

same group enhanced the surface roughness of flax fiber using plasma treatment and recommended it as an efficient reinforcement for polymer composite [86]. de Farias et al. [87] employed plasma treatment to alter the surface property of coir fiber. Plasma-treated fiber possesses a rough surface with low amount of surface impurities (lignin, waxes, and cellulose), which was confirmed by FTIR and SEM analysis (Figure 7.20). As a result, the treated composite possesses better mechanical parameters than the untreated composite (Figure 7.21).

Gibeop et al. [88] compared the performance of plasma-treated jute/PLA composite with alkali-treated composite. The etching effect of plasma makes the surface of fiber rougher and thus provides a better mechanical interlocking with the matrix. Mechanical analysis shows that plasma-treated fiber possesses superior mechanical properties (tensile strength, Young's modulus, and flexural strength) than the alkali-treated ones. Liu and Cheng [89] reported low-temperature plasma treatment on ramie fiber. Their studies show that plasma exposure time and output power play a major role in the surface property of the fiber. The fiber treated with plasma (200 W) for three minutes exhibits high surface energy, contact area, and wettability and hence displays a high tensile strength (253 MPa). Sinha and Panigrahi [90] investigated the influence of plasma treatment time on the mechanical properties of jute fiber. The authors reported that jute fiber exposed to 10 minutes of plasma treatment possesses high mechanical performances and flexural strength. However, high treatment time (15 minutes) decreases the mechanical parameter, which could be due to the damaging effect of plasma on the structure of the fiber (Figure 7.22).

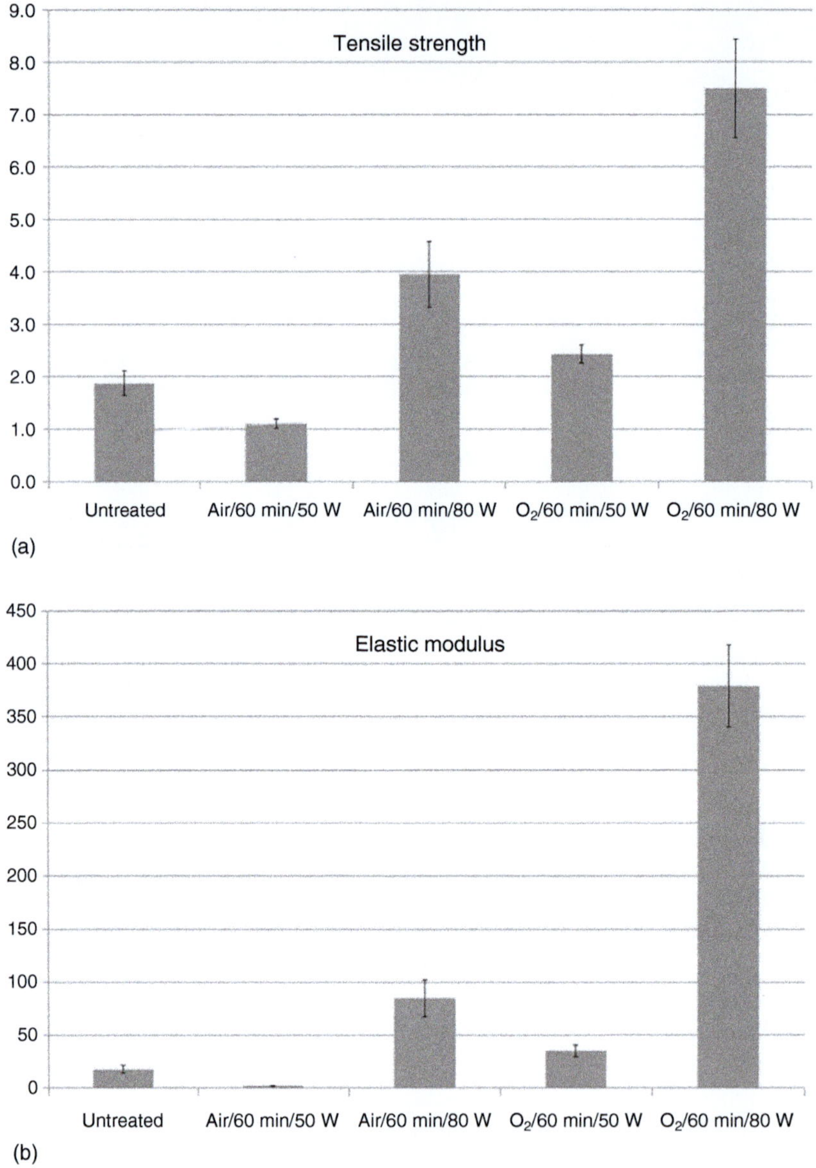


**Figure 7.21** (a) Tensile strength and (b) elastic modulus of untreated and plasma-treated fiber. Source: From de Farias et al. [87]. © 2017, Elsevier.

### 7.5.6 Corona Treatment

This is one of the widely used techniques to enhance the surface properties of natural fibers. This method attracted attention from researchers due to its effectiveness, low energy requirement, and less expensive nature. In this technique, a high voltage ionizes the air and generates plasma with a blue color hue (corona). Upon exposure

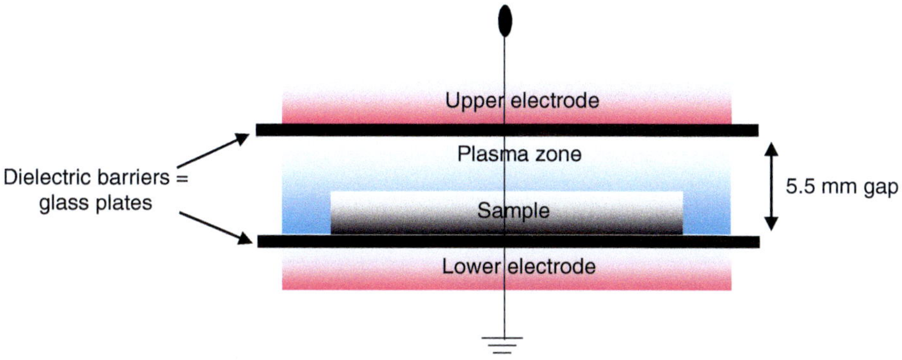


**Figure 7.22** Schematic representation of corona treatment.

to corona, the surface of the fiber gets activated by the generation of oxygen free radical and facilitates its adhesion to the polymer [91].

Ragoubi et al. [92] employed corona treatment to improve the mechanical and thermal properties of natural fiber/polymer composite. The good performance of corona-treated fiber composite is attributed to the corona-induced surface modification such as oxidation and etching. Belgacem et al. [93] modified the surface of cellulose fiber by corona treatment and used it to fabricate a polyproline composite. The corona-treated fiber exhibits better mechanical properties than untreated composites. Based on the experimental studies, the authors conclude that the corona treatment energy and specific fiber/polypropylene combination highly influence the overall performance of the composite. On the contrary, Gassan et al. [94] report a detrimental effect of corona treatment on the mechanical properties of *Aloe vera* fiber. The authors claimed that corona treatment reduces the crystallinity of the fiber by disturbing the ordered structure of lignocellulosic of the fiber. As crystallinity is related to the overall mechanical performance of the fiber, a fiber with less crystallinity will yield a low-strength composite.

### 7.5.7 Ozone Treatment

Ozone ($O_3$) is an allotrope of oxygen with powerful oxidizing ability. Ozone treatment is used to improve the surface properties of fibers. This treatment method tends to increase the surface roughness of fibers by destroying the lignin part of the fiber [95, 96]. Thus, ozone treatment improves the surface roughness and thus the fiber could adhere more strongly with the polymer matrix. Recently, this method has received significant attention from researchers as it is safe, environmentally friendly, and requires less energy. Rajendran Royan et al. [97] reported that UV/$O_3$-treated rice husk fiber possesses a rough surface with good mechanical properties. The improvement in surface morphology of UV/$O_3$-treated fiber is ascribed to the decomposition of lignin from the fiber surface while the high tensile strength is attributed to the retention of particle rigidity due to the mild treatment conditions. Lette et al. [98] noted a significant improvement in the flexural property of wood–polymer composite reinforced with ozone-treated bamboo fiber. This is attributed to the fact that ozone treatment improves the compatibility between the fiber and the matrix. A similar result was reported by Chtourou et al. [99] for polyethylene/pulp composite when treated with oxygen–fluorine gas.

## 7.6 Summary

Owing to their salient features such as abundance, nontoxicity, biodegradability, and high specific strength to weight ratio, natural fibers are considered as a better alternative for synthetic fibers to reinforce polymer composites. However, high hydrophilicity of natural fibers creates incompatibility with the polymer matrix and thus leads to poor mechanical interlocking of the fiber with the matrix. As a result, natural fiber-reinforced polymer composites are mechanically inferior to synthetic fiber composites. This limitation of natural fibers could be improved by several chemical (alkali treatment, silane treatment, maleated coupling, acetylation, permanganate peroxide treatment, benzoylation, isocyanate treatment, etc.) and physical treatments (plasma, corona, ozone treatment, etc.). A dramatic increase in surface roughness and hydrophobicity was reported for natural fibers treated with various surface treatments. Thus, surface treatment enhances the interface adhesion between the fiber and the polymer matrix and significantly improves the mechanical parameters of natural fiber-reinforced composites.

## References


**1** Delicano, J.A. (2018). A review on abaca fiber reinforced composites. *Compos. Interfaces* 25: 1039–1066.

**2** Joshi, S.V., Drzal, L.T., Mohanty, A.K., and Arora, S. (2004). Are natural fiber composites environmentally superior to glass fiber reinforced composites? *Composites Part A* 35: 371–376.

**3** Monteiro, S.N., Lopes, F.P.D., Ferreira, A.S., and Nascimento, D.C.O. (2009). Natural-fiber polymer-matrix composites: cheaper, tougher, and environmentally friendly. *JOM* 61: 17–22.

**4** Ahmad, F., Choi, H.S., and Park, M.K. (2015). A review: natural fiber composites selection in view of mechanical, light weight, and economic properties. *Macromol. Mater. Eng.* 300: 10–24.

**5** Saba, N., Tahir, P., and Jawaid, M. (2014). A review on potentiality of nano filler/natural fiber filled polymer hybrid composites. *Polymers* 6: 2247–2273.

**6** Mueller, D.H. and Krobjilowski, A. (2016). New discovery in the properties of composites reinforced with natural fibers. *J. Ind. Text.* 33: 111–130.

**7** Radoor, S., Karayil, J., Rangappa, S.M. et al. (2020). A review on the extraction of pineapple, sisal and abaca fibers and their use as reinforcement in polymer matrix. *Express Polym. Lett.* 14: 309–335.

**8** Li, X., Tabil, L.G., and Panigrahi, S. (2007). Chemical treatments of natural fiber for use in natural fiber-reinforced composites: a review. *J. Polym. Environ.* 15: 25–33.

**9** Singha, A.S. and Thakur, V.K. (2008). Mechanical properties of natural fibre reinforced polymer composites. *Bull. Mater. Sci.* 31: 791–799.

**10** Sarikaya, E., Çallioğlu, H., and Demirel, H. (2019). Production of epoxy composites reinforced by different natural fibers and their mechanical properties. *Composites Part B* 167: 461–466.

**11** Balla, V.K., Kate, K.H., Satyavolu, J. et al. (2019). Additive manufacturing of natural fiber reinforced polymer composites: processing and prospects. *Composites Part B* 174: 1–29.

**12** Cheung, H.-y., Ho, M.-p., Lau, K.-t. et al. (2009). Natural fibre-reinforced composites for bioengineering and environmental engineering applications. *Composites Part B* 40: 655–663.

**13** Komuraiah, A., Kumar, N.S., and Prasad, B.D. (2014). Chemical composition of natural fibers and its influence on their mechanical properties. *Mech. Compos. Mater.* 50: 359–376.

**14** Ali, A., Shaker, K., Nawab, Y. et al. (2016). Hydrophobic treatment of natural fibers and their composites – a review. *J. Ind. Text.* 47: 2153–2183.

**15** Faruk, O., Bledzki, A.K., Fink, H.-P., and Sain, M. (2012). Biocomposites reinforced with natural fibers: 2000–2010. *Prog. Polym. Sci.* 37: 1552–1596.

**16** Sgriccia, N., Hawley, M.C., and Misra, M. (2008). Characterization of natural fiber surfaces and natural fiber composites. *Composites Part A* 39: 1632–1637.

**17** Shukor, F., Hassan, A., Saiful Islam, M. et al. (2014). Effect of ammonium polyphosphate on flame retardancy, thermal stability and mechanical properties of alkali treated kenaf fiber filled PLA biocomposites. *Mater. Des. (1980–2015)* 54: 425–429.

**18** Sinha, A.K., Narang, H.K., and Bhattacharya, S. (2017). Effect of alkali treatment on surface morphology of abaca fibre. *Mater. Today: Proc.* 4: 8993–8996.

**19** Valadez-Gonzalez, A., Cervantes-Uc, J.M., Olayo, R., and Herrera-Franco, P.J. (1999). Effect of fiber surface treatment on the fiber–matrix bond strength of natural fiber reinforced composites. *Composites Part B* 30: 309–320.

**20** Cai, M., Takagi, H., Nakagaito, A.N. et al. (2015). Influence of alkali treatment on internal microstructure and tensile properties of abaca fibers. *Ind. Crops Prod.* 65: 27–35.

**21** Bachtiar, D., Sapuan, S.M., and Hamdan, M.M. (2008). The effect of alkaline treatment on tensile properties of sugar palm fibre reinforced epoxy composites. *Mater. Des.* 29: 1285–1290.

**22** Edeerozey, A.M.M., Akil, H.M., Azhar, A.B., and Ariffin, M.I.Z. (2007). Chemical modification of kenaf fibers. *Mater. Lett.* 61: 2023–2025.

**23** Goda, K., Sreekala, M.S., Gomes, A. et al. (2006). Improvement of plant based natural fibers for toughening green composites – effect of load application during mercerization of ramie fibers. *Composites Part A* 37: 2213–2220.

**24** Fiore, V., Di Bella, G., and Valenza, A. (2015). The effect of alkaline treatment on mechanical properties of kenaf fibers and their epoxy composites. *Composites Part B* 68: 14–21.

**25** Mahjoub, R., Yatim, J.M., Mohd Sam, A.R., and Hashemi, S.H. (2014). Tensile properties of kenaf fiber due to various conditions of chemical fiber surface modifications. *Constr. Build. Mater.* 55: 103–113.

**26** Venkateshwaran, N., Perumal, A.E., and Arunsundaranayagam, D. (2013). Fiber surface treatment and its effect on mechanical and visco-elastic behaviour of banana/epoxy composite. *Mater. Des.* 47: 151–159.

**27** Ibrahim, M.I., Hassan, M.Z., Dolah, R. et al. (2018). Tensile behaviour for mercerization of single kenaf fiber. *Malaysian J. Fundam. Appl. Sci.* 14: 437–439.

**28** Wong, K.J., Yousif, B.F., and Low, K.O. (2016). The effects of alkali treatment on the interfacial adhesion of bamboo fibres. *Proc. Inst. Mech. Eng., Part L* 224: 139–148.

**29** Siddika, S., Mansura, F., Hasan, M., and Hassan, A. (2014). Effect of reinforcement and chemical treatment of fiber on the properties of jute-coir fiber reinforced hybrid polypropylene composites. *Fibers Polym.* 15: 1023–1028.

**30** Xie, Y., Hill, C.A.S., Xiao, Z. et al. (2010). Silane coupling agents used for natural fiber/polymer composites: a review. *Composites Part A* 41: 806–819.

**31** Thakur, V.K., Singha, A.S., Kaur, I. et al. (2010). Silane functionalization of *Saccaharum cilliare* fibers: thermal, morphological, and physicochemical study. *Int. J. Polym. Anal. Charact.* 15: 397–414.

**32** Singha, A.S. and Thakur, V.K. (2009). Morphological, thermal, and physicochemical characterization of surface modified Pinus fibers. *Int. J. Polym. Anal. Charact.* 14: 271–289.

**33** Valadez-Gonzalez, A., Cervantes-Uc, J.M., Olayo, R., and Herrera-Franco, P.J. (1999). Chemical modification of henequén fibers with an organosilane coupling agent. *Composites Part B* 30: 321–331.

**34** Herrera-Franco, P.J. and Valadez-González, A. (2004). Mechanical properties of continuous natural fibre-reinforced polymer composites. *Composites Part A* 35: 339–345.

**35** Herrera-Franco, P.J. and Valadez-González, A. (2005). A study of the mechanical properties of short natural-fiber reinforced composites. *Composites Part B* 36: 597–608.

**36** Sepe, R., Bollino, F., Boccarusso, L., and Caputo, F. (2018). Influence of chemical treatments on mechanical properties of hemp fiber reinforced composites. *Composites Part B* 133: 210–217.

**37** Jappes, J.T.W. and Siva, I. (2011). Studies on the influence of silane treatment on mechanical properties of coconut sheath-reinforced polyester composite. *Polym.-Plast. Technol. Eng.* 50: 1600–1605.

**38** Liu, Y., Lv, X., Bao, J. et al. (2019). Characterization of silane treated and untreated natural cellulosic fibre from corn stalk waste as potential reinforcement in polymer composites. *Carbohydr. Polym.* 218: 179–187.

**39** Rowell, R.M. (2004). Acetylation of natural fibers to improve performance. *Mol. Cryst. Liq. Cryst.* 418: 153–164.

**40** Rana, A.K., Basak, R.K., Mitra, B.C. et al. (1997). Studies of acetylation of jute using simplified procedure and its characterization. *J. Appl. Polym. Sci.* 64: 1517–1523.

**41** Zaman, H.U. and Khan, R.A. (2019). Acetylation used for natural fiber/polymer composites. *J. Thermoplast. Compos. Mater.* 1: 1–21.

**42** Bledzki, A.K., Mamun, A.A., Lucka-Gabor, M., and Gutowski, V.S. (2008). The effects of acetylation on properties of flax fibre and its polypropylene composites. *Express Polym. Lett.* 2: 413–422.

**43** Senthilraja, R., Sarala, R., Antony, A.G., and Seshadhri, S. (2020). Effect of acetylation technique on mechanical behavior and durability of palm fibre vinyl-ester composites. *Mater. Today: Proc.* 21: 634–637.

**44** Chen, J., Su, M., Ye, J. et al. (2014). All-straw-fiber composites: benzylated straw as matrix and additional straw fiber reinforced composites. *Polym. Compos.* 35: 419–426.

**45** Chen, C., Cho, M., Kim, B.-W. et al. (2012). Thermo plasticization and characterization of kenaf fiber by benzylation. *J. Ind. Eng. Chem.* 18: 1107–1111.

**46** Nayak, S. and Mohanty, J.R. (2018). Influence of chemical treatment on tensile strength, water absorption, surface morphology, and thermal analysis of areca sheath fibers. *J. Nat. Fibers* 16: 589–599.

**47** Nair, K.C.M. and Thomas, S. (2003). Effect of interface modification on the mechanical properties of polystyrene-sisal fiber composites. *Polym. Compos.* 24: 332–343.

**48** Majid, R.A., Ismail, H., and Taib, R.M. (2016). Benzoyl chloride treatment of kenaf core powder: the effects on mechanical and morphological properties of PVC/ENR/kenaf core powder composites. *Procedia Chem.* 19: 803–809.

**49** Yadav, A. and Gupta, M.K. (2019). Development and characterization of jute composites for sustainable product: effect of chemical treatments and polymer coating. *Mater. Res. Express* 7: 1–13.

**50** Sari, N.H., Sanjay, M.R., Arpitha, G.R. et al. (2019). Synthesis and properties of pandanwangi fiber reinforced polyethylene composites: evaluation of dicumyl peroxide (DCP) effect. *Compos. Commun.* 15: 53–57.

**51** Joseph, K., Thomas, S., and Pavithran, C. (1996). Effect of chemical treatment on the tensile properties of short sisal fibre-reinforced polyethylene composites. *Polymer* 37: 5139–5149.

**52** Ahmad, E.E.M. and Luyt, A.S. (2012). Effects of organic peroxide and polymer chain structure on morphology and thermal properties of sisal fibre reinforced polyethylene composites. *Composites Part A* 43: 703–710.

**53** Mokoena, M.A., Djoković, V., and Luyt, A.S. (2004). Composites of linear low density polyethylene and short sisal fibres: the effects of peroxide treatment. *J. Mater. Sci.* 39: 3403–3412.

**54** Sapieha, S., Allard, P., and Zang, Y.H. (1990). Dicumyl peroxide-modified cellulose/LLDPE composites. *J. Appl. Polym. Sci.* 41: 2039–2048.

**55** Madhu, P., Sanjay, M.R., Jawaid, M. et al. (2020). A new study on effect of various chemical treatments on agave Americana fiber for composite reinforcement: Physico-chemical, thermal, mechanical and morphological properties. *Polym. Test.* 85: 1–7.

**56** Malunka, M.E., Luyt, A.S., and Krump, H. (2006). Preparation and characterization of EVA–sisal fiber composites. *J. Appl. Polym. Sci.* 100: 1607–1617.

**57** Rytlewski, P., Moraczewski, K., Malinowski, R., and Żenkiewicz, M. (2014). Assessment of dicumyl peroxide ability to improve adhesion between polylactide and flax or hemp fibres. *Compos. Interfaces* 21: 671–683.

**58** Keener, T.J., Stuart, R.K., and Brown, T.K. (2004). Maleated coupling agents for natural fibre composites. *Composites Part A* 35: 357–362.

**59** Mohanty, S., Verma, S.K., Nayak, S.K., and Tripathy, S.S. (2004). Influence of fiber treatment on the performance of sisal-polypropylene composites. *J. Appl. Polym. Sci.* 94: 1336–1345.

**60** Mohanty, S., Nayak, S.K., Verma, S.K., and Tripathy, S.S. (2016). Effect of MAPP as a coupling agent on the performance of jute–PP composites. *J. Reinf. Plast. Compos.* 23: 625–637.

**61** Kakou, C.A., Arrakhiz, F.Z., Trokourey, A. et al. (2014). Influence of coupling agent content on the properties of high density polyethylene composites reinforced with oil palm fibers. *Mater. Des.* 63: 641–649.

**62** Mutjé, P., Vallejos, M.E., Gironès, J. et al. (2006). Effect of maleated polypropylene as coupling agent for polypropylene composites reinforced with hemp strands. *J. Appl. Polym. Sci.* 102: 833–840.

**63** Wong, K.H., Mohammed, D.S., Pickering, S.J., and Brooks, R. (2012). Effect of coupling agents on reinforcing potential of recycled carbon fibre for polypropylene composite. *Compos. Sci. Technol.* 72: 835–844.

**64** Baykus, O., Mutlu, A., and Doğan, M. (2015). The effect of pre-impregnation with maleated coupling agents on mechanical and water absorption properties of jute fabric reinforced polypropylene and polyethylene biocomposites. *J. Compos. Mater.* 50: 257–267.

**65** George, J., Sreekala, M.S., and Thomas, S. (2001). A review on interface modification and characterization of natural fiber reinforced plastic composites. *Polym. Eng. Sci.* 41: 1471–1485.

**66** Ashori, A. and Nourbakhsh, A. (2009). Polypropylene cellulose-based composites: the effect of bagasse reinforcement and polybutadiene isocyanate treatment on the mechanical properties. *J. Appl. Polym. Sci.* 111: 1684–1689.

**67** Karmarkar, A., Chauhan, S.S., Modak, J.M., and Chanda, M. (2007). Mechanical properties of wood–fiber reinforced polypropylene composites: effect of a novel compatibilizer with isocyanate functional group. *Composites Part A* 38: 227–233.

**68** Raj, R.G., Kokta, B.V., and Daneault, C. (1989). Effect of chemical treatment of fibers on the mechanical properties of polyethylene-wood fiber composites. *J. Adhes. Sci. Technol.* 3: 55–64.

**69** Nourbakhsh, A., Kokta, B.V., Ashori, A., and Jahan-Latibari, A. (2008). Effect of a novel coupling agent, polybutadiene isocyanate, on mechanical properties of wood-fiber polypropylene composites. *J. Reinf. Plast. Compos.* 27: 1679–1687.

**70** Joseph, P.V., Joseph, K., and Thomas, S. (2012). Short sisal fiber reinforced polypropylene composites: the role of interface modification on ultimate properties. *Compos. Interfaces* 9: 171–205.

**71** De Rosa, I.M., Kenny, J.M., Maniruzzaman, M. et al. (2011). Effect of chemical treatments on the mechanical and thermal behaviour of okra (*Abelmoschus esculentus*) fibres. *Compos. Sci. Technol.* 71: 246–254.

**72** Dhanalakshmi, S., Ramadevi, P., and Basavaraju, B. (2015). Influence of chemical treatments on flexural strength of areca fiber reinforced epoxy composites. *Chem. Sci. Trans.* 4: 409–418.

**73** Dhanalakshmi, S., Ramadevi, P., and Basavaraju, B. (2015). Effect of chemical treatments on tensile strength of areca fiber reinforced natural rubber composites. *J. Appl. Chem.* 8: 43–52.

**74** Hasany, S.F., Rejaband, M.R.M., Bachtiar, D., and Mohammed, A.A. (2017). Effect of potassium permanganate on tensile properties of sugar palm fibre reinforced thermoplastic polyurethane. *Indian J. Sci. Technol.* 10: 1–5.

**75** Zaman, H.U., Khan, M.A., Khan, R.A. et al. (2010). Role of potassium permanganate and urea on the improvement of the mechanical properties of jute polypropylene composites. *Fibers Polym.* 11: 455–463.

**76** Patra, A., Bisoyi, D.K., Manda, P.K., and Singh, A.K. (2013). Electrical and mechanical properties of the potassium permanganate treated short sisal fiber reinforced epoxy composite in correlation to the macromolecular structure of the reinforced fiber. *J. Appl. Polym. Sci.* 128: 1011–1019.

**77** Khan, J.A., Khan, M.A., and Islam, R. (2013). Mechanical, thermal and degradation properties of jute fabric – reinforced polypropylene composites: effect of potassium permanganate as oxidizing agent. *Polym. Compos.* 34: 671–680.

**78** Kiattipanich, N., Kreua-ongarjnukool, N., Pongpayoon, T., and Phalakornkule, C. (2007). Properties of polypropylene composites reinforced with stearic acid treated sugarcane fiber. *J. Polym. Eng.* 27: 411–428.

**79** Nekkaa, S., Chebira, F., and Haddaoui, N. (2006). Effect of fiber treatment on the mechanical and rheological properties of polypropylene/broom fiber spartium junceum composites. *J. Eng. Appl. Sci.* 3: 278–283.

**80** Kalaprasad, G., Francis, B., Thomas, S. et al. (2004). Effect of fibre length and chemical modifications on the tensile properties of intimately mixed short sisal/glass hybrid fibre reinforced low density polyethylene composites. *Polym. Int.* 53: 1624–1638.

**81** Enriquez, J.K.E.D.V., Macalino Santiago, P.J., Funcion Ong, T., and Chakraborty, S. (2009). Fabrication and characterization of high-density polyethylene – coconut coir composites with stearic acid as compatibilizer. *J. Thermoplast. Compos. Mater.* 23: 361–373.

**82** Torres, F.G. and Cubillas, M.L. (2005). Study of the interfacial properties of natural fibre reinforced polyethylene. *Polym. Test.* 24: 694–698.

**83** Sun, D. and Stylios, G.K. (2016). Investigating the plasma modification of natural fiber fabrics-the effect on fabric surface and mechanical properties. *Text. Res. J.* 75: 639–644.

**84** Kan, C.-w. and Lam, Y.-l. (2015). The effect of plasma treatment on water absorption properties of silk fabrics. *Fibers Polym.* 16: 1705–1714.

**85** Bozaci, E., Sever, K., Sarikanat, M. et al. (2013). Effects of the atmospheric plasma treatments on surface and mechanical properties of flax fiber and adhesion between fiber–matrix for composite materials. *Composites Part B* 45: 565–572.

**86** Bozaci, E., Sever, K., Demir, A. et al. (2010). Effect of the atmospheric plasma treatment parameters on surface and mechanical properties of jute fabric. *Fibers Polym.* 10: 781–786.

**87** de Farias, J.G.G., Cavalcante, R.C., Canabarro, B.R. et al. (2017). Surface lignin removal on coir fibers by plasma treatment for improved adhesion in thermoplastic starch composites. *Carbohydr. Polym.* 165: 429–436.

**88** Gibeop, N., Lee, D.W., Prasad, C.V. et al. (2013). Effect of plasma treatment on mechanical properties of jute fiber/poly (lactic acid) biodegradable composites. *Adv. Compos. Mater* 22: 389–399.

**89** Liu, X. and Cheng, L. (2016). Influence of plasma treatment on properties of ramie fiber and the reinforced composites. *J. Adhes. Sci. Technol.* 31: 1723–1734.

**90** Sinha, E. and Panigrahi, S. (2009). Effect of plasma treatment on structure, wettability of jute fiber and flexural strength of its composite. *J. Compos. Mater.* 43: 1791–1802.

**91** Hassani, F.O., Merbahi, N., Oushabi, A. et al. (2020). Effects of corona discharge treatment on surface and mechanical properties of aloe vera fibers. *Mater. Today: Proc.* 24: 46–51.

**92** Ragoubi, M., George, B., Molina, S. et al. (2012). Effect of corona discharge treatment on mechanical and thermal properties of composites based on miscanthus fibres and polylactic acid or polypropylene matrix. *Composites Part A* 43: 675–685.

**93** Belgacem, M.N., Bataille, P., and Sapieha, S. (1994). Effect of corona modification on the mechanical properties of polypropylene/cellulose composites. *J. Appl. Polym. Sci.* 53: 379–385.

**94** Gassan, J. (2000). Effects of corona discharge and UV treatment on the properties of jute-fibre epoxy composites. *Compos. Sci. Technol.* 60: 2857–2863.

**95** Hu, S.G., Jou, C.H., and Yang, M.C. (2003). Antibacterial and biodegradable properties of polyhydroxyalkanoates grafted with chitosan and chitooligosaccharides via ozone treatment. *J. Appl. Polym. Sci.* 88: 2797–2803.

**96** Avinc, O., Eren, H.A., Uysal, P., and Wilding, M. (2012). The effects of ozone treatment on soybean fibers. *Ozone Sci. Eng.* 34: 143–150.

**97** Zhao, F., Rajendran Royan, N.R., Sulong, A.B. et al. (2018). UV/$O_3$ treatment as a surface modification of rice husk towards preparation of novel biocomposites. *PLoS One* 13: 1–17.

**98** Lette, M.J., Takasaki, A., Okabe, T. (2016). Processing and characterization of ozone treated bamboo fiber/phenolic resin wood-polymer composite. The 9th International Conference on Green Composites (ICGC9): Kobe University Centennial Hall, Kobe University, Kobe, Japan.

**99** Chtourou, H., Riedl, B., and Kokta, B.V. (1997). Strength properties of wood-PE composites: influence of pulp ratio and pretreatment of PE fibers. *Tappi J.* 80: 141–151.

# 8

# Mechanical and Tribological Characteristics of Industrial Waste and Agro Waste Based Hybrid Composites

*Vigneswaran Shanmugam[1,2], Uthayakumar Marimuthu[1], Veerasimman Arumugaprabu[1], Sundarakannan Rajendran[1], and Rajendran Deepak Joel Johnson[2]*

[1]*Kalasalingam Academy of Research and Education, Faculty of Mechanical Engineering, Krishnankoil 626 126, India*
[2]*Saveetha School of Engineering, Saveetha Institute of Medical and Technical Sciences, Department of Mechanical Engineering, Chennai, Tamilnadu, India*

## 8.1 Introduction

Erosion loss due to solid particle impact produces a high rate of damage in materials. Erosion wear produces an adverse effect on the material surface. When a material surface is exposed to the continuous impact of particles, it will result in a change in mass; either the mass increases due to the clinging of particles or there is mass loss due to material removal. Various parameters constitute the erosion loss including erodent impact angle, velocity of erodent, particle flow, and type of erodent [1]. Besides, material morphology also has a significant effect on erosion resistance. In a recent research on polymer erosion, Jena et al. [2] investigated the influence of erosion parameters in correlation with material behavior on erosion loss. The results proved the influence of the erosion variables on material loss in a cenosphere-filled bamboo fiber composite. The most critical variables affecting the erosion rate were the velocity and impact angle of the abrasive particles. Effectual fiber reinforcement in the composite matrix significantly influences the composite's erosion resistance [3]. Deepak et al. [4] reported that fiber properties show minimum contribution in affecting erosion behavior. Yet the fiber contribution cannot be ignored since it has significant interaction with other erosion variables. Although there are many reports in the literature on the effect of fiber reinforcement on erosion behavior of the composites, only a limited number of studies discuss the effect of treated fiber reinforcement on erosion behavior.

From a thorough literature analysis, the relation between material morphology and erosion loss was understood. Maximum erosion for the brittle material occurs at or near 90° impact angle; for ductile material, maximum erosion occurs at a lower impact angle, less than 30°, but for fiber-reinforced polymers maximum erosion occurs at 60° because of their semi-ductile behavior. In fiber-reinforced polymer

*Mechanical and Dynamic Properties of Biocomposites,* First Edition. Edited by Senthilkumar Krishnasamy, Rajini Nagarajan, Senthil Muthu Kumar Thiagamani, and Suchart Siengchin.

composites, semi-ductile behavior results in brittle failure at or near 90° and ductile failure at 30° or less [5]. Erosion test on hybrid composites filled with boric acid and glass fiber showed maximum erosion at 30° impact angle, which is similar to the failure in ductile material [6]. In a particle-added hybrid composite, the maximum erosion loss occurs at a lower impact angle, as reported by Bagci [7].

The residue collected during alumina segregation in the Bayer method is known as red mud. Owing to increasing production, these wastes continue to generate severe environmental pollution, and their disposal costs are also high. The adverse effects of red mud have persuaded researchers to find a way to make use of red mud. Arumugaprabu et al. [8] examined the effect of red mud on natural-fiber-reinforced polymers. The investigation results reported the effectiveness of red mud addition in the polymer matrix. Rachchh et al. [9] studied red-mud-reinforced polymer composites and reported that the high hardness offered by red mud reinforcement would help in increasing the erosion resistance. The wear performance of the composites was increased considerably with red mud addition and red mud particle, which were highly suitable for application in erosion environment [10]. Increasing the red mud percentage reduces the strength of the composites due to improper bonding features [11]. This can be avoided by increasing the bonding characteristics in the hybrid composites. Bonding properties of composites can be increased with chemical treatment [12]. Chemically treated fiber reinforcement in polymer composites could increase erosion resistance [13]. The solid waste red mud reinforcement was found to be more effective against erosion. Therefore, the present work aims to examine the effect of red mud on composite erosion behavior. The findings of the analysis compare the erosion resistance of filled and unfilled hybrid composites with untreated and treated fiber.

## 8.2 Materials and Methods

Different treated jute fiber and red mud hybrid composites were fabricated through a press molding process. The matrix used is polyester matrix. Polyester matrix is preferred owing to its fast curing time and good particle dispersion ability. Red mud is reinforced at four different weight percentages (0%, 10%, 20%, and 30%) with jute fiber reinforcement of 40 wt.%. The fiber is treated with alkali (5% sodium hydroxide) and silane (2% triethoxy(ethyl)silane). The fiber structural variation was analyzed through X-ray diffraction (XRD) test carried out using X-ray diffractometer (ECO D8, Bruker, Germany). Details of the fabricated composite are shown in Table 8.1. Mechanical tests such as hardness 9 (ASTM D-2240) and tensile (ASTM D-3039) tests were performed. The mechanical test was performed on five samples, and an average value was reported.

Fabricated composites were investigated by solid particle erosion test as per ASTM G76. The erosion tester of DUCOM was used to conduct the experiments, which consists of the dry air supplying compressor, erodent feeder unit, and erosion chamber. The erodent from the feeder unit was mixed well with the dry compressed air and passed through a nozzle of 50 mm length and 1.5 mm diameter made of

**Table 8.1** Details of fabricated composite.

| S. No. | Composition | Designation |
|---|---|---|
| 1 | 60 wt.% polyester + 40 wt.% untreated fiber + 0 wt.% red mud | A |
| 2 | 50 wt.% polyester + 40 wt.% untreated fiber + 10 wt.% red mud | B |
| 3 | 40 wt.% polyester + 40 wt.% untreated fiber + 20 wt.% red mud | C |
| 4 | 30 wt.% polyester + 40 wt.% untreated fiber + 30 wt.% red mud | D |
| 5 | 60 wt.% polyester + 40 wt.% NaOH treated fiber + 0 wt.% red mud | E |
| 6 | 50 wt.% polyester + 40 wt.% NaOH treated fiber + 10 wt.% red mud | F |
| 7 | 40 wt.% polyester + 40 wt.% NaOH treated fiber + 20 wt.% red mud | G |
| 8 | 30 wt.% polyester + 40 wt.% NaOH treated fiber + 30 wt.% red mud | H |
| 9 | 60 wt.% polyester + 40 wt.% silane treated fiber + 0 wt.% red mud | I |
| 10 | 50 wt.% polyester + 40 wt.% silane treated fiber + 10 wt.% red mud | J |
| 11 | 40 wt.% polyester + 40 wt.% silane treated fiber + 20 wt.% red mud | K |
| 12 | 30 wt.% polyester + 40 wt.% silane treated fiber + 30 wt.% red mud | L |

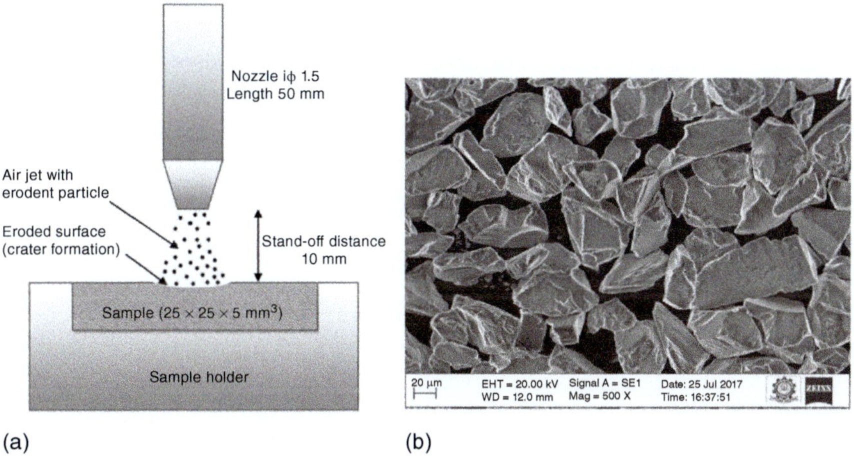


**Figure 8.1** (a) Schematic representation of the erosion process. (b) Morphology of erodent particle.

tungsten carbide material. Figure 8.1a shows a schematic view of the experimental setup inside the erosion chamber. The work sample is placed in the sample holder having an stand-off-distance (SoD) of 10 mm. The erodent used in the present investigation is aluminum oxide of an average size of 50 μm whose morphology is shown in Figure 8.1b.

The hard-abrasive particles coming out of the nozzle strike the sample, which leads to erosion loss. The erosion test was done for 10 minutes by varying impact angles (30°, 45°, 60°, and 90°) at a constant erodent velocity of 100 m/s. The velocity of the erodent particle striking the sample was found by the double-disc method. The erodent was delivered at a rate of 3.3 g/m. Since mass loss is a major concern, the weight of the sample was measured before and after the erosion test. To achieve quality measurement, each sample was tested thrice and the average weight loss was calculated. To determine the erosion rate, Eq. (8.1) is used:

$$\text{Erosion rate} = \frac{\text{Weight loss}}{\text{discharge} \times \text{time}}(\text{g/g}) \tag{8.1}$$

### 8.2.1 Scanning Electron Microscopy (SEM)

Inspection of the failure mechanism in the mechanical- and erosion-tested composites was performed using scanning electron microscopy (SEM) (EVO18, CARL ZEISS) operating at 25 kV. The fractured surfaces of the tensile-tested composites were scanned, and in erosion-tested samples the eroded surface was scanned.

## 8.3 Result and Discussion

### 8.3.1 Effect of Chemical Treatment on Fiber

The treated and untreated jute fiber XRD patterns are illustrated in Figure 8.2. In all the three XRD patterns, two peaks were observed, which are the defined XRD peaks for the natural fibers as reported by researchers [14, 15]. The two peaks in the XRD represent cellulose I and IV of the natural fiber [16]. Chemical treatment changed the crystallinity of the fiber. The jute fiber crystallinity index (CI) was determined using the Segal analytical method (Eq. (8.2)) [17]:

$$\text{CI} = \frac{I_{002} - I_{\text{am}}}{I_{002}} \times 100 \tag{8.2}$$

where $I_{002}$ is the maximum intensity of the crystalline peak and $I_{\text{am}}$ is the minimum intensity of the amorphous material calculated from the XRD pattern shown in Figure 8.2. Table 8.2 presents the CI value calculated through the XRD pattern.

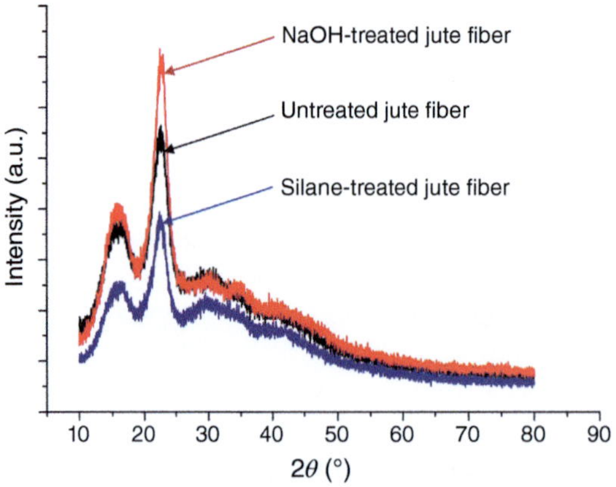


**Figure 8.2** XRD pattern of fiber.

**Table 8.2** Crystallinity property of fiber.

| Fiber | Crystallinity (%) |
|---|---|
| Untreated jute fiber | 39.8 |
| NaOH-treated jute fiber | 45.2 |
| Silane-treated jute | 43.7 |

On NaOH treatment, the crystallinity of the fiber increases when compared with the untreated fiber. This is due to the reaction of NaOH with the cellulose present in the fiber and removal of non-cellulose substances. This enhances the packing of the cellulose chain and increases the crystallinity. This aids in the development of good bonding with the matrix and increases the mechanical strength. Silane-treated fibers showed the opposite trend. The crystallinity of the silane-treated fiber is lower than that of the NaOH-treated fiber and higher than that of the untreated fiber. The reaction of silane with the cellulose present in the fiber surface results in the formation of hydrogen cellulose chains, which modifies the crystalline nature of the fiber surface to amorphous. Further diffusion of the silane coupling agent by developing the amorphous region in the surface and reacting with the crystalline cellulose produces more amorphous cellulose.

### 8.3.2 Mechanical Behavior

The results of the tensile and hardness test are shown in Figure 8.3. The outcomes reveal the enhancement in the properties because of the fiber treatment. In both the treated and untreated composites, increment in the red mud decreases the tensile strength of the composites. The hardness of the composites increases with red mud increment. Both the tensile and hardness properties of treated fiber composites are higher than those of the untreated fiber composites. Notably, silane-treated composites exhibit superior characteristics than the untreated fiber and NaOH-treated fiber composites. It is concluded that silane treatment of jute fiber greatly influences the tensile strength of the composites. However, chemical treatment showed only a marginal variation in the hardness characteristics.

The mechanical test results show that increase in the red mud particle inclusion reduces the tensile characteristics and improves the hardness. Failure analysis on the fracture surfaces revealed the two main aspects of the failure mechanism. One is the development of poor interlocking between the fiber, red mud, and matrix. At this condition, the matrix, fiber, and red mud fail to transfer the tensile stress. Although the bonding was improved on chemical treatment, the tensile strength was lower in high red mud added composites. The second one is the agglomeration of red mud. At high red mud addition, the particles stick together and aggregate. This develops a particle-rich area, which shows negative results under mechanical loading. On tensile load, these regions fail first due to ineffective load-carrying

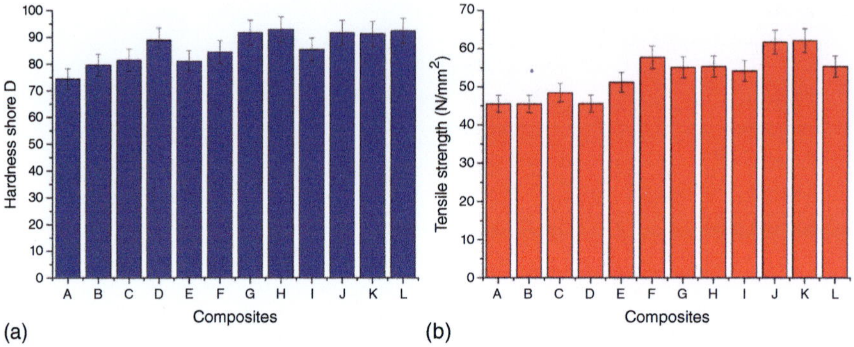


**Figure 8.3** Hardness and tensile strength of composites (a) Hardness (b) Tensile strength.

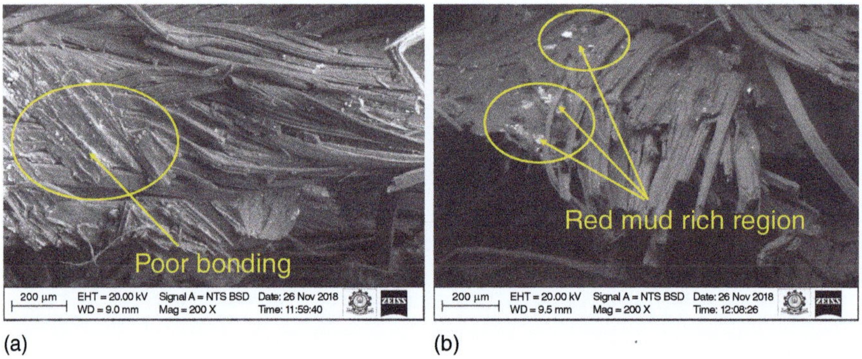


**Figure 8.4** SEM image of tensile-tested composites (a) Poor bonding (b) Red mud agglomeration.

ability, which significantly reduces the tensile strength. This is evident in SEM images of the fracture's specimen analysis (Figure 8.4).

Wettability of the fiber composites is a function of fiber characteristics. On chemical treatment, the fiber develops two different interlocking mechanisms: structural bonding and chemical bonding. On alkaline NaOH treatment, the contaminants in the fiber are removed, which develops a rougher surface. This rougher surface provides effective structural bonding with the matrix [16]. Treatment of fiber with coupling agents such as silane results in chemical bonding with the matrix. Silane settles on the fiber surface and reacts with the matrix, developing a strong covalent bond [17, 18].

From the findings, it can be concluded that on chemical treatment only the tensile strength of the composites is improved. The hardness of the composites is not influenced in a significant way on chemical treatment. This is because of the variation in stress under tensile and impact strength. Under tensile stress, the fiber tends to pull out, debond, and break, which is highly dependent on fiber–matrix interlocking, whereas in the hardness test compression stress acts on the composite surface. Compression stress will be distributed uniformly within the composite's structure irrespective of fiber–matrix bonding.

### 8.3.3 Erosion Behavior

The erosion mechanism at four impact angles at an erodent velocity of 100 m/s is shown in Figure 8.5. The yellow marking in the figure represents the erosion exposure area, and it is understood that under erosion attack on the material surface, a crater of 10 mm diameter is developed. This is because of the divergence in the air jet. The concentration of erosion attack varies with the impact angle. In all the four eroded surfaces at the high concentration regions, the fibers were exposed and visible owing to repeated erodent impact in that region. The woven fiber is visible in Figure 8.5, which denotes the high impact region.

#### 8.3.3.1 Effect of Fiber Treatment on Erosion Rate

The experimental results showed the effect of fiber treatment against erosion. Figure 8.6 shows the erosion test result between the unfilled treated and untreated composites at different impact angles. At all experimental conditions, erosion loss of the treated composites was found to be the least. Similar results were noted on the red-mud-filled jute composites.

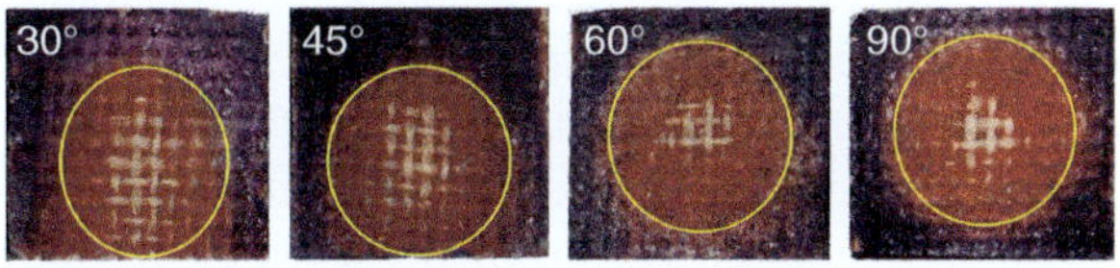


**Figure 8.5** Erosion exposure region at different impact angles.

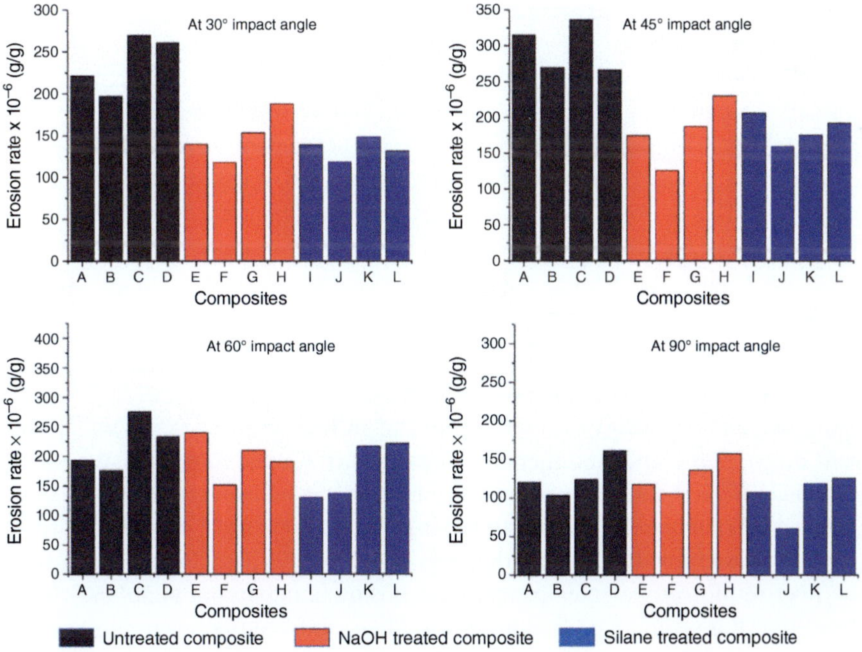


**Figure 8.6** Erosion rate at different impact angles.

Chemical treatment modifies the fiber structure and composition. This modification increases the wettability of the fiber with the matrix by producing a strong interfacial bonding mechanism [19]. On alkaline treatment, the contaminant present in the fibers is removed. Biological contents such as lignin, cellulose, and other waxy substances in the fiber are also removed. This makes the fiber surface rougher and produces good interlocking with the matrix. Because of these reasons, the treated fiber reinforcement develops a good interface between the matrix and the red mud. As a result, the surface and interfacial properties of the composites are enhanced considerably. But on silane treatment, the fiber develops enhanced bonding features due to chemical bonding.

#### 8.3.3.2 Effect of Red Mud Addition on Erosion Rate

In both the treated and untreated composites, addition of red mud leads to good resistance against the erosion. From the erosion rate analysis, it is found that the filled composites display higher wear loss on particle impact than the unfilled composites. This is because of the development of good surface properties in the filled composite. From Figure 8.6 it is noted that on increasing red mud addition, the erosion loss also increases. The addition of 10% red mud shows consistent erosion loss than in the unfilled composites. When the high accelerated erodent meets the composite surface, the kinetic energy of the erodent particle is converted to impact energy. The impact energy of erodent on the composite surface is the main reason for material removal. Red mud particles in the composite matrix act as an absorber of erodent impact energy. This reduces the chance of material removal and deformation occurs in the form of plastic deformation.

But on further increment, the erosion rate was found to increase consistently irrespective of fiber treatment. Increment in red mud weight percentage reduced the fiber matrix wettability, leading to the formation of poor surface interface. This was observed in 30% red-mud-filled composites. Owing to the poor surface interface, the erodent on impact easily removes the surface layer and starts attacking the fiber more quickly. The red mud was found to accumulate (Figure 8.7a), developing a hill-like structure (Figure 8.7b) on the composite surface. At this area the bonding is poor and on erodent impact, the hill structure collapses, developing micro-craters, cracks, and pores.

#### 8.3.3.3 Effect of Impact Angle on Erosion Rate

Material morphology also has some significant influence on erosion loss. In the present investigation, maximum erosion loss is noted for the composite at a lower impact angle. This is due to the variation in erosion mechanism for brittle and ductile material. Notably, maximum material loss in brittle material occurs at normal impact angle, and for ductile material, it is at a lower impact angle. But in case of the fiber-reinforced polymer composites, erosion is maximum at 30°–60°. This is because of the semi-ductile nature of the fiber-reinforced polymer composites. The variation in failure mechanism of the materials with impact angle is shown in Figure 8.8.

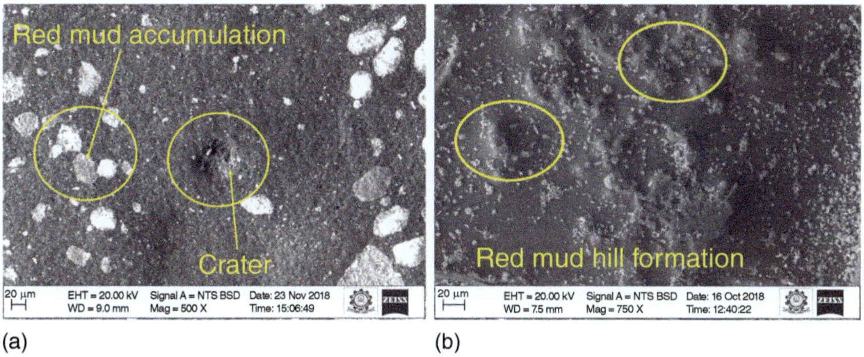


**Figure 8.7** (a) Crater development due to erodent impact. (b) Red mud hill formation.

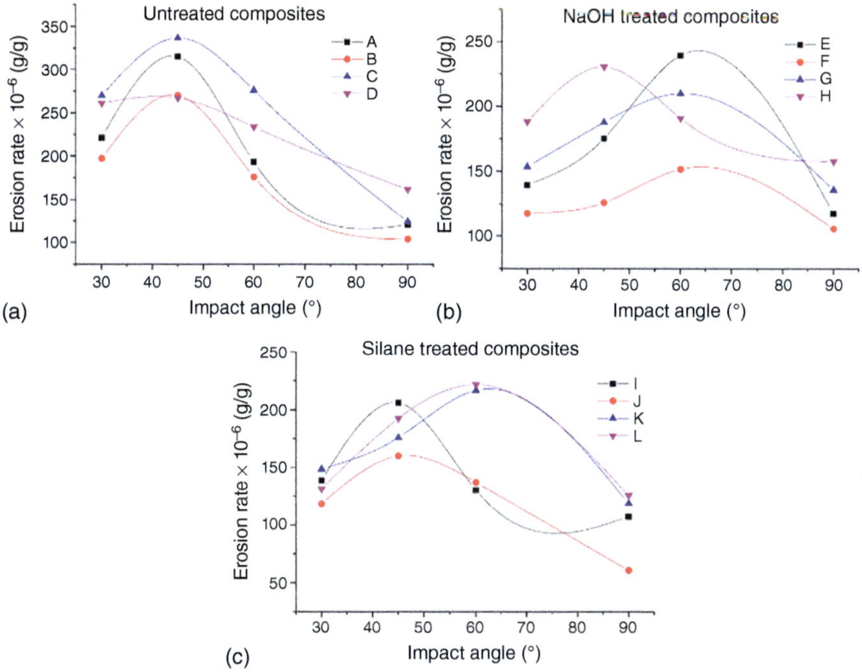


**Figure 8.8** Erosion behavior at different impact angles of treated and untreated composites (a) untreated composites, (b) NaOH treated composites (c) silane treated composites.

Maximum erosion was noted in untreated composites at an angle of impact of 45°. This is found to be varied in the composites reinforced with chemically treated fiber. This is owing to the variation of the composites' morphological properties.

In both the chemical-treated composites, maximum erosion rate occurs at 45° and 60° impact angle. This variation is due to the bonding between the fiber and matrix. Although the composite exhibits maximum loss at various impact angles, all the composites are found to be semi-ductile since maximum erosion occurs at a lower impact angle. At lower impact angle, the surface is highly characterized by

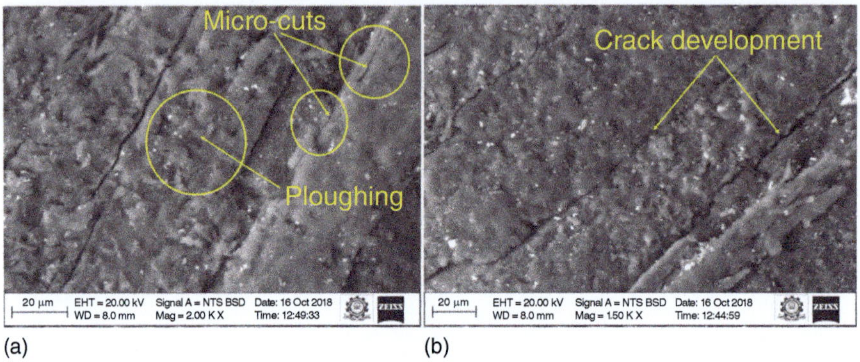


**Figure 8.9** (a) Micro cuts and ploughing (b) Crack developement.

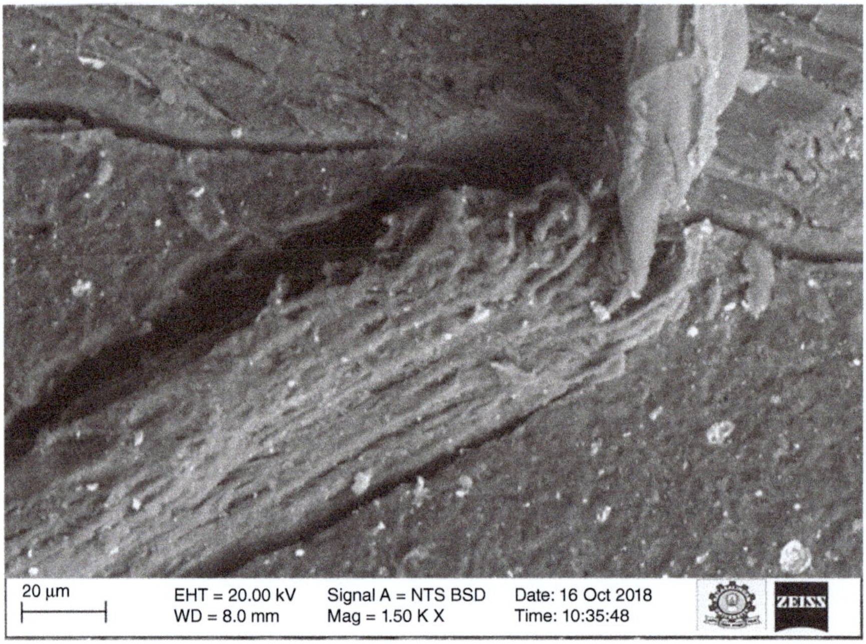


**Figure 8.10** Fiber damage due to erodent impact.

micro-cuts and severe ploughing (Figure 8.9a), whereas in the case of higher impact angle the surface has craters and cracks (Figure 8.9b). More material loss occurs at low angles since the sharp cutting edge of the erodent ploughs the composite surface and chip of the surface material. On continuous impact, the surface material is completely removed, which exposes the fiber surface (Figure 8.10). The exposed fibers are easily damaged and fragmented by the hitting erodent. In contrast, in the case of treated fiber reinforcement, the bond between the matrix and fiber resists the surface damage against erosion.

Minimum erosion rate for all composites was reported at an angle of impact of 90°. This is because of the ductile characteristics of the fiber composite. The surface of the composite is covered with the brittle matrix material where the surface exhibits semi-ductility on fiber reinforcement. At 90° impact angle, the erodent hits the composite surface where the impact energy is absorbed by the matrix material and transferred through to the fiber reinforcement. This does not produce a big impact on the wear loss. But on continuous impact, the surface undergoes plastic deformation and formation of craters and micro-cracks in the composite surface.

## 8.4 Conclusion

A hybrid composite having treated/untreated jute fiber and red mud reinforcement is successfully fabricated. Chemical treatment does not show a significant impact on the hardness strength but red mud addition showed a major influence in increasing the hardness. With red mud and treated fiber reinforcement, the tensile strength of the composites was improved. Maximum tensile strength was noted on 20% red mud addition and silane-treated fiber-reinforced composites. The chemical treatment of fibers increased the erosion resistance characteristics of the composites. Chemically treated fiber reinforcement varied the morphology of the hybrid composites, which is observed through the variation in the maximum erosion rate. Compared with NaOH treatment, silane treatment of jute fiber resulted in effective erosion resistance characteristics on the hybrid composite. Although the chemical treatment led to good erosion resistance, at 30% red mud reinforcement the matrix interface with fiber was poor due to agglomeration, which subsequently reduced the erosion resistance. It can be concluded that reinforcement with chemically treated fiber is an effective method for increasing the erosion resistance of fiber composites.

## References


1 Deliwala, A.A., Peter, M.R., and Yerramalli, C.S. (2018). A multiple particle impact model for prediction of erosion in carbon-fiber reinforced composites. *Wear* 406: 185–193.

2 Jena, H., Pradhan, A.K., and Pandit, M.K. (2018). Study of solid particle erosion wear behavior of bamboo fiber reinforced polymer composite with cenosphere filler. *Adv. Polym. Tech.* 37: 761–769.

3 Das, G. and Biswas, S. Erosion wear behavior of coir fiber-reinforced epoxy composites filled with $Al_2O_3$ filler. *J. Ind. Text.* 47: 472–488.

4 Deepak, R., Arumugaprabu, V., Uthayakumar, M. et al. (2018). Erosion performance studies on sansevieria cylindrica reinforced vinylester composite. *Mater. Res. Express* 3: 035309.

5 Patnaik, A., Alok, S., Navin, C. et al. (2010). Solid particle erosion wear characteristics of fiber and particulate filled polymer composites: a review. *Wear* 268: 249–263.

6 Bagci, M. and Imrek, H. (2011). Solid particle erosion behaviour of glass fibre reinforced boric acid filled epoxy resin composites. *Tribol. Int.* 44: 1704–1710.

7 Bagci, M. (2016). Determination of solid particle erosion with Taguchi optimization approach of hybrid composite systems. *Tribol. Int.* 94: 336–345.
8 Arumugaprabu, V., Uthayakumar, M., Manikandan, V. et al. (2014). Influence of redmud on the mechanical, damping and chemical resistance properties of banana/polyester hybrid composites. *Mater. Des.* 64: 270–279.
9 Rachchh, N.V., Misra, R.K., and Roychowdhary, D.G. (2015). Effect of red mud filler on mechanical and buckling characteristics of coir fibre-reinforced polymer composite. *Iran. Polym. J.* 24: 253–265.
10 Biswas, S. and Satapathy, A. (2009). Tribo-performance analysis of red mud filled glass-epoxy composites using Taguchi experimental design. *Mater. Des.* 30: 2841–2853.
11 Biswas, S. and Satapathy, A. (2010). A comparative study on erosion characteristics of red mud filled bamboo–epoxy and glass–epoxy composites. *Mater. Des.* 31: 1752–1767.
12 Li, X., Tabil, L.G., and Satyanarayan, P. (2007). Chemical treatments of natural fiber for use in natural fiber-reinforced composites: a review. *J. Polym. Environ.* 15: 25–33.
13 Vigneshwaran, S., Uthayakumar, M., and Arumugaprabu, V. (2017). A review on erosion studies of fiber-reinforced polymer composites. *J. Reinf. Plast. Compos.* 36: 1019–1027.
14 Cai, M., Takagi, H., Nakagaito, A.N. et al. (2016). Effect of alkali treatment on interfacial bonding in abaca fiber-reinforced composites. *Composites Part A* 90: 589–597.
15 Sreenivasan, V., Somasundaram, S., Ravindran, D. et al. (2011). Microstructural, physico-chemical and mechanical characterisation of Sansevieria cylindrica fibres–an exploratory investigation. *Mater. Des.* 32: 453–461.
16 Jebadurai, S., Garette, E.R., Sreenivasan, V., and Binoj, S. (2019). Comprehensive characterization of natural cellulosic fiber from *Coccinia grandis* stem. *Carbohydr. Polym.* 207: 675–683.
17 Rong, M.Z., Zhang, M.Q., Liu, Y. et al. (2001). The effect of fiber treatment on the mechanical properties of unidirectional sisal-reinforced epoxy composites. *Compos. Sci. Technol.* 61: 1437–1447.
18 Xie, Y., Hill, C.A.S., Xiao, Z. et al. (2010). Silane coupling agents used for natural fiber/polymer composites: a review. *Composites Part A* 41: 806–819.
19 Vigneshwaran, S., Sundarakannan, R., John, K.M. et al. (2020). Recent advancement in the natural fiber polymer composites: a comprehensive review. *J. Cleaner Prod.* 277: 124109.

# 9

# Dynamic Properties of Kenaf-Fiber-Reinforced Composites

*Rashed Al Mizan[1], Nur N. Akter[2], and Mohammad I. Iqbal[3]*

[1]*Ahsanullah University of Science and Technology, Department of Textile Engineering, 141 & 142, Love Road, Dhaka, 1208, Bangladesh*
[2]*BGMEA University of Fashion and Technology, Department of Apparel Manufacturing Technology, Nishatnagar, Turag, Dhaka, 1230, Bangladesh*
[3]*BGMEA University of Fashion and Technology, Department of Textile Engineering, Nishatnagar, Turag, Dhaka, 1230, Bangladesh*

## 9.1 Introduction

Natural fibre reinforced composite is becoming a high matter of concern to the industrialized and researcher [1, 2]. The well-studied form of this is polymer matrix composites (PMCs), which has expanded day by day because of their unique characteristics such as lightweight, nontoxicity, minimum production and processing cost, and biodegradability [3–5]. The utilization of natural fibers, collected from inexhaustible resources used in producing thermoplastic matrix composites, as a reinforcing agent provides some ecological advantages such as disposability and raw material usage [6, 7]. The main purpose of PMC manufacturing is to create a new material with some extraordinary properties that are not found in its components individually. Kenaf fiber (KF) has the potential as a reinforcement filler in the manufacture of PMC. There are mainly two types of kenaf fibers (KFs), bast and core. The outermost layer of kenaf is known as bast and the inner part is called core. Because of its toughness properties, the bast fiber is used in plastic, textile, and fiberglass technology applications by blending. On the other hand, the core is very malleable and vuggy, and hence compatible for use in plastic as organic filler. Moreover, kenaf fiber is used for reinforcing diversified products, as it is not hazardous to the environment.

To analyze the mechanical properties of materials under various process parameters such as frequency of loading, stress, and temperature, the Dynamic Mechanical Analysis (DMA) technique is used. A small sinusoidal time-varying oscillating force is employed using this technique and the mechanical response of the materials is divided into two parts, namely, an elastic part and a viscous part; if the polymer composites are considered as viscoelastic material, they have both elastic and viscous nature. The ratio of the loss modulus to the storage modulus is determined

*Mechanical and Dynamic Properties of Biocomposites*, First Edition. Edited by Senthilkumar Krishnasamy, Rajini Nagarajan, Senthil Muthu Kumar Thiagamani, and Suchart Siengchin.

by the viscous component and denoted by tan $\delta$ [8]. Dynamic environments are a popular technique to determine the sensitivity of a material during processing [9].

Ku et al. [10] have reported that bleaching of hemp strands has great effects in upgrading interfacial holding. To improve the mechanical properties of the sisal fiber polyester composites such as tensile strength, flexural strength, impact strengths, electrostatic charge, dielectric constant, thermal conductivity, and stability, sisal fibers are treated with benzoylation, mercerization, permanganate, polymethyl methacrylate, and micellar polymerization treatments for enhancing the interfacial adhesion [11–14]. While the damping factor, storage modulus, and loss modulus increased with treatment, they diminished with increasing quantity of fiber. The mechanical properties of pultruded kenaf fiber composites over a range of temperatures under differing frequencies of oscillating load were observed in an experiment [15]. Aziz and Ansell [6] have contemplated composites of alkalized polyester–kenaf and polyester–hemp. To determine the unrivaled properties of composites with treated fibers, TGA (thermogravimetric analysis), DMA, and DSC (differential scanning calorimetry) have been conducted. A survey of the ongoing research on DMTA (dynamic mechanical and thermal analysis) has been conducted by Costa et al. [16]. The several layers of sandwich PMCs of the jute/carbon/polyester composites were examined with various layer arrangement. Here the sample having an outside carbon layer shows an expansion away and loss modulus without a significant impact on $T_g$ of the composite [17]. Another study reviewed the viscoelastic temperature and glass transition temperature of other natural fabric composites that can substitute kevlar fiber composites for protection applications and noticed that fiber quantity has influence on the abovementioned properties [18]. DMA of glass/epoxy blended with banana fiber nanoparticles was conducted, and it was concluded that obstacle versatility of atomic chains improved the properties of the material by adding 8% of banana particles by weight of resin. Decrease in hysteresis and increase in versatility attributes were also found [19].

This chapter highlights the overall concept and some insight into the dynamics characteristics of kenaf-fiber-reinforced composite, where the DMA, TGA, vibration-damping acoustic properties, and acoustic properties are broadly discussed in the context of KF reinforcement; these properties are analyzed considering hybridization of other fibers for reinforcement.

## 9.2 Manufacturing Techniques for Kenaf-Fiber-Reinforced Composites

This section reviews the methods of manufacturing PMCs with kenaf fiber reinforcement. However, in most cases the manufacturing route used for kenaf fiber composites is merely incidental to the results obtained, with minimal contribution to knowledge of processing. For the bulk manufacturing of short-fiber-reinforced thermoplastics or thermoset composites, the widely used processes are describing below.

The mass production of short-fiber-reinforced thermoplastics and thermoset composites are fabricated by hot pressing [20] and injection molding [21]. The assembling of continuous fiber-reinforced composites has been reviewed by Astrom [21–25]. Spray and hand lamination processes are widely chosen because of the moderately cheap raw materials; the procedure can utilize generally unskilled labor, and the shape equipment are not required to be vacuum tight for low-performance components. In any case, these processes produce composites with low fiber volume fractions and a significant number of voids (porosity). For high-performance composites, vacuum packing of wet-lay-up or pre-impregnated (prepreg) materials can include 1 bar of pressure for consolidation, while for the best performance, an autoclave can add further outside pressure to the bag [26–29]. Elite composites can likewise be produced by utilizing compression molding in a (normally hydraulic) press between a coordinated (male and female) mold tool set with heated platens or integral heated tooling [21, 30]. The molds are normally made of steel to withstand the high shutting forces. For thermosetting matrix materials, the least difficult type of the procedure includes loading the open mold cavity with a shaped arrangement of the reinforcement to be utilized. The preform might be gathered in the instrument, or a particular shape, or sprayed offline. Liquid resin is added to the mold cavity; at this point the press is shut and the part cured under pressure. An elective methodology is the utilization of glass fiber reinforcement sheet molding compounds (SMCs) or bulk molding compounds (BMCs) materials, both dependent on the catalyzed polyester resins. Materials are formulated to give a quick cure, and molding cycle durations of a couple of moments are conceivable at process temperatures of around 150 °C. For thermoplastic composites, the standard commercial compression molding route includes the preheating of a pile of pre-impregnated sheets outside the mold apparatus, commonly in an oven or by infrared lamps. This procedure is sometimes alluded to as "stamping" and is generally utilized for the large-scale manufacturing of glass mat-reinforced thermoplastic parts for automotive applications. Pre-impregnation can be difficult and tedious; as another option, thin sheets of reinforcement and polymer can be "film stacked." When softened, they are quickly moved to the coordinated mold instrument, and at the same time shaped and cooled. The process duration is resolved by heat transfer rates, and in addition by the capacity of the matrix to stream and accomplish sufficient impregnation and consolidation. Resin transfer molding (RTM) [21, 28, 31, 32] is ideal for large-scale manufacturing of complex shaped composites of small or medium size. In RTM, dry fibers are loaded in a mold, and before the resin gets fixed to create a solid component, it streams into the fabric stack. RTM is suitable for moderately small parts; the mold closure forces become exorbitant as part sizes increase.

## 9.3 Characterization

The performance of materials is generally governed by their dynamic elements qualities, for example, storage and loss moduli (G), storage modulus ($E$), loss modulus ($E$), thermogravimetric properties, vibration damping, and sound properties. These attributes are imperative for assessing material capacity at extreme and critical conditions. Recently, various processes have been introduced to completely characterize the kenaf-fiber-reinforced composite behavior [15, 33–35]. However, the most acceptable mechanisms of testing are given below.

### 9.3.1 Dynamic Mechanical Analysis (DMA)

The dynamic mechanical characteristics of composites were analyzed using a dynamic mechanical analyzer. The dimensions of the specimen are length, 30 mm; width, 6.3 mm; and thickness, 0.13 mm. At variable heating rates and frequency levels, measurements were taken from −135 to 250 °C temperature.

### 9.3.2 Thermogravimetric Analysis (TGA)

At variable time and temperature, a thermogravimetric analyzer was used for the TGA of the composite under controlled conditions. The sample was heated at 10 °C/min heating rate, ranging from 25 to 800 °C temperature, with a nitrogen gas flow rate of 50 ml/min. In the sample pan, 8–10 mg of the sample was heated and the recorded information is shown as TG and DTG, where TG indicates the weight loss and DTG derivative the thermogravimetric weight loss rate with temperature variation.

### 9.3.3 Vibration-Damping Testing

According to ASTM E756-98, vibration testing was performed with a cantilevered setup mode on 10–25 mm sample clamping distance. Sample vibration process was finished using a finger with an estimated separation of 2 cm. The vibration was estimated utilizing a LutronV8200 transducer mounted at a distance of 10–15 mm from the free end of the sample. Signals from the transducer were enhanced by a digital storage oscilloscope (DSO) and viewed on a PC display (the vibration testing scheme can be found in Figure 9.1). The testing information was calculated and processed using equations to yield the vibration-damping factor value. The vibration-damping factor is a damping an incentive as a non-dimensional number, which indicates the measure of energy dissipation to confine the vibration amplitude [18].

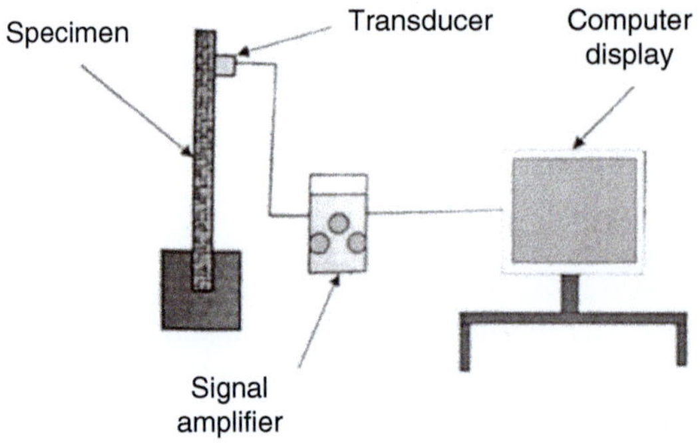


**Figure 9.1** Vibration testing scheme. Source: Adapted from Ismail et al. [36].

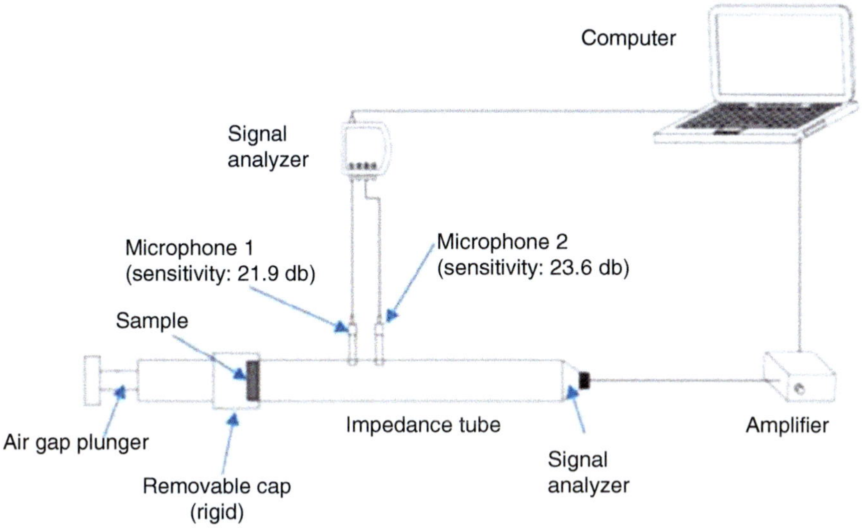


**Figure 9.2** Experimental setup for acoustic testing. Source: Ismail et al. [36]. Liscenced under (CC BY4.0).

### 9.3.4 Acoustic Properties:

The sound coefficient of the composite was measured according to ISO 10534-2 using the impendence tube. The sample size was maintained as per the standard. Figure 9.2 indicates the schematic of the test.

Initially, the sample was set in a removable cap on one side of the tube and the other side was attached to an amplifier. After setting the specimen, a blank noise was fed into the tube to provide a consistent sound energy per constant bandwidth/hertz. There are two acoustic microphones used to track the incident sound and the sound signal resulting from the composite. By using an analyzer, the signal was recorded after 10 minutes and the data were analyzed using the attached computer to produce the required spectra to convey the function of transfer. Moreover, the coefficient of absorption can also be determined by the same process. Varying sound frequency range was used to determine the sound ingestion coefficient.

## 9.4 Overview of the Dynamics Properties of Kenaf-Fiber-Reinforced Composite

Strict control is needed to manufacture kenaf-fiber-reinforced polymer composites with excellent dynamics properties. Moreover, the selection of a suitable polymeric matrix is crucial for such properties in the kenaf-fiber-reinforced polymer composite. Thus, this section reviews some related dynamics properties and composite structure relationship in a step by step manner.

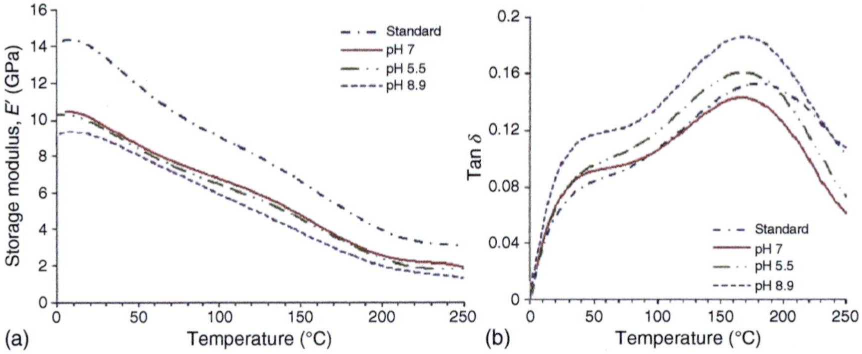


**Figure 9.3** Graph for the effect of various solutions of kenaf-fiber-reinforced composite. (a) Variations of storage modulus, *E*. (b) Variations of tan $\delta$ for 24 weeks. Source: From Akil et al. [38]. © 2011, Elsevier.

### 9.4.1 Dynamic Mechanical Properties (DMA)

This test is a delicate measuring process that reveals the mechanical attributes of materials by detecting the property variation with temperature and for various stress and frequency levels of loading. In DMA, a sinusoidal time-differing oscillating force is applied to the sample to determine the two processing components [8]. One is the elastic part to determine the energy storage capacity of the material to which it is applied, called storage modulus ($E'$), and the other is the viscous component that defines the energy dissipate capacity of the material, known as the loss modulus ($E''$). All the fibrous composite materials are viscoelastic material, having both elastic and viscous qualities. Also, the ratio between the loss modulus and the storage modulus is denoted by tan $\delta$. This is also called the dynamic loss factor or mechanical loss factor and varies proportionally with the intensity of adhesion between the fiber and the matrix. The weak adhesion between the fiber and matrix gives higher tan $\delta$ values and vice versa [15]. Nishino et al. [37] conducted DMA on the kenaf-fiber-reinforced polylactic acid (PLLA) composites. They observed that the developed composites confirmed a better $E'$ than the individual constituents. Moreover, such high storage modulus continued up to the melting point of PLLA (161 °C). Moreover, the immobility of the polymer matrix gives relatively high thermal stability. They also observed a shift in the glass transition of PLLA in tan $\delta$ peak spectra, which indicates a good interaction between PLLA resin and kenaf fiber by ensuring the reinforcement. In another study on DMA, Nosbi et al. [15] immersed kenaf-fiber-reinforced composites in various pH solutions, leading to a fall in the dynamic mechanical properties. It can be summarized from their study that the immersion in various solutions has a devastating effect on the dynamic mechanical characteristics of kenaf-fiber-reinforced composites in terms of $E'$, $E''$, and their ratio tan $\delta$, as illustrated in Figure 9.3a,b.

The highest measurable reduction of kenaf-fiber-reinforced composite occurred when it was immersed in sea water (pH 8.9), followed by refined water (pH 7) and rainwater (pH 5.5). The reduction pattern recorded for tan $\delta$ after the immersion process is attributed to the ductility increase for all samples tested due to the movement restriction of the polymer molecules. When the incorporation of reinforcing fibers restricts the mobility of polymer molecules under deformation, the tan $\delta$ values of the composites are reduced. Therefore, storage modulus values are raised and between the stress and the strain, the viscoelastic lag lessens [34, 35]. Meanwhile, for the immersed kenaf-fiber-reinforced composite storage modulus values were decreased due to the matrix damage, weakening the fiber–matrix interfacial adhesion and bond strength. The ability of the material to bear external stress depends on the van der Waals bonding and hydrogen bonding, and on the behaviors of the chemical bonding of chains in the polymer molecular structure [35]. Hydrolysis, plasticization, and water penetration into the polymer cause rupture the combinations of chemical and bonding of the matrix. On account of encountering stress, when a more prominent strain is induced, it would in a roundabout way lead to a decrease in the storage modulus. In addition, it has been explained that the storage modulus of these materials is lower because of the expanded mobility of polymer chains in composites with poor adhesion compared to weak interfacial bonding [39].

Furthermore, some researchers [40] used a co-rotating twin screw extruder to measure the DMA characteristics depending on the processing parameter of kenaf fiber/high-density polyethylene (HDPE) composites at low processing temperature (LPT) and high processing temperature (HPT). They established the dynamic mechanical properties at variable extrusion parameters. The storage modulus ($E''$) curves are shown in Figure 9.4 for composites at LPT and HPT.

In general, as the temperature increases, the softening of the polymer matrix leads to a decrease in the storage modulus; this concept is also supported in Figure 9.4 [41, 42]. From Figure 9.4a, it is clear that the higher fiber content in the LPT composite makes a negative impression on the storage modulus, irrespective of whether the temperature is glassy (−130 °C) or higher. However, the composites having lower fiber content (3.4–8.5%) show an enhancement in reinforcement with increased fiber loading. Moreover, at LPT the composites with 3.4 and 8.5 wt.% fiber have a higher $E''$ associated to the pure HDPE, but the 17.5 wt.% fiber composites display a lower

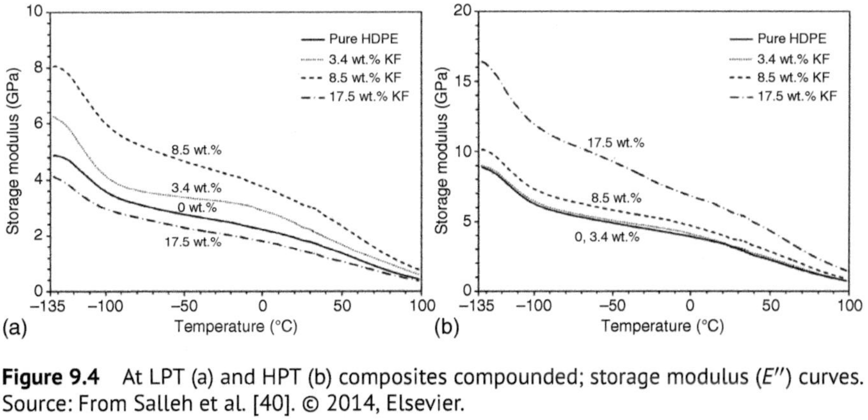


**Figure 9.4** At LPT (a) and HPT (b) composites compounded; storage modulus ($E''$) curves. Source: From Salleh et al. [40]. © 2014, Elsevier.

value than pure HDPE. This is because the LPT results in low wetting of the fiber with the polymer matrix in the extruder, thus reducing the stress-bearing capability from the fiber to the polymer matrix, leading to a drop in the storage modulus [34, 41, 42].

In contrast, in the case of HPT the scenario is nearly altered; that is, $E''$ increases with higher fiber quantity, as seen in Figure 9.4b. As expected, the composite containing high fiber (17.5 wt.%) content shows a higher storage modulus than other composites and pure HDPE, irrespective of whether the temperature is glassy (−130 °C) or higher. At HPT the fiber–polymer matrix adhesion is imparted, which gives a good reinforcement effect leading to transfer of the stress from the host polymer to the fiber [42, 43]. In this case, there is no change in properties at very low reinforcement (3.4 wt.%), where the fiber may have performed as an hollow, subsequently declining reinforcing usefulncss [44].

The loss modulus for the composite at LPT and HPT is presented in Figure 9.5. Figure 9.5a reveals that there are three transition peaks (a, b, and c) related to the pure HDPE and kenaf-fiber-reinforced composite. At LPT, the range of 38.5–66.5 °C is related to the long linear chain motion in the crystalline region of the HDPE. Likewise, the range of −32.0 to −26.0 °C is for the segmental motion of methyl slackening in the amorphous region of the HDPE, and finally the range of −120 and −117 °C is ascribed to long linear chain crankshaft slackening in the amorphous region of the HDPE chain.

The same behavior is also observed in [45], through the β and γ transition for HDPE were at −45 and −107 °C (±1 °C) respectively, due to segmental activities of (-CH3) groups in the amorphous phase. At LPT, there is a change toward lower temperature of about 5.5 and 6 °C compared to the pure HDPE with the composites of 8.5 and 17.5 wt.% fiber. This indicates the existence of certain actions, which have perhaps encouraged the softening of the matrix, leading to the free mobility of polymer chains in the composite [46]. On the other hand, the loss modulus at HPT of composites shows similar transition peaks (a, b, and c) for high fiber loading (17.5 wt.%) while the other composites with low fiber loading show only two transition peaks (a and c) (Figure 9.5b). This indicates a better uniformity of fiber achieved with high

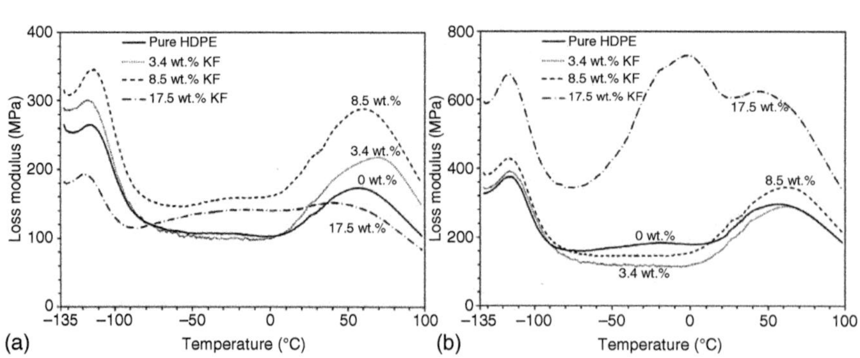


**Figure 9.5** Composites compounded at (a) LPT and (b) HPT; loss modulus graph. Source: From Salleh et al. [40]. © 2014, Elsevier.

fiber content in the composite than that with low fiber content at HPT. However, the highest magnitude of loss modulus contrasted with the pure HDPE found in 17.5 wt.% fiber composites [46].

In case of tan $\delta_{max}$ at LPT pure HDPE offered the highest value than other composites and the 3.4 wt.% fiber loading composite displayed the maximal tan $\delta_{max}$ than other fiber loadings. The damping factor represents the flaws in the elasticity of a polymer [47]; thus at LPT the high fiber loading gives a nonhomogeneous cohesion between the matrix and the kenaf fiber, leading to an increase in the molecular mobility of the composite as well as in the pure polymer. However, at HPT the scenario with tan $\delta_{max}$ values is the reverse, where the magnitude of tan $\delta_{max}$ decreases with increasing fiber content. This obviously suggests the positive effect of HPT in fiber to polymer matrix adhesion and thus the reinforcement effect imparted by the fibers to composite. Also, this is due to the limitation of mobility of the polymer matrix, which was produced by the enhancement in fiber spreading in the composites [48].

Another type of composite, kenaf epoxy composite, was also studied by Rajnish Kumar et al. [49], where they tried to find the effect of fiber length on the dynamic mechanical characteristics at fixed fiber loading. The storage modulus ($E'$) of kenaf–epoxy composite is shown in Figure 9.6a, with varied fiber length (5–50 mm) at fixed 25 wt.% content of fibers.

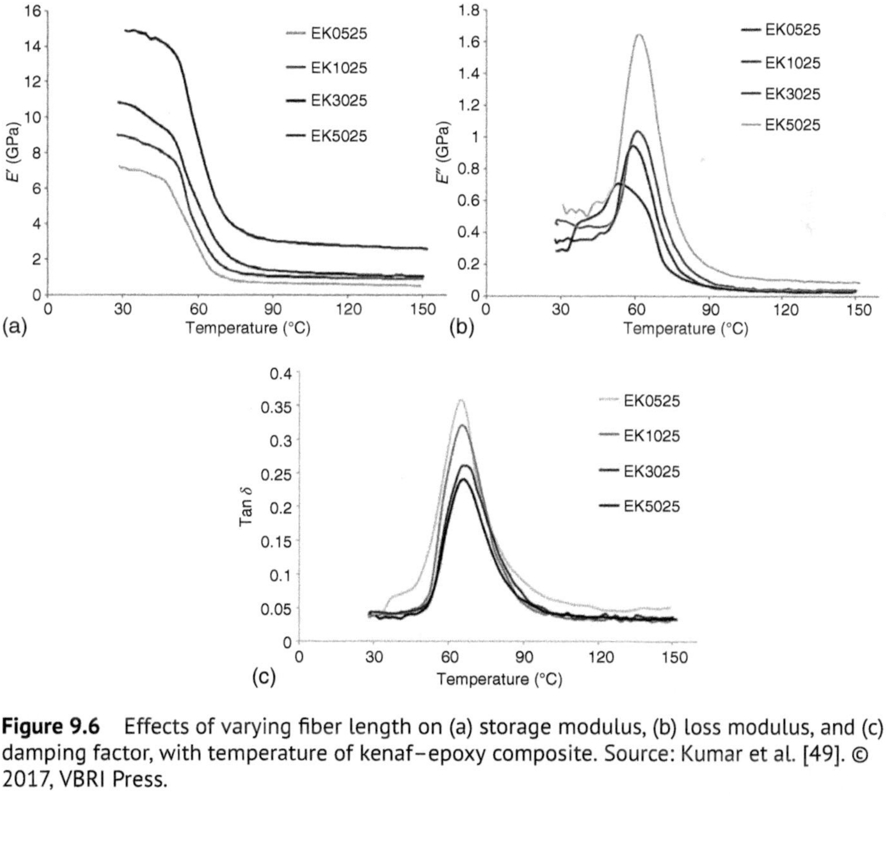


**Figure 9.6** Effects of varying fiber length on (a) storage modulus, (b) loss modulus, and (c) damping factor, with temperature of kenaf–epoxy composite. Source: Kumar et al. [49]. © 2017, VBRI Press.

On comparing various composites, higher $E'$ values were found for longer fibers at low temperature. Because of loss in firmness of the matrix the storage modulus drops at higher temperature. An acute decrease in $E'$ value was seen in the region of the glass transition temperature ($T_g$), showing that the material experienced a glass/rubbery transition stage. There is a direct relation between $E'$ and the interface holding [50]. The outcomes show better interface bonding with the expanding fiber length in the composite because of viable stress transformation in the fiber and the matrix. Figure 9.6b depicts the change in loss modulus $E''$ with composite temperatures at varying fiber lengths (5–50 mm) at steady 25 wt.% loading of fibers. Loss modulus $E''$ increased in the plastic area, and afterward began diminishing with increasing temperature in the rubbery area. Loss modulus $E''$ likewise demonstrated comparative patterns as if there should be an occurrence of storage modulus $E'$. All things considered, the $T_g$ value obtained from $E''$ indicated an expanding pattern with increase in the length of fibers in the composite, demonstrating better capacity of composites to disseminate energy during the distortions. Also, Figure 9.6c shows the varying tan $\delta$ values with temperature of composites with the difference in fiber length (5–50 mm) at consistent 25 wt.% loading of strands. Increase in fiber length caused decreased tan $\delta$ peak height, indicating lower damping and better adhesion between the fiber and the matrix.

### 9.4.2 TGA Analysis of Composites:

The mass of the sample can be obtained as a function of temperature by the TGA measurement. The materials mass changes are reasonably identified with thermal stability usually occurring during decomposition, sublimation, chemical reaction, evaporation, and magnetic or electrical transformation [33]. In order to assess the thermal attributes of kenaf-filled chitosan biocomposites, both DSC and TGA were investigated. In DSC investigation, it was observed that all samples demonstrated an expansive endothermic peak at the first heating scan, which was related with water hydration. In the interim, it demonstrated a diminishing endothermic temperature by increasing kenaf dust content in the chitosan film for the subsequent heating scan. Researchers asserted that this turn was credited to hydrogen bond formation [33]. However, there are no correlations between enthalpy values ($\Delta H$) and the amount of kenaf fiber. Meanwhile, TGA reveals that there is no prominent change in the chitosan film's thermal stability with the addition of kenaf dust. Lee et al. [51] observes the impacts of fiber loading in KF-reinforced polypropylene (PP) composites on thermal properties. In common, it is observed that at 383 °C, the fiber reinforced thermoplastics composites show an initial degradation. Thermal decomposition occurred in two stages – at a lower temperature and at a higher temperature, when the fiber and filler split. Every single composite showed superior mass residue over the pure PP polymer. However, if the quantity of kenaf fiber increased in composite making, it showed higher mass loss gradually. In another study, it was also found that at the onset temperature, gradual increment of the cellulose and hemicellulose substances caused higher mass loss of the composites [38, 48]. For example, the 10 wt.% kenaf fiber test showed a mass loss of 8.49%. This

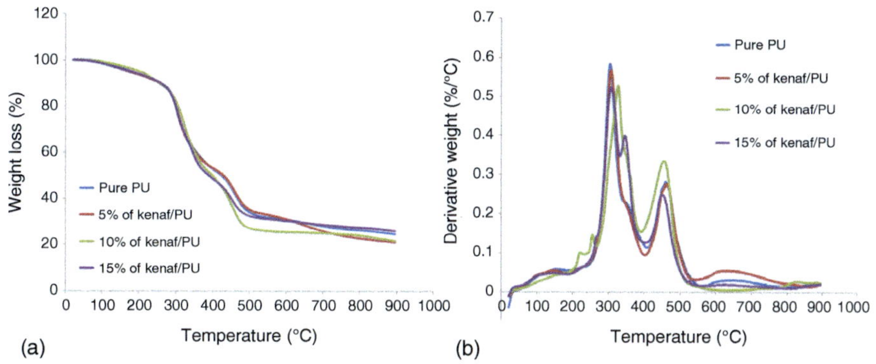


**Figure 9.7** Graphical representation of different fiber content loaded kenaf/PU biocomposite's (a) TGA and (b) differential thermal analysis (DTA). Source: Adapted from Athmalingam et al. [52].

increased to 9.664%, 10.03%, and 24.09% for 15 , 20, and 25 wt.% samples respectively. Therefore, increasing fiber quantity induced a lower onset temperature as the kenaf fiber showed higher mass loss due to lower thermal stability than the PP matrix [51].

Furthermore, a similar trend of degradation was found in other studies [51], where higher fiber content led to deterioration of mass loss according to the temperature profile. In Figure 9.7a,b, TGA and DTG curves of pure polyurethane (PU) and 5, 10, and 15 wt.% fiber loaded kenaf/PU biocomposites are shown.

The mass loss for each composite occurred between 200 and 400 °C, with the extreme loss at 300 °C relating to a weight reduction of about half. At 253.33 °C, the initial mass loss for 10 wt.% kenaf/PU biocomposite is much lower compared to that for 5 wt.% kenaf/PU, 15 wt.% kenaf/PU, and pure PU. The main causes of mass loss are the hemicellulose deterioration from kenaf and disturbance of urethane bond. Furthermore, with increasing temperature, progressive decomposition of raw materials caused mass loss continuously. This indicated a decrease in thermal stability with increasing kenaf fiber in the matrix; thus, it is ideal to state that the decrease in weight reduction was the aftereffect of thermal stability improvement [53]. It is additionally proved from earlier studies that generally natural fibers with hemicellulose have low thermal stability compared to others [54]. Pure PU decomposes sometime later in contrast with kenaf/PU composites because of the early disintegration of kenaf in the composition. The highest weight reduction of approximately 70% occurred at around 450 °C with subsequent decomposition happening in the range of 350–500 °C. This degradation technique shows that the consolidation of kenaf fiber does not influence the pure PU decomposition behavior. In Figure 9.7b, PU/kenaf composite's DTG curve is delineated. The temperature at which extreme weight reduction occurred can be identified from the DTG bends. It may be observed that the central peaks of pure polyurethane composites have higher values in contrast with kenaf composites. From these outcomes, it is clear that the level of kenaf composition in polyurethane influenced the composites decomposition temperature and thermal stability. Because of good interfacial adhesion between the fiber and the matrix, degradation temperature of composites increased in contrast with pure PU.

Finally, it is outlined that fiber content, the matrix used, and fiber–matrix adhesion influence the thermal stability of kenaf-fiber-reinforced composites.

In addition, it was observed that the 54.9% and 265% higher damping values contrasted with kenaf- and bamboo-based composites respectively. Thus, it is demonstrated that at the optimum weight rate, bamboo – 30% and kenaf – 70%, hybrid composites consolidate the advantages of their individual constituents; hence some prevalent properties can be achieved by these hybrid composites [55].

### 9.4.3 Acoustic Properties

Generally, delicate, malleable, or permeable materials (such as fabrics) serve as great acoustic separators – absorbing the most sound, while thick, hard, invulnerable materials (for example, metals) reflect sound the most. How well a material retains sound is measured by the powerful retention zone of the materials. An experiment was done on the acoustic characteristics of kenaf-fiber-based composites, and based on the result it is seen that the composites retain sound less effectively as sound assimilation of the composites is lower than 0.5. Air flow resistivity, porosity, viscoelasticity, density, thickness, and tortuosity are some factors on which sound retention of the composite relies [45]. It could be used to improve the sound ingestion of the composites by the air gap between the sample and the inflexible composites wall [54]. Figure 9.8 shows the sound absorption coefficient with various air gap thicknesses.

Figure 9.8a depicts the impact of air gap of kenaf composite on sound assimilation. In the frequency range of 500–3000 Hz sound retention of kenaf composites is enhanced. The peak of the most extreme sound ingestion increased, and due to

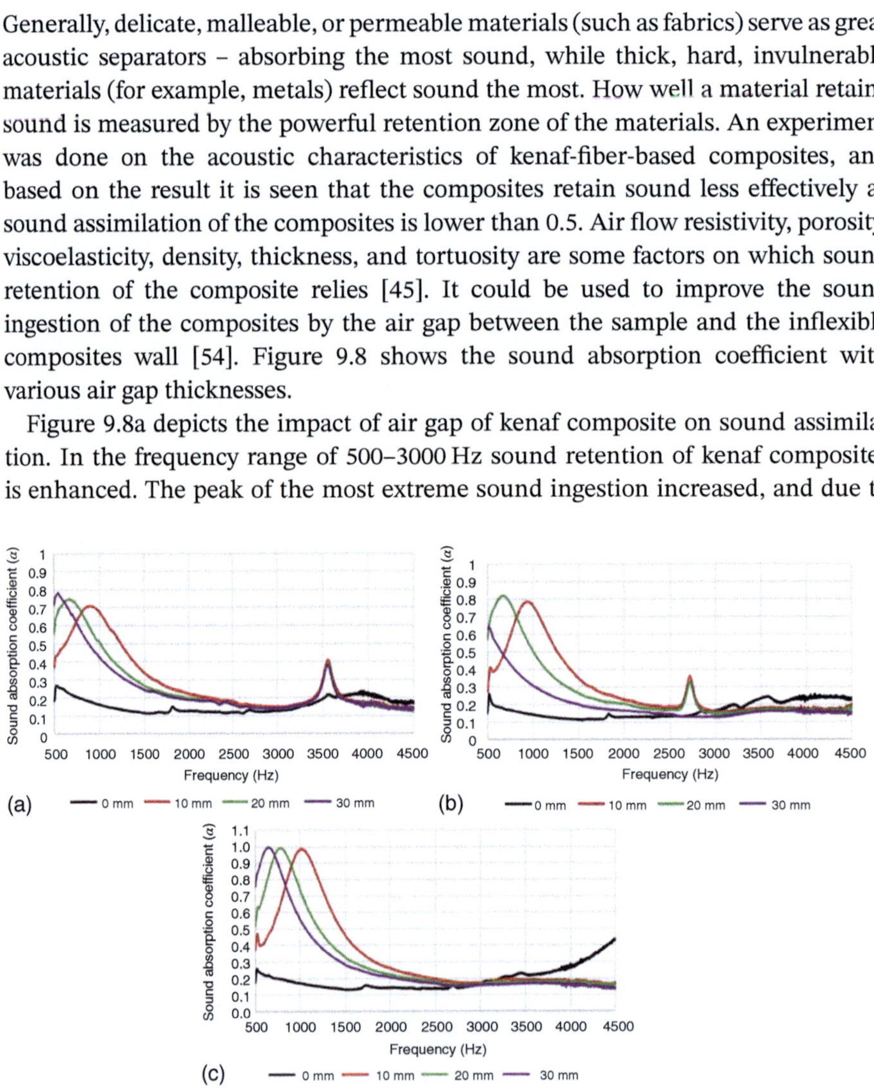


**Figure 9.8** Graph of sound absorption coefficient with various air gap thicknesses: (a) kenaf, (b) kenaf/bamboo (30/70), and (c) kenaf/bamboo (50/50). Source: Ismail et al. [36]. Liscenced under (CC BY 4.0).

the air gap increase it moved to a lower frequency. Figure 9.8b,c shows the sound ingestion of hybrid composites in relation to the air gap. It is found that compared with the other two hybrid composites, the proportion of half kenaf and half bamboo showed superior sound assimilation with the highest peak. The most extreme peak sound retention of the hybrid composites moved to bring down the frequency with increasing air gap thickness [54, 56].

## 9.5 Conclusion

In the field of composite materials manufacturing, kenaf fibers have an enormous advantage for use as a substitute for human-made fibers; for example, because of low manufacturing cost, kenaf fiber is modest over human-made fibers. Besides, KF is lighter in weight because of its low density. Regardless of its great characteristics, KF is not absolutely faultless. Various methods have been suggested by researchers to overcome its inadequacies; for example, surface treatment is one of them. The modifications provide incredible outcomes regarding better attributes when KF is incorporated in composites, yet the financial viewpoint ought to be constantly remembered. A few techniques are currently known to enhance the bond inside composites to improve the dynamic properties, for instance, improving the adhesion to the material by chemical bonding by applying sodium hydroxide, silane, or other combined treatments of those chemicals. Depending on the form of the KF, there is an assortment of treatments for KF. A concise discussion of the dynamic properties of kenaf fiber reinforced composites is provided alongside a study of past experiments. Earlier findings are provided to enable researchers to find out the benefits and prospects of kenaf fiber as an alternative to supplant traditional methods and materials used in composites manufacturing. Processing technique and the fundamental properties of kenaf-fiber-reinforced composite are well described and examined. Fundamentally, to completely recognize the general execution of the composite with critical aspect it is said that the dynamics analysis is mandatory characterization. This chapter has thus attempted to do it in a sequential order. Subsequently, kenaf-fiber-reinforced composite manufacturing can assist with producing employment in both provincial and urban areas, notwithstanding assisting with decreasing waste, and providing a healthier environment. However, looking at future demands and for large-scale end products, more crucial studies on manufacturing processes are required for product commercialization.

## References


1. Iqbal, M.I., Kabir, S.F., Rakib, M.A.N., Rashid, M.M., and Sikdar, P.P. (2018). Characterization of Waste Bamboo Strips Underscoring Node Effects. *J. Sustainable Constr. Mater. Technol.* 3 (1): 163–173.
2. Iqbal, M.I. and Al Mizan, R. (2021). The Effects of Culm Nodes on Bamboo Fiber Properties. In *Bamboo Fiber Composites* (pp. 175–187). Singapore: Springer.

3 Iqbal, M.I., Mizan, R., and Khan, A.N. (2020). An approach to search the potentiality of node in bamboo strip reinforced composites. *J. Met. Mater. Miner.* 30: 11–19.
4 Iqbal, M.I., Al Mizan, R., and Khan, A.N. (2020). *Biofibers and Biopolymers for Biocomposites*, 159–169. Springer.
5 Yanga, R. et al. (2018). The Effect of Carbon Fibre Reinforced on Polyamide1012. *J. Fiber Bioeng. Inform.* 11 (1): 41–47.
6 Aziz, S.H. and Ansell, M.P. (2004). The effect of alkalization and fibre alignment on the mechanical and thermal properties of kenaf and hemp bast fibre composites: Part 1 – polyester resin matrix. *Compos. Sci. Technol.* 64: 1219–1230.
7 Hasan, K.F. et al. (2016). Dynamic mechanical behavior & analysis of the jute-glass fiber reinforced polyester hybrid composites. *Am. J. Appl. Phys* 1 (1): 1–12.
8 Misganew, M., Rao, D.N., Reddy, K.R.N., and Avvari, M. (2019). *International Conference on Advances of Science and Technology*, 684–694. Springer.
9 Mohanty, S. and Nayak, S.K. (2007). Rheological characterization of HDPE/sisal fiber composites. *Polym. Eng. Sci.* 47: 1634–1642.
10 Ku, H., Wang, H., Pattarachaiyakoop, N., and Trada, M. (2011). A review on the tensile properties of natural fiber reinforced polymer composites. *Composites Part B* 42: 856–873.
11 Towo, A.N. and Ansell, M.P. (2008). Fatigue evaluation and dynamic mechanical thermal analysis of sisal fibre–thermosetting resin composites. *Compos. Sci. Technol.* 68: 925–932.
12 Sreekumar, P.A. et al. (2008). Effect of chemical treatment on dynamic mechanical properties of sisal fiber-reinforced polyester composites fabricated by resin transfer molding. *Compos. Interfaces* 15 (2–3): 263–279.
13 Sreekumar, P. et al. (2009). Dynamic mechanical properties of sisal fiber reinforced polyester composites fabricated by resin transfer molding. *Polym. Compos.* 30: 768–775.
14 Oladele, I., Omotoyinbo, J., and Adewara, J. (2010). Investigating the effect of chemical treatment on the constituents and tensile properties of sisal fibre. *J. Miner. Mater. Charact. Eng.* 9: 569.
15 Nosbi, N., Akil, H.M., Ishak, Z.M., and Bakar, A.A. (2010). Degradation of compressive properties of pultruded kenaf fiber reinforced composites after immersion in various solutions. *Mater. Des.* 31: 4960–4964.
16 Costa, C., Fonseca, A., Serra, A., and Coelho, J. (2016). Dynamic mechanical thermal analysis of polymer composites reinforced with natural fibers. *Polym. Rev.* 56: 362–383.
17 Sezgin, H., Berkalp, O., Mishra, R., and Militky, J. (2016). Investigation of dynamic mechanical properties of jute/carbon reinforced composites. *Composites* 14: 19.
18 Neves Monteiro, S. et al. (2018). Fique fabric: a promising reinforcement for polymer composites. *Polymers* 10: 246.
19 Nagendra, P.S., Prasad, V., and Ramji, K. (2017). A study on dynamic mechanical analysis of natural nano banana particle filled polymer matrix composites. *Mater. Today: Proc.* 4: 9081–9086.

**20** Nicolais, L., Borzacchiello, A., and Lee, S.M. (2012). *Wiley Encyclopedia of Composites*. Wiley.

**21** Summerscales, J. and Grove, S. (2014). *Natural Fibre Composites*, 176–215. Elsevier.

**22** Gutowski, T. (1997). *Advanced Composites Manufacturing*. New York: Wiley.

**23** Davé, R.S. and Loos, A.C. (2000). *Processing of Composites*. Munich: Hanser Publishers.

**24** Campbell, F.C. Jr. (2003). *Manufacturing Processes for Advanced Composites*. Elsevier.

**25** Strong, A.B. (2008). *Fundamentals of Composites Manufacturing: Materials, Methods and Applications*. Society of Manufacturing Engineers.

**26** Ciriscioli, P.R. and Springer, G.S. (2004). *Smart Autoclave Cure of Composites*. RSL.

**27** Noakes, K. (2008). *Successful Composite Techniques: A Practical Introduction to the Use of Modern Composite Materials*. Crowood Press.

**28** Summerscales, J. (2016). *Marine Applications of Advanced Fibre-Reinforced Composites*, 19–55. Elsevier.

**29** Davé, R.S., Henne, M., Louis, B., and Ermanni, P. (2011). Processing: autoclave modeling. In: *Wiley Encyclopedia of Composites* (ed. L. Nicolais), 1–13. Wiley.

**30** Nairn, J., Talreja, R., and Manson, J. (2000). *Comprehensive Composite Materials*, vol. 2, 403–432. Elsevier Science.

**31** Shenoi, R.A., Wellicome, J.F., and Dyer, I. (1993). *Composite Materials in Maritime Structures: Fundamental Aspects*, vol. 1, 1. Cambridge University Press.

**32** Mahrholz, T., Stängle, J., and Sinapius, M. (2009). Quantitation of the reinforcement effect of silica nanoparticles in epoxy resins used in liquid composite moulding processes. *Composites Part A* 40: 235–243.

**33** Julkapli, N.M. and Akil, H.M. (2010). Thermal properties of kenaf-filled chitosan biocomposites. *Polym.-Plast. Technol. Eng.* 49: 147–153.

**34** Rana, A., Mitra, B., and Banerjee, A. (1999). Short jute fiber-reinforced polypropylene composites: dynamic mechanical study. *J. Appl. Polym. Sci.* 71: 531–539.

**35** Gu, H. (2009). Dynamic mechanical analysis of the seawater treated glass/polyester composites. *Mater. Des.* 30: 2774–2777.

**36** Ismail, A.S., Jawaid, M., and Naveen, J. (2019). Void content, tensile, vibration and acoustic properties of kenaf/bamboo fiber reinforced epoxy hybrid composites. *Materials* 12: 2094.

**37** Nishino, T., Hirao, K., Kotera, M. et al. (2003). Kenaf reinforced biodegradable composite. *Compos. Sci. Technol.* 63: 1281–1286.

**38** Akil, H. et al. (2011). Kenaf fiber reinforced composites: a review. *Mater. Des.* 32: 4107–4121.

**39** Gerard, J., Perret, P., and Chabert, B. (1990). *Controlled Interphases in Composite Materials*, 449–456. Springer.

**40** Salleh, F.M., Hassan, A., Yahya, R., and Azzahari, A.D. (2014). Effects of extrusion temperature on the rheological, dynamic mechanical and tensile properties of kenaf fiber/HDPE composites. *Composites Part B* 58: 259–266.

**41** Liang, Z., Pan, P., Zhu, B. et al. (2010). Mechanical and thermal properties of poly(butylene succinate)/plant fiber biodegradable composite. *J. Appl. Polym. Sci.* 115: 3559–3567.

**42** Huda, M.S., Drzal, L.T., Mohanty, A.K., and Misra, M. (2006). Chopped glass and recycled newspaper as reinforcement fibers in injection molded poly(lactic acid) (PLA) composites: a comparative study. *Compos. Sci. Technol.* 66: 1813–1824.

**43** Hassan, A., Rahman, N.A., and Yahya, R. (2011). Extrusion and injection-molding of glass fiber/MAPP/polypropylene: effect of coupling agent on DSC, DMA, and mechanical properties. *J. Reinf. Plast. Compos.* 30: 1223–1232.

**44** Hassan, A. et al. (2011). Interfacial shear strength and tensile properties of injection-molded, short- and long-glass fiber-reinforced polyamide 6,6 composites. *J. Reinf. Plast. Compos.* 30: 1233–1242.

**45** Khanna, Y.P., Turi, E.A., Taylor, T.J. et al. (1985). Dynamic mechanical relaxations in polyethylene. *Macromolecules* 18: 1302–1309.

**46** Tajvidi, M., Falk, R.H., and Hermanson, J.C. (2006). Effect of natural fibers on thermal and mechanical properties of natural fiber polypropylene composites studied by dynamic mechanical analysis. *J. Appl. Polym. Sci.* 101: 4341–4349.

**47** Bleach, N., Nazhat, S., Tanner, K. et al. (2002). Effect of filler content on mechanical and dynamic mechanical properties of particulate biphasic calcium phosphate—polylactide composites. *Biomaterials* 23: 1579–1585.

**48** Kumar, K.S., Siva, I., Rajini, N. et al. (2016). Layering pattern effects on vibrational behavior of coconut sheath/banana fiber hybrid composites. *Mater. Des.* 90: 795–803.

**49** Kumar, R., Hashmi, S., Nimanpure, S., and Naik, A. (2017). Enhanced dynamic mechanical properties of kenaf epoxy composites. *Adv. Mater. Proc.*

**50** Sreekala, M., George, J., Kumaran, M., and Thomas, S. (2002). The mechanical performance of hybrid phenol-formaldehyde-based composites reinforced with glass and oil palm fibres. *Compos. Sci. Technol.* 62: 339–353.

**51** Lee, C., Sapuan, S., and Hassan, M. (2017). Mechanical and thermal properties of kenaf fiber reinforced polypropylene/magnesium hydroxide composites. *J. Eng. Fibers Fabr.* 12, https://doi.org/10.1177/155892501701200206.

**52** Athmalingam, M. and Vicki, W. (2018). IOP Publishing Ltd. Investigation on thermal properties of kenaf fibre reinforced polyurethane bio-composites. *IOP Conf. Ser. Mater. Sci. Eng.* 303 (1): 012011.

**53** Sarifuddin, N., Ismail, H., and Ahmad, Z. (2015). Studies of properties and characteristics of low-density polyethylene/thermoplastic sago starch-reinforced kenaf core fiber composites. *J. Thermoplast. Compos. Mater.* 28: 445–460.

**54** Ogbomo, S.M., Ayre, B., Webber, C.L., and D'Souza, N.A. (2014). Effect of kenaf fiber age on PLLA composite properties. *Polym. Compos.* 35: 915–924.

**55** Sathishkumar, T., Naveen, J.A., and Satheeshkumar, S. (2014). Hybrid fiber reinforced polymer composites – a review. *J. Reinf. Plast. Compos.* 33: 454–471.

**56** Koizumi, T., Tsujiuchi, N., and Adachi, A. (2002). The development of sound absorbing materials using natural bamboo fibers. *WIT Trans. Built Environ.* 59.

# 10

# Effect of Micro-Dry-Leaves Filler and Al-SiC Reinforcement on the Thermomechanical Properties of Epoxy Composites

*Mohit Hemath[1], Govindrajulu Hemath Kumar[2], Varadhappan Arul Mozhi Selvan[3], Mavinkere R. Sanjay[1], and Suchart Siengchin[1]*

[1] *King Mongkut's University of Technology, Materials and Production Engineering Department, Natural Composite Research Group, The Sirindhorn International Thai-German Graduate School of Engineering, 1518 Pracharat 1 Road, Wongsawang, Bangsue, Bangkok 10800, Thailand*
[2] *Composite Research Center, Department of Mechanical Engineering, EVR Street, Ambattur, Chennai 600053, India*
[3] *National Institute of Technology, Department of Mechanical Engineering, Tanjore Main Road, National Highway 83, Tiruchirappalli, Tamilnadu 620015, India*

## 10.1 Introduction

Bio-based composites form a developing field of interest in the area of polymer composite materials because of the opportunity to achieve materials that are cost-efficient and entirely biodegradable [1, 2]. On the other hand, the use of marine or agricultural biowastes presents an encouraging trend in advanced materials that acquire higher sustainability from both the ecological and economic viewpoints. Furthermore, an essential parameter for bio-based composites possesses bad mechanical characteristics when applied to the applications because of its brittle nature [3].

The expectations on plant cellulose fiber incorporated laminates are rising continuously because of the lower weight, low cost, higher resistance to chemicals, and higher mechanical characteristics [4]. Hybridization in bio-based laminates is performed by the addition of plant fibers, fillers, and synthetic textile fibers. In the bulk of the synthetic/plant hybridization, fiber glass was applied with plant-based fibers [5]. Presently, synthetic fibers have been replaced with plant fibers, which possess biodegradable characteristics with eco-friendly nature. The mechanical characteristics of these laminates are also comparable with synthetic nonrenewable fibers [6, 7]. The mechanical characteristics of plant fiber-reinforced laminates are influenced by parameters such as material of the matrix, the fiber utilized, its orientation, length of the fiber, method of production, and surface treatment. Matrix materials such as polyester, polypropylene, phenol, and epoxy are applied generally. Surface treatment is also a primary parameter in enhancing the characteristics of the plant fiber. Acrylonitrile grafting, silane treatment, acetylation, alkali, acrylation, sodium chloride salt solution, and benzoylation are the different kinds of treatment processes

*Mechanical and Dynamic Properties of Biocomposites*, First Edition. Edited by Senthilkumar Krishnasamy, Rajini Nagarajan, Senthil Muthu Kumar Thiagamani, and Suchart Siengchin.

applied in plant fibers [8–10]. Among these, alkali and sodium chloride salt solution treatment is utilized for the higher mechanical characteristics accomplished [10, 11]. The sisal fiber treated with 5% of alkali displayed higher mechanical characteristics because of interactions between the fiber and the matrix [12].

Different types of manufacturing methods were applied to fabricate plant-based laminates. Compression molding and hand lay-up are the commercially applied techniques in the field because of low cost and simplicity. The hybrid kenaf/banana fiber-reinforced in epoxy laminates displayed the highest mechanical characteristics under 45° inclinations than other horizontal and vertical orientations [13]. Epoxy laminates reinforced with coir fiber manufactured with the help of hand lay-up technique displayed an improvement of 76% in tensile characteristics; similarly, impact and flexural characteristics improved by 10% and 29% with the effect of fiber length of 30 mm [14]. In plant fiber laminates, the fillers were also reinforced to enhance the mechanical characteristics by decreasing the fiber pullouts and gaps. Nanoclay, coconut shell nanopowders, and red mud particles were applied in enhancing the characteristics [15–17]. The addition of coconut shell nanopowder in coir/kenaf hybrid laminates improved the interfacial characteristics and improved impact, flexural, and tensile characteristics of the laminates [15]. The addition of nanoclay (1–5%) in sisal fiber laminates improved the bonding within the fibers and polymer matrix, and also enhanced the thermal, impact, and tensile characteristics [17].

In recent years, carbon nanomaterials such as nanographite, carbon nanotubes, and graphene nanoparticles attracted much interest from researchers, because of extraordinary electrical, mechanical, and thermal characteristics [18–20]. It is well-established that nanofiller carbon nanotubes in one-dimensional nanostructure have been broadly investigated with their efficiency determined as filler in polymer-based composites [21–23]. Furthermore, the higher price of carbon nanotubes compared with other carbon-based nanomaterials has restricted their utilization in different industrial researches. Another critical limitation is the potential for toxicity from carbon nanotubes. On the other hand, the nano-graphene is an efficient reinforcement in polymer composites, due to its excellent electrical, thermal, and mechanical characteristics and two-dimensional structure. Moreover, nano-graphene can be an efficient filler selection because of the lower price and uniform dispersion in any type of polymer matrix offering a higher interfacial bonding within the polymer material [24, 25]. As explained above, many scientists have investigated the graphene nanoparticles reinforced laminates in different conditions. The size scale influence of nano-graphene particles on the thermal and mechanical characteristics of epoxy laminates has also been investigated and it has been depicted that all particle sizes of nano-graphene can significantly influence the materials characteristics, enhancing the structural properties. Moreover, the higher surface area has further enhanced the crack bridging and toughness properties [26]. Hadden et al. [27] documented their multiscale modeling outcomes with the experimental one, and graphene nanoparticles and carbon fiber-reinforced epoxy laminates were fabricated as materials in their investigation. The experimental outcomes and developed model exhibited that the graphene nanoparticles entirely affect the tensile characteristics of the composites. The mechanical properties of basalt fiber-incorporated graphene nanoparticles/epoxy laminates have been investigated [28], and a particular content of graphene nanoparticles has led to improvement in flexural, impact, and impact strength of epoxy/basalt laminate.

Literature survey reveals that limited research dealing with particles-incorporated hybrid polymer laminates has been reported. To fill this research gap, particularly for hybrid filler-incorporated polymer laminates, in light of the current investigation, it is proposed to fabricate hybrid laminates incorporated with inorganic nanoparticles to comprehend the mechanical and viscoelastic characteristics of new hybrid composite material manufactured. This investigation reports the outcomes of a comprehensive investigation on the fabrication and characterization of hybrid micro-dry-leaves fiber and aluminum silicon carbide (Al-SiC) incorporated epoxy laminates. The novelty of this investigation comes from the combined influence of micro-dry-leaves fiber and Al-SiC hybridization on the mechanical and viscoelastic characteristics in terms of tensile, flexural, impact, storage, loss modulus, and damping factors characteristics.

## 10.2 Materials and Methods

### 10.2.1 Materials

Aluminum (Al – 10 nm) particles were procured from Ganapathy Colours, Parrys, Chennai, India, and they were utilized as the primary toughening microparticles in the epoxy polymer, whereas the silicon carbide (SiC) was purchased from Carborundum Universal, Kochi, India, with a particle size of 150 μm.

A lower viscosity epoxy resin, which includes two parts as resin (LY556) and hardener (HY951), procured from Sakthi Fiber Glass Inc., Chennai, India, at a proportion ratio of 10 : 1 by weight, was applied as polymer matrix.

Around 1 kg of dry leaves as agricultural wastes were collected from nearby trees of different types from Trichy, Tamilnadu, India, and crushed using an industrial grinder, and then subjected to separation through a micro-sieve of 120 μm (average size).

### 10.2.2 Production of Al-SiC Nanoparticles

The process of mechanical alloying has been performed in horizontal type of high energy ball mill equipment. To avoid severe welding, 3 wt.% of silica gel was mixed with the Al and SiC mixture during the process of milling. The milling time, balls to powder weight ratio, and milling speed were chosen as 180 minutes, 10, and $200 \pm 5$ rpm respectively. An interim period of 1200 seconds for every one hour was offered to avoid overheating. After the mechanical alloying process, the nanoparticles of Al-SiC were separated from a 55 nm (average size) sieve and utilized as a reinforcement material to the polymer matrix.

### 10.2.3 Fabrication of Epoxy Composites

An ultrasonicator probe, UP400St, Hielscher, Germany, has been applied to disperse both micro-dry-leaves and Al-SiC nanofillers in the epoxy resin. The 5 wt.% of micro-dry-leaves and Al-SiC nanofillers, and epoxy polymer (LY 556) were weighed carefully and placed in a cylindrical beaker of 1000 ml. Two types of epoxy composites have been fabricated, one reinforced with dry leaves (ERD), and another with both micro-dry-leaves and Al-SiC nanoparticles (ERDA). Either the single or multiple particles were dispersed in epoxy resin at the higher sonotrode power of 400 W and 24 kHz of frequency, as defined by the manufacturer. After the ultrasonication, the hardener is included with the epoxy polymer mixture using stirrer for 15 minutes to get uniform dispersion within the matrix. Finally, the mixture was poured in the mold cavity and shifted to a vacuum chamber to remove the entrapped air and void contents. Both the fabricated composite specimens were post-cured at $80 \pm 3\,°C$ for four hours in an electric furnace to eliminate the intrinsic moisture content [29].

### 10.2.4 Epoxy Composite Characterization

#### 10.2.4.1 Porosity, Density, and Volume Fraction

The porosity and volume fraction of filler reinforcement for all epoxy samples were determined as per the ASTM D 792 standard. The samples were sectioned into 15 mm × 15 mm and dried in an electric furnace for 24 hours under 100 °C before checking the weight. The volume of the samples was calculated after removing from the furnace. The density of the epoxy composite samples was measured using the Archimedes technique and the difference between the measured density and theoretical density showed the porosity level of the epoxy composites. The filler volume fraction signifies the balance weight measured after degrading the laminates entirely with 600 °C and the volume fraction was calculated as per the ASTM D3171 standard.

#### 10.2.4.2 Tensile Properties

Uniaxial tensile test was performed with the help of displacement regulated Tinius Olsen University Tensile Testing machine, H-50 kN capacity, USA, under room temperature to determine the influence of incorporation of micro-dry-leaves fiber and Al-SiC nanoparticles on the tensile characteristics of the fabricated epoxy laminate samples as per the ASTM D 3039/D3039M-17 standard. An axial extensometer was applied to evaluate the axial displacement quantities, and the cross-head speed of the machine was 1 mm/min.

#### 10.2.4.3 Flexural Properties

The three-point flexural test samples were fabricated and measured as per the ASTM D790-15 standard from Tinius Olsen University Tensile Testing machine, H-50 kN capacity, USA, under room temperature to determine the flexural characteristics of the fabricated epoxy laminates. The samples were 100 mm in length and 12.5 mm in width respectively. The flexural test was conducted till the epoxy composite sample failure, and its stress–strain curves were reported for comparison.

#### 10.2.4.4 Impact Strength

The impact strength experimentation was conducted as per the ASTM D256 standard on an Izod impact testing equipment, Bagga International Scientific Instruments, India, which includes a 10.4 N-m hammer with 20 J energy, and the size of the specimen was 62.5 mm × 12.5 mm × 5 mm with a notch of 2.5 mm.

#### 10.2.4.5 Dynamic Mechanical Analysis (DMA)

The viscoelastic properties (dynamic mechanical analysis [DMA]) were observed for measuring the dependent temperature characteristics such as loss, storage modulus, and damping factor, according to ASTM D4065-01 standard. A three-point bend test with a force of 150 N was applied for the examination. The sample measurement of 50 mm × 12 mm × 5 mm was experimented at nitrogen environment from 30 to 250 °C with 3 °C/min of heating rate. A fixed frequency of 1 Hz and strain rate of 0.2% were employed in this test.

#### 10.2.4.6 Morphological Properties

The fracture surface microstructure of both single and multiple filler-reinforced epoxy composites was examined by considering a scanning electron microscope (SEM) with an accelerating voltage of 15 kV. Before the SEM observations, the epoxy composite specimens were sputtered with gold particles to convert the material as conducting under the electron beam modes.

## 10.3 Results and Discussion

### 10.3.1 Quality of Fabrication and Volume Fraction of Epoxy Composites

The density, thickness, filler volume fraction, and porosity values of ER, ERD, and ERDA composites are given in Table 10.1. The thickness of these epoxy laminates was varied between 2.8 and 3.7 mm. The difference in thickness of the epoxy laminates was attributed to the particle size and volume fraction of the fillers. Furthermore, the viscoelastic and damping characteristics of the epoxy laminates could not be modified significantly with the different thicknesses [30]. However, the mechanical characteristics could not be completely based on the little deviations in the laminate's thickness [31]. Necessarily, the ultrasonication time and frequency of 120 minutes and 24 kHz were constantly controlled for all ultrasonication-assisted hand lay-up techniques to continuously determine the volume fraction of filler and polymer matrix [31].

Consequently, the densities of ERD and ERDA laminates were 977.90 and 1854.20 kg/m$^3$ respectively. Therefore, the ERD composites have a lower density than the ERDA composites, because of lower density of dry leaves fiber when compared with Al-SiC nanoparticles. The porosity of ER, ERD, and ERDA composites

**Table 10.1** Fabrication qualities of the epoxy composite samples.

| | Density in kg/m$^3$ | Porosity in % | Thickness in mm | Filler volume fraction in % |
|---|---|---|---|---|
| ER | 1270 | 8.17 | 2.86 | 0 |
| ERD | 977.90 | 8.49 | 3.12 | 15.57 |
| ERDA | 1854.20 | 7.66 | 3.48 | 7.78/7.78 |

are 8.17%, 8.49%, and 7.66% respectively. Dry leaves fiber reinforcement improved the porosity of the epoxy laminates since plant fibers primarily contain lumen and cellulose fiber, which are hydrophilic in nature [32]. Lastly, the volume fractions of fillers of the ERD and ERDA composites were 15.57% (dry leaves fiber) and 7.78% (dry leaves fiber and Al-SiC) respectively. The difference between these volume fractions of fillers is primarily based on the ultrasonication process and geometry of the micro-dry-leaves fiber/Al-SiC nanoparticles applied during the fabrication process.

### 10.3.2 Tensile Characteristics

The tensile strength test is an essential experiment to identify the strength of the composite materials in the mode of elasticity. The tensile stress–strain graphs and tensile characteristics (strength and modulus) are shown in Figure 10.1a,b respectively. The addition of Al-SiC nanoparticles in dry leaves fiber epoxy composites improved the strength to the highest value, but the strength was lower with single filler (dry leaves fiber) reinforcement.

Furthermore, it is clearly observed that the addition of Al-SiC nanoparticles in dry leaves epoxy laminate could significantly influence the strength. In this context, the ERDA composites have higher rate of strain (elongation) than the dry leaves fiber epoxy composites. The tensile strength for dry leaves fiber and dry leaves fiber/Al-SiC epoxy composites are 14.67 and 22.52 MPa respectively. After the reinforcement of dry leaves fiber and both dry leaves fiber/Al-SiC nanoparticles, the tensile strength of epoxy composites improved by 24.74% and 57.48% respectively when compared with the neat epoxy resin. The reason may be attributed to the uniform dispersion of particles within the polymer, which leads to stronger interfacial bonding between the polymer and the fillers (either single or multiple fillers). Figure 10.1b presents the tensile modulus of ERD and ERDA composite samples. The tensile modulus of ERD and ERDA were 0.91 and 1.02 GPa respectively. The lowest and highest tensile modulus are assigned to the ERD and ERDA composites, respectively. The tensile modulus of ERD composites was 12.08% lower than that of ERDA composites.

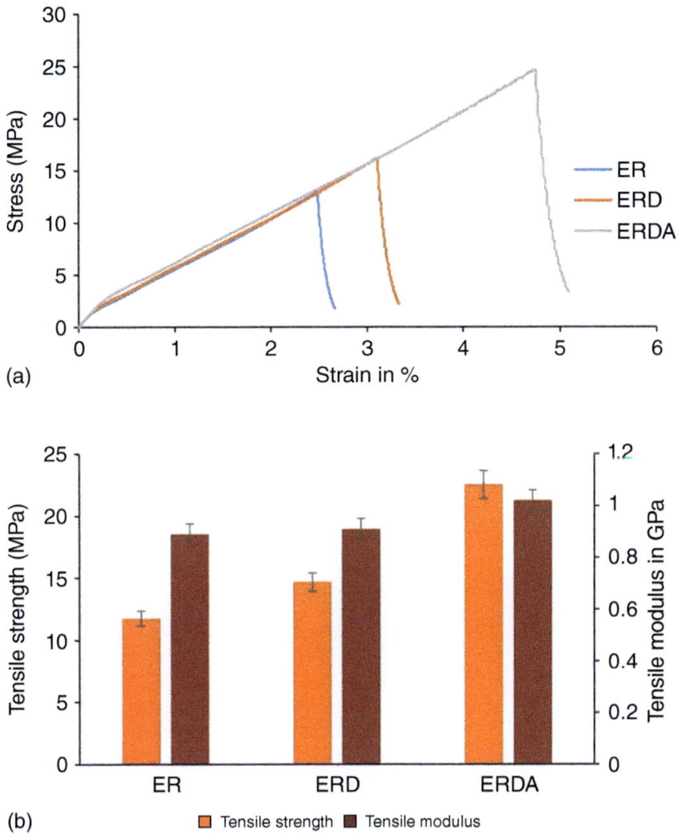


**Figure 10.1** Dry leaves fiber/Al-SiC-reinforced epoxy composites: (a) tensile stress–strain and (b) tensile properties.

### 10.3.3 Flexural Characteristics

Figure 10.2 exhibits the flexural experiments of ER, ERD, and ERDA composite samples. The stress–strain graphs of ER, ERD, and ERDA composites collected from three-point bending test are shown in Figure 10.2a. It is evident that the ER, ERD, and ERDA composites graphs exhibit one peak, which signifies that both the single filler and multiple filler reinforcement on epoxy composites are not failed at the similar time [34]. The reason may be attributed to the delamination and flexural load that the samples undergo [33]. Figure 10.2b shows that the flexural strength of ER, ERD, and ERDA composites were 29.73, 3092, and 44.17 MPa respectively. Significant modifications can be seen in epoxy composites in the appearance of dry leaves fibers; the flexural strength of ERD and ERDA composites were improved by 4% and 48.57% respectively when compared with the neat epoxy polymer. Figure 10.2b also depicts

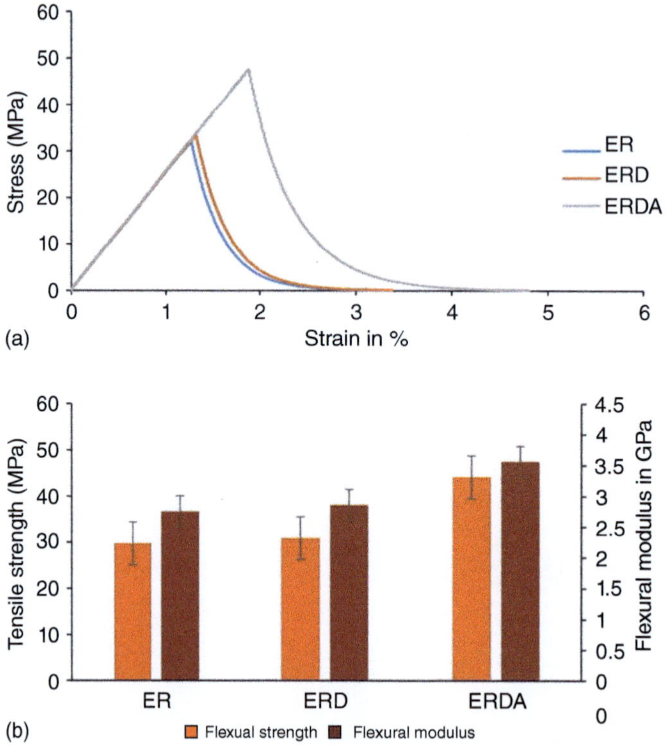


**Figure 10.2** Dry leaves fiber/Al-SiC-reinforced epoxy composites: (a) flexural stress–strain and (b) flexural properties. Source: Based on Chensong [33].

the flexural modulus of ER, ERD, and ERDA composites, which are 2.75, 2.86, and 3.56 GPa respectively, which follows the trend of flexural strength. The ERDA composites have a higher flexural modulus, followed by ERD composites. The flexural modulus of ERD composite was 4% higher than that of the neat epoxy polymer. Further, the reinforcement of Al-SiC nanoparticles in ERDA composites resulted in flexural modulus of 29.45% and 24.47%, which is higher when compared with neat polymer and ERD composites respectively. The reason may also be attributed to the uniform dispersion of single or multiple fillers within the matrix, which tends to improve the flexural characteristics of the ERD and ERDA composites.

### 10.3.4 Impact Characteristics

The outcome of the impact tests for ER, ERD, and ERDA composites is shown in Figure 10.3. The impact strengths of ER, ERD, and ERDA composites were 0.35, 0.36, and 0.51 $kJ/m^2$, respectively. The outcomes showed that the ERDA composites had higher impact strength when compared with ERD composites and the neat polymer, because of Al-SiC capability to withstand under impact strength [29]. Both neat polymer and ERD composite have lower impact strength; due to higher brittleness and relatively lower strain, they do not absorb much impact energy. In the

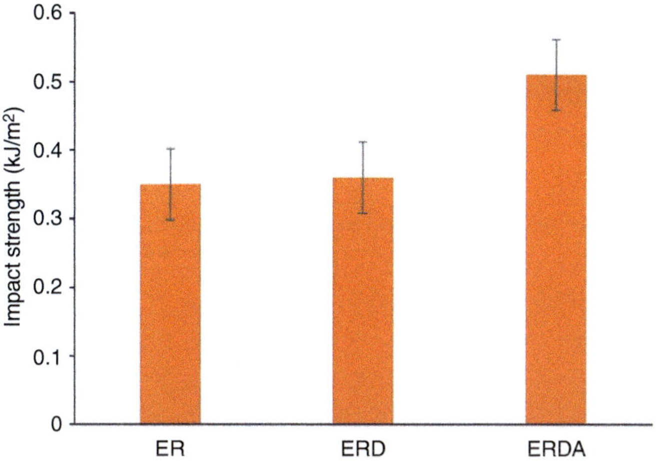


**Figure 10.3** Impact strength of dry leaves fiber/Al-SiC-reinforced epoxy composites.

case of ERD composites, the impact strength was lower because of the lower impact strength of dry leaves fiber or natural fabrics [31, 35].

### 10.3.5 Dynamic Mechanical Analysis

The reinforcement of dry leaves fibers and Al-SiC nanoparticles in the epoxy laminate under different temperatures and frequencies is determined using DMA experimental work. The outcomes of loss, storage modulus, and damping factors were evaluated. The variation of temperature effect for ER, ERD, and ERD composites is shown in Figure 10.4a–c.

#### 10.3.5.1 Storage Modulus

The storage modulus confronts the molecular relaxation and stiffness for all fabricated composite specimen Figure 10.4a exhibits the variation of storage modulus significantly for pure ER, and the developed composites reinforced with dry leaves fiber and dry leaves fiber/Al-SiC nanoparticles. The variation in storage modulus quantities is examined with temperature for ERD and ERDA composites. In all the three conditions, the values of storage modulus decrease continuously with rise in temperature.

From the outcome, it is also observed that the storage modulus remains wider in the glassy region, as the composition is completely closed in the lower range of temperature and the value of storage modulus sustains in the higher range before invading the glass transition region. Furthermore, in 55–100 °C range, which represents the leather to rubbery region, the modulus decreases drastically. When the temperature rises, movement in the molecular chain of the ER, ERD, and ERDA composites takes place suddenly; this may be because of breaking of the molecular linkages. Then slightly, the storage modulus value reduces in the rubbery region. But there is no significant modification in the rubbery region for the ERD and ERDA

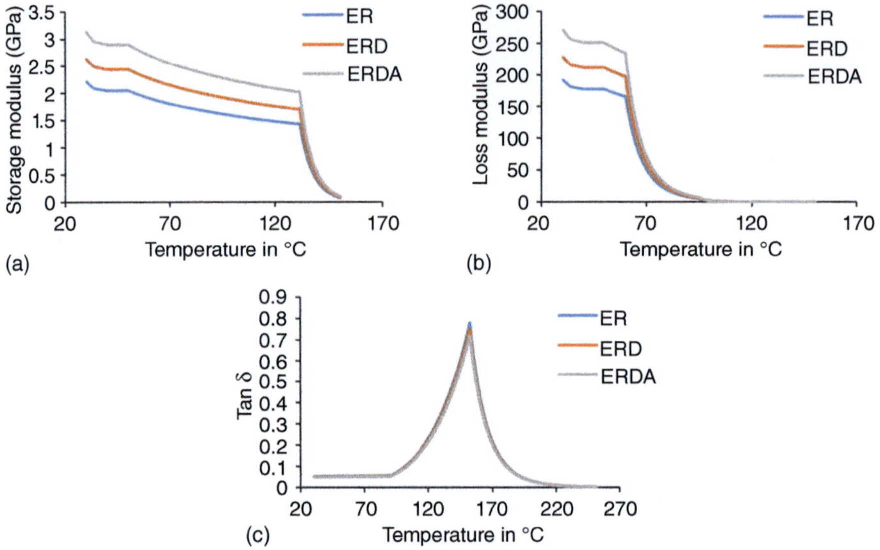


**Figure 10.4** Dynamic mechanical properties of dry leaves fiber/Al-SiC-reinforced epoxy composites: (a) storage modulus, (b) loss modulus, and (c) damping factor.

composites examined. The storage modulus graph shows the effect of reinforcement of Al-SiC nanoparticles in the dry leaves fiber-reinforced epoxy polymer laminate. The storage modulus value maximizes with the addition of Al-SiC nanoparticles, which tends to improve the interfacial adhesion within the epoxy polymer and dry leaves fiber/Al-SiC nanoparticles [36, 37]. The storage modulus for pure epoxy is nearly 2.21 GPa, which has been enhanced to the highest value of nearly 3.13 GPa for the ERDA composite sample, thus establishing a significant enhancement in storage modulus of about 41.62% with the reinforcement of Al-SiC nanoparticles and micro-dry-leaves fibers. The higher value of storage modulus of ERDA composite also followed relatively higher thermomechanical characteristics when compared with pure epoxy polymer.

#### 10.3.5.2 Loss Modulus

The values of loss modulus of the dynamic mechanical analysis outcomes with respect to temperature for the ER, ERD, and ERDA composites are presented in Figure 10.4b. These curves also present the same trend as the value of storage modulus for single and multiple filler reinforcement in polymers. The value of loss modulus improves with the addition of Al-SiC nanoparticles. In all the three conditions, the loss modulus achieved a maximum point for the largest depletion of mechanical energy and decreased with rising temperature, due to the free movement of polymeric networks, fascinating the value of loss modulus for ERDA composites when compared with ERD and pure epoxy polymer composites [38]. Pure epoxy and dry leaves fibers reinforcement alone showed improvement in the value of loss modulus. Furthermore, the loss modulus graph declines at 100 °C temperature. It can be described that in both the conditions of ERD and ERDA composites, the loss and storage modulus enhanced with the incorporation of Al-SiC nanoparticles.

#### 10.3.5.3 Damping Factor

The damping factor for single (dry leaves fiber) and multiple fillers (dry leaves fiber/Al-SiC nanoparticles) reinforcement on epoxy polymer is illustrated in Figure 10.4c. The neat epoxy polymer exhibited higher damping factor value and the ERDA composite showed lower damping factor among all the fabricated epoxy composite specimens. The damping factor value improves with temperature increment and it attains the highest level in the transition region, followed by reduction in the rubbery region for ERD and ERDA composites. Hence, with the incorporation of Al-SiC nanoparticles, the damping factor peak becomes broader, which presents a higher relaxation time for molecules due to the movement of interior polymeric network chains. A higher cross-linking density is introduced significantly for the ERD and ERDA composites, because of good interfacial bonding. The outcomes shown are also in line with the recent literature findings [29, 37, 39–41].

### 10.3.6 Morphological Characteristics

Figure 10.5 exhibits the fracture morphology of the tensile sample, which is examined to verify the satisfactory bonding between the polymer matrix and filler reinforcement. Usually, the interfacial adhesion within the polymer matrix and fillers reinforcement is weak and the fillers are dragged out. In the ERD composite fracture, the fillers initially deteriorated and were dragged out as the epoxy matrix became damaged (Figure 10.5a). In the fracture surface of ERDA composites, the fillers were damaged without being dragged out, signifying that the interfacial adhesion within the polymer and fillers was excellent (Figure 10.5b), which leads to higher mechanical and dynamic mechanical properties when compared with the neat epoxy polymer.

## 10.4 Conclusion

The present investigation is focused on reinforcing composites with waste dry leaves fiber, which is adequately feasible. The aim is to reduce specific issues related to the utilization of plant fibers in epoxy resin. Micro-sized dry leaves fiber and Al-SiC nanoparticles are applied for producing a new type of composites for improving the thermomechanical and mechanical characteristics. After a detailed investigation on the newly developed ERD and ERDA composites, successive conclusions are compiled below:

1. The dry leaves fiber and Al-SiC nanoparticles-reinforced epoxy (ERDA) composites showed improved mechanical and viscoelastic characteristics; the strong interfacial bonding between the matrix and the Al-SiC nanoparticles also supported the development of the conductive path within the polymers.

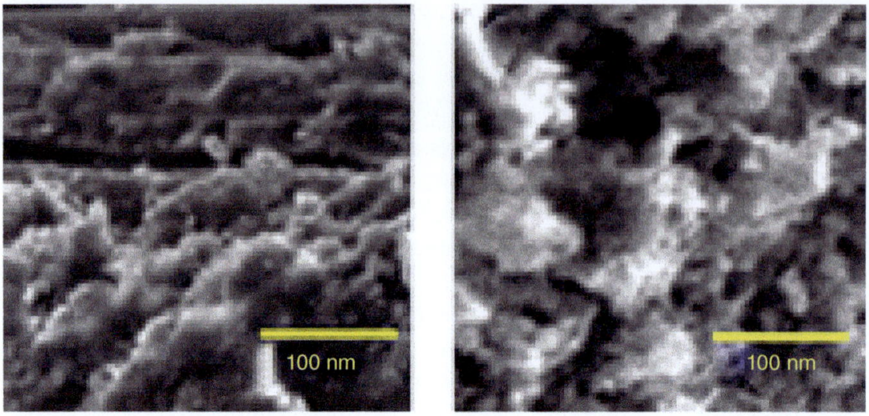


(a) (b)

**Figure 10.5** Fracture SEM micrographs of dry leaves fiber/Al-SiC-reinforced epoxy composites: (a) ERD and (b) ERDA.

2. Reinforcement with Al-SiC nanoparticles played a vital role in reducing the generation of crack, which decreased the brittleness characteristics of the polymer composites.
3. Reinforcement with hybrid particles enhanced the synergistic effect in the epoxy laminate, which led to improving the loss modulus, storage modulus, and damping parameter and also supported to increment the temperature range of the polymeric material.
4. The mechanical characteristics such as impact strength, flexural and tensile strength, and loss modulus exhibited higher values for ERDA hybrid laminates.

## References

**1** Benito-Gonzalez, I., Lopez-Rubio, A., and Martínez-Sanz, M. (2017). Potential of lignocellulosic fractions from *Posidonia oceanica* to improve barrier and mechanical properties of bio-based packaging materials. *Int. J. Biol. Macromol.* 118: 542–551.

**2** Liu, W., Drzal, L.T., Mohanty, A.K., and Misra, M. (2007). Influence of processing methods and fiber length on physical properties of kenaf fiber reinforced soy based biocomposites. *Composites Part B* 38: 352–359.

**3** Scaffaro, R., Maio, A., Gulino, E.F., and Megna, B. (2019). Structure-property relationship of PLA-Opuntia Ficus Indica biocomposites. *Composites Part B* 167: 199–206.

**4** Wu, C., Yen, F., and Wang, C. (2011). Polyester/natural fiber biocomposites: preparation, characterization, and biodegradability. *Polym. Bull.* 67: 1605–1619.

**5** Shah, D.U. (2013). Developing plant fibre composites for structural applications by optimizing composite parameters: a critical review. *J. Mater. Sci.* 48: 6083–6107.

**6** Li, S., Ball, B., Donner, E. et al. (2018). Mechanical properties of green canola meal composites and reinforcement with cellulose fibers. *Polym. Bull.* 76: 1257–1275.
**7** Patel, J.P. and Parsania, P.H. (2016). Fabrication and comparative mechanical, electrical and water absorption characteristic properties of multifunctional epoxy resin of bisphenol-C and commercial epoxy-treated and -untreated jute fiber-reinforced composites. *Polym. Bull.* 74: 485–504.
**8** Sanjay, M.R., Siengchin, S., Parameswaranpillai, J. et al. (2019). A comprehensive review of techniques for natural fibers as reinforcement in composites: preparation, processing, and characterization. *Carbohydr. Polym.* 207: 108–121.
**9** Mohit, H. and Selvan, V.A.M. (2018). A comprehensive review on surface modification, structure interface, and bonding mechanism of plant cellulose fiber reinforced polymer based composites. *Compos. Interfaces* 25 (5–7): 629–667.
**10** Mohit, H. and Selvan, V.A.M. (2019). Thermo-mechanical properties of sodium chloride and alkali treated sugarcane bagasse fiber. *Ind. J. Fiber Text. Res.* 44 (3): 286–293.
**11** Rahman, R., Hamdan, S., Jayamani, E. et al. (2018). Tert-butyl catechol/alkaline-treated kenaf/jute polyethylene hybrid composites: impact on physico-mechanical, thermal and morphological properties. *Polym. Bull.* 76: 763–784.
**12** Annie, S., Boudenne, A., Ibos, L. et al. (2008). Effect of fiber loading and chemical treatments on thermophysical properties of banana fiber/polypropylene commingled composite materials. *Composites Part A* 39: 1582–1588.
**13** Kesavan, P.S.R., Ramnath, B., and Vijaya, C.V. (2015). Effect of fiber orientation and stacking sequence on mechanical and thermal characteristics of banana-kenaf hybrid epoxy composite. *Silicon* 9: 577–585.
**14** Biswas, S., Kindo, S., and Patnaik, A. (2011). Effect of fiber length on mechanical behavior of coir fiber reinforced epoxy composites. *Fibers Polym.* 12: 73–78.
**15** Abdul Khalil, H.P.S., Masri, M., Saurabh Chaturbhuj, K. et al. (2017). Incorporation of coconut shell based nanoparticles in kenaf/coconut fibres reinforced vinyl ester composites. *Mater. Res. Express* 4: 1–20.
**16** Prabu, V.A., Kumaran, S.T., and Uthayakumar, M. (2017). Influence of redmud particle hybridization in banana/sisal and sisal/glass composites. *Part. Sci. Technol.* 36: 402–407.
**17** Ibrahim, I.D., Jamiru, T., Sadiku, R.E. et al. (2017). Dependency of the mechanical properties of sisal fiber reinforced recycled polypropylene composites on fiber surface treatment, fiber content and nanoclay. *J. Polym. Environ.* 25: 427–434.
**18** Shin, M.K., Lee, B., Kim, S.H. et al. (2012). Synergistic toughening of composite fibres by self-alignment of reduced graphene oxide and carbon nanotubes. *Nat. Commun.* 3: 650.
**19** Plyushch, A., Macutkevic, J., Kuzhir, P. et al. (2016). Electromagnetic properties of graphene nanoplatelets/epoxy composites. *Compos. Sci. Technol.* 128: 75–83.
**20** Ahmad, S.R., Xue, C., and Young, R.J. (2017). The mechanisms of reinforcement of polypropylene by graphene nanoplatelets. *Mater. Sci. Eng., B* 216: 2–9.

**21** Mittal, G., Dhand, V., Rhee, K.Y. et al. (2015). A review on carbon nanotubes and graphene as fillers in reinforced polymer nanocomposites. *J. Ind. Eng. Chem.* 21: 11–25.

**22** Zhao, D., Jiang, Y., Ding, Y. et al. (2018). Polymer/carbon nanotubes nanocomposites: relationship between interfacial adhesion and performance of nanocomposites. *J. Mater. Sci.* 53: 10160–10172.

**23** Li, Y., Wang, S., Wang, Q., and Xing, M. (2018). Enhancement of fracture properties of polymer composites reinforced by carbon nanotubes: a molecular dynamics study. *Carbon* 129: 504–509.

**24** King, J.A., Klimek, D.R., Miskioglu, I., and Odegard, G.M. (2013). Mechanical properties of graphene nanoplatelet/epoxy composites. *J. Appl. Polym. Sci.* 128: 4217–4223.

**25** Feng, C., Kitipornchai, S., and Yang, J. (2017). Nonlinear bending of polymer nanocomposite beams reinforced with non-uniformly distributed graphene platelets (GPLs). *Composites Part B* 110: 132–140.

**26** Chatterjee, S., Nafezarefi, F., Tai, N.H. et al. (2012). Size and synergy effects of nanofillers hybrids including graphene nanoplatelets and carbon nanotubes in mechanical properties of epoxy composites. *Carbon* 50: 5380–5386.

**27** Hadden, C.M., Klimek-McDonald, D.R., Pineda, E.J. et al. (2015). Mechanical properties of graphene nanoplatelet/carbon fiber/epoxy hybrid composites: multiscale modeling and experiments. *Carbon* 95: 100–112.

**28** Bulut, M. (2017). Mechanical characterization of Basalt/epoxy composite laminates containing graphene nanopellets. *Composites Part B* 122: 71–78.

**29** Mohit, H. and Selvan, V.A.M. (2020). Effect of Al-SiC nanoparticles and cellulose fiber dispersion on the thermomechanical and corrosion characteristics of polymer nanocomposites. *Polym. Compos.* 41 (5): 1–22.

**30** Crane, R.M. and Gillespie, J.W. Jr., (1991). Characterization of the vibration damping loss factor of glass and graphite fiber composites. *Compos. Sci. Technol.* 40: 355–375.

**31** Kim, C. and Song, J. (2020). Effect of hybrid reinforcement on the mechanical properties of vinyl ester green composites. *Fibers Polym.* 21 (2): 428–436.

**32** Prabhakar, M.N. and Song, J.-I. (2018). Fabrication and characterisation of starch/chitosan/flax fabric green flame-retardant composites. *Int. J. Biol. Macromol.* 119: 1335–1343.

**33** Chensong, D. (2016). Uncertainties in flexural strength of carbon/glass fibre reinforced hybrid epoxy composites. *Composites Part B* 98: 176–181.

**34** Cao, M., Zhao, Y., Gu, B.H. et al. (2019). Progressive failure of inter-woven carbon-Dyneema fabric reinforced hybrid composites. *Compos. Struct.* 211: 175–186.

**35** Kumar, C.N., Prabhakar, M.N., and Song, J.-I. (2019). Effect of interface in hybrid reinforcement of flax/glass on mechanical properties of vinyl ester composites. *Polym. Test.* 73: 404–411.

**36** Prolongo, S.G., Moriche, R., Jiménez-Suárez, A., and SánchezMand Ureña, A. (2014). Advantages and disadvantages of the addition of graphene nanoplatelets to epoxy resins. *Eur. Polym. J.* 61: 206–214.

**37** Gouda, K., Bhowmik, S., and Das, B. (2020). Thermomechanical behavior of graphene nanoplatelets and bamboo micro filler incorporated epoxy hybrid composites. *Mater. Res. Express* 7: 015328.
**38** Saba, N., Paridah, M.T., Abdan, K., and Ibrahim, N.A. (2016). Dynamic mechanical properties of oil palm nano filler/kenaf/epoxy hybrid nanocomposites. *Constr. Build. Mater.* 124: 133–138.
**39** Wang, F., Drzal, L.T., Qin, Y., and Huang, Z. (2015). Mechanical properties and thermal conductivity of graphene nanoplatelet/epoxy composites. *J. Mater. Sci.* 50: 1082–1093.
**40** Chandrasekaran, S., Seidel, C., and Schulte, K. (2013). Preparation and characterization of graphite nano-platelet (GNP)/epoxy nanocomposite: mechanical, electrical and thermal properties. *Eur. Polym. J.* 49: 3878–3888.
**41** Jawaid, M., Khalil, H.A., Hassan, A. et al. (2013). Effect of jute fibre loading on tensile and dynamic mechanical properties of oil palm epoxy composites. *Composites Part B* 45: 619–624.

# 11

# Effect of Fillers on Natural Fiber–Polymer Composite: An Overview of Physical and Mechanical Properties

*Annamalai Saravanakumaar, Arunachalam Senthilkumar, and Balasundaram Muthu Chozha Rajan*

*Sethu Institute of Technology, Department of Mechanical Engineering, Virudhunagar 626115, Tamil Nadu, India*

## 11.1 Introduction

Nowadays, many researchers are working toward biodegradable materials that are ready to substitute synthetic materials due to the limited resources that come from petroleum-based polymers [1]. In addition to their use as composite reinforcement such as fillers, particles, fibers, and whickers, natural fibers are used to enhance the mechanical properties of composites [2]. A lot of work has been done using natural fibers with different natural fillers that increase the composite performance [3]. Synthetic filler is also used to improve the performance of the composite laminates. However, this is not recommended because it is not environmentally friendly. Several residues available from agriculture and forestry are made use of as primary resources of renewable energy in various industries [4]. Indirect agricultural resources such as waste materials (waste dust and waste paper) provide economic benefits and are also environmentally friendly to use as fillers [5]. Using natural fibers as reinforcements, the composite laminates are affected by reduced fiber–matrix interaction, lower durability, and low water resistance. The reduced bonding between the fiber and the polymeric matrix affects their properties: mechanical, physical, dynamic, chemical, etc. [6].

Nowadays, researchers use fillers to enhance the properties of composites [7]. The fillers can be obtained from different parts of the plants: bark, pulp, bagasse, stem, and cereal straw. The properties of the natural fiber fillers greatly change with age and the part of the plant from which they are extracted [8]. The properties also vary with varying the chemical modification methods, such as mechanical combing and water retting. [9]. The final result of incorporating fillers in composites varies with the specimen size, the filler particle size, and the volume ratio of the filler. The reinforcement properties of any composite are based on the uniform distribution of the matrix, orientation of the fiber, aspect ratio, and the adhesion property [10].

Researchers reported that the natural fiber composites reinforced with fillers in the structure showed superior performance in mechanical properties than

*Mechanical and Dynamic Properties of Biocomposites*, First Edition. Edited by Senthilkumar Krishnasamy, Rajini Nagarajan, Senthil Muthu Kumar Thiagamani, and Suchart Siengchin.

non-filler-reinforced natural fiber composites [11]. Addition of fillers reduces the water intake property and increases the mechanical properties of the composites.

This review article offers a broad overview of recent developments in the research of adding fillers in the natural fiber–polymer matrix composites. It also throws light on the effect of adding fillers on the physical and mechanical properties of natural fiber–polymer composites.

## 11.2 Influence of Cellulose Micro-filler on the Flax, Pineapple Fiber-Reinforced Epoxy Matrix Composites

Sumesh et al. [12] evaluated the mechanical performance of pineapple leaf/flax fiber/cellulose micro-filler (CMF)-reinforced hybrid composites. In this research, the hybrid composites are fabricated with a combination of pineapple leaf fiber (PALF) and flax fiber (phenol formaldehyde, PF) at 30 and 35 wt.%. Besides, CMF was added (1–3 wt.%) to enhance their properties. The overall results were derived using Table 11.1, in which the composites were found to be enhanced by incorporating the CMF. The inclusion of 2% CMF with 30 wt.% of PF improved the overall mechanical properties when compared with other combinations. It was observed that the overall mechanical properties were found to be improved at 35 wt.% composite specimen; it was ascribed to good fiber–matrix adhesion properties [12].

## 11.3 Influence of Sugarcane Bagasse Filler on the Cardanol Polymer Matrix Composites

Balaji et al. [13] evaluated the effect of sugarcane bagasse (SCB) fiber filler in the cardanol polymer composite specimens. Composite specimens were prepared using

**Table 11.1** Mechanical properties of PF/CMF hybrid fiber composites.

| Hybrid composites | Cellulose microfiber (wt.%) | Tensile strength (MPa) | Flexural strength (MPa) | Impact strength (J/m) |
|---|---|---|---|---|
| 30 wt.% PF | 0 | 23.04 | 49.3 | 64.53 |
| | 1 | 27.53 | 58.7 | 72.5 |
| | 2 | 34.5 | 66.2 | 94.98 |
| | 3 | 33 | 60.9 | 80.24 |
| 35 wt.% PF | 0 | 24.8 | 51.4 | 66.25 |
| | 1 | 27.9 | 57.1 | 71.89 |
| | 2 | 35.8 | 67.8 | 85.85 |
| | 3 | 36.1 | 68.93 | 96.13 |

Source: Adapted from Sumesh et al. [12].

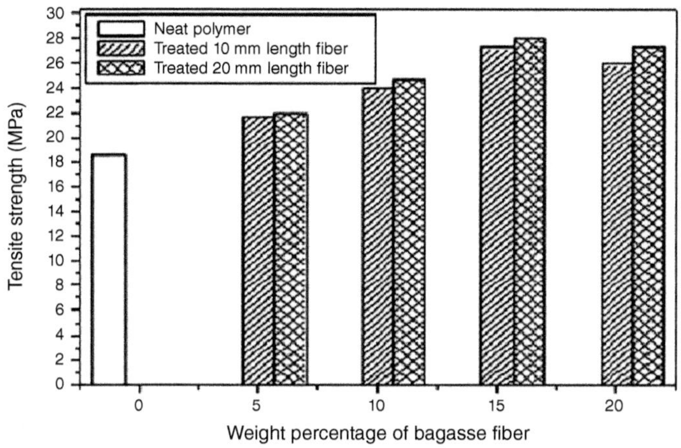


**Figure 11.1** Effect of fiber content on tensile strength of bagasse composite. Source: From Balaji et al. [13].

bagasse filler with varying percentage of filler (0, 5, 10, 15, and 20 wt.%). Tensile, flexural, and hardness tests were conducted with different fiber contents and different fiber lengths and their properties compared with the neat polymer. For the tensile and flexural properties, an increasing trend occurred up to 15 wt.% of fiber. From Figure 11.1, it was identified that the tensile strength increased from 18.2 MPa to a maximum of 27.3 MPa for 15 wt.% in 10 mm fiber length and to 28 MPa for 15 wt.% in 20 mm fiber length. It can be noticed that the tensile strength increases with increase in fiber content only within 15 wt.%. Beyond 15 wt.% there was a decreasing trend in tensile strength.

A higher hardness value of 45 HRB (Rockwell Hardness measured on the B scale) was achieved for 20 wt.% bagasse fiber in 10 mm fiber length and 46 HRB was achieved for 20 wt.% bagasse fiber in 20 mm fiber length. It indicates that the hardness value increases with an increase in the weight percentage of fiber.

## 11.4 Influence of Sugarcane Bagasse Filler on the Natural Rubber Composites

De Paiva et al. [14] examined the mechanical properties of sugarcane bagasses fiber/sandal/natural rubber composites. It was observed that the dispersion of filler and alkali treatment improved their mechanical properties.

From Table 11.2, it can be observed that the composite specimen with 10 phr of treated sugarcane bagasse ($SCB_T$) had a tensile strength value of 12.89 MPa, which is higher than that of all types of composites. The lowest value was achieved at 3.72 MPa for the 40 phr of untreated sugarcane bagasse ($SCB_U$) natural rubber (NR) composite. It was also identified that when adding SCB in the composite specimen, the tensile strength of both (NR/$SCB_U$ and NR/$SCB_T$) specimens got reduced because of less filler adhesiveness to the matrix [14].

**Table 11.2** Tensile strength of natural rubber, and raw and treated bagasse fiber composite.

| Composite specimen | Tensile strength (MPa) | |
|---|---|---|
| | $NR/SCB_T$ | $NR/SCB_U$ |
| NR | — | 12.35 |
| 10 | 12.89 | 11.41 |
| 20 | 11.13 | 7.11 |
| 30 | 9.19 | 4.75 |
| 40 | 7.36 | 3.72 |

Source: Adapted from De Paiva et al. [14].

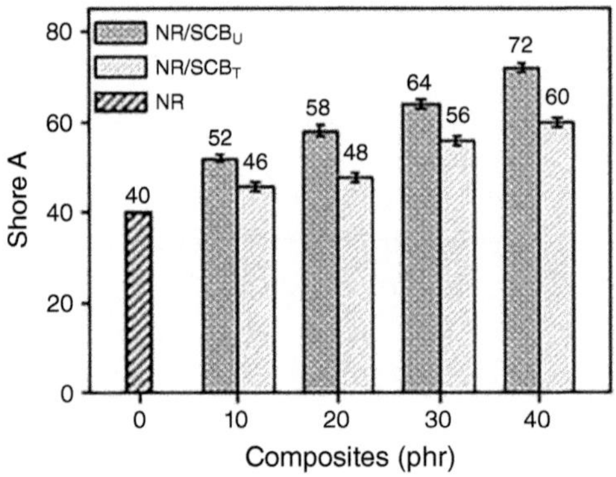


**Figure 11.2** Effect of fiber content on hardness of bagasse composite. Source: From De Paiva et al. [14].

From Figure 11.2 it is seen that the hardness value (Shore A) of the composite specimens for both $NR/SCB_U$ and $NR/SCB_T$ increases with increase in the SCB filler. This behavior is because of the fibrous characteristics of SCB.

## 11.5 Influence of Fly Ash on Wood Fiber Geopolymer Composites

Furtos et al. [15] measured the mechanical properties of wood fiber geopolymer composites by adding 5 wt.% of sand as a filler. The wood fiber composites were prepared with a variable percentage of 5–35 wt.% with an increment of 5 wt.%. The compressive strength (CS) and compressive modulus (CM) of the samples were found from Eq. (11.1) and (11.2).

$$CS_{cylindrical} = F/r^2 \tag{11.1}$$

$$CS_{cubic} = F/a^2 \tag{11.2}$$

where $F$ = applied load; $r$ = radius of the sample; and $a$ = length of a side of the sample.

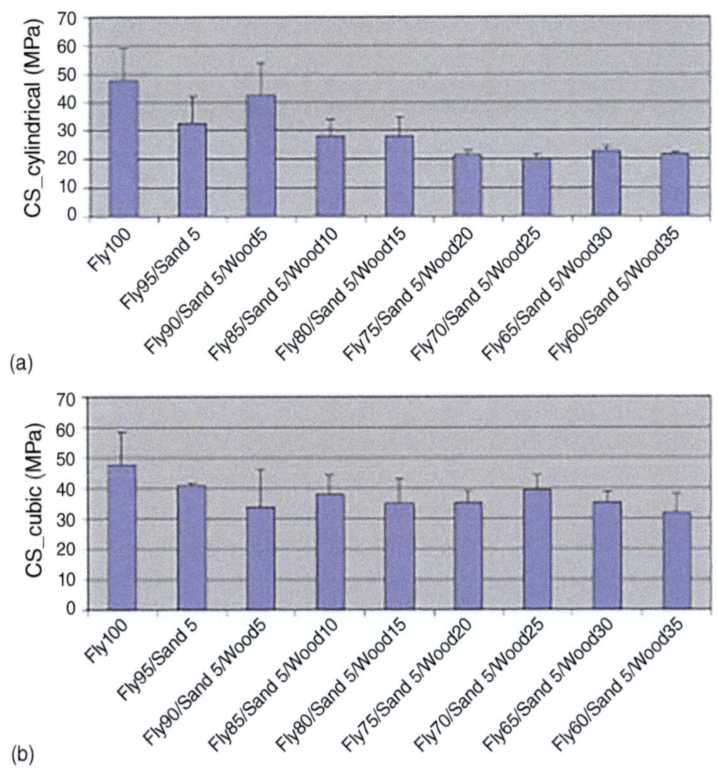


**Figure 11.3** Mechanical properties (a) $CS_{cylindrical}$ and (b) $CS_{cubic}$. Source: From Furtos et al. [15].

Figure 11.3 showed that the fly 100 configuration from both $CS_{cylindrical}$ and $CS_{cubic}$ composites showed a better CS than the rest of the composites. Further, reduction of fly ash percentage decreases the CS [15].

## 11.6 Influence of Eggshell Powder/Nanoclay Filler on the Jute Fiber Polyester Composites

Ganesan et al. 2018 [16] evaluated the mechanical properties of jute fiber composite with eggshell powder and nanoclay (NC) filler using the polyester matrix. The specimens were prepared with polyester as matrix and NaOH treated jute mat (NJM), Un treated jute mat (UJM), egg shell powder (ESP), and NC as filler. Four types of combinations that are tried for this experiment are listed below:

(i) (NJM/UJM)/polyester composite
(ii) (NJM/UJM)/NC/polyester composite
(iii) (NJM/UJM)/ESP/polyester composite
(iv) (NJM/UJM)/ESP/NC/polyester composite

**Table 11.3** Mechanical strength of different combinations of jute fiber composites.

| Combination | Tensile strength (MPa) | | Flexural strength (MPa) | | Impact strength (J) | |
|---|---|---|---|---|---|---|
| | UJM | NJM | UJM | NJM | UJM | NJM |
| Neat | 16.37 | 27.65 | 8.77 | 33.61 | 2.26 | 2.37 |
| ESP | 17.93 | 27.4 | 20.22 | 36.3 | 2.42 | 2.56 |
| NC | 17.88 | 28.35 | 17.94 | 37.12 | 2.32 | 2.55 |
| ESP (1.5 wt.%) + NC (1.5 wt.%) | 18.22 | 29.5 | 17.52 | 39.52 | 2.10 | 3.06 |

Source: Adapted from Ganesan et al. [16].

The mechanical properties of NJM and the UJM composites have been compared and their results are shown in Table 11.3. It was noted that the addition of fillers improved the tensile strength of jute fiber with or without treatment. Comparing the treated and untreated fibers, the treated fibers had higher tensile strength. It is because of the surface bonding of fibers with the polyester matrix. The untreated fibers had poor surface bonding due to the nonremoval of impurities. With respect to flexural strength the NJM composites showed the highest value of 39.52 MPa when compared with UJM composites (17.52 MPa). This was because the filler composition resists crack formation and propagation during the bending load.

The filler size is also a major factor for the impact strength of both NJM and UJM composite specimens. In the raw condition, the fiber composite produced with filler and without filler showed a better and more or less equal impact strength when compared with the treated composite specimens with or without added fillers. The impact strength of UJM with hybrid filler was very less because of the large filler size and the agglomeration [16].

## 11.7 Influence of *Portunus sanguinolentus* Shell Powder on the Jute Fiber–Epoxy Composite

Kumaran et al. [17] studied the effective usage of *Portunus sanguinolentus* shell powder (solid waste in the seafood industry). The traditional hand lay-up method was used to produce two types of composites, one with four layers of jute fiber fabric filled with 10 wt.% untreated *Portunus sanguinolentus* sheel filler (CF-2) and another with chemically treated 10 wt.% *Portunus sanguinolentus* sheel filler (CF-3). The results were compared with the four layers of jute fiber fabric composite (CF-1). From Figure 11.4, it was identified that the tensile strength of CF-3 composite was higher (55 N/mm$^2$) when compared to the tensile strength of other composite specimens CF-2 (42 N/mm$^2$) and CF-1 (31 N/mm$^2$). It was because of the excellent fiber and filler interaction.

The value of density in the CF-3 composite is higher than (1.19 g/cc) that of the other two composites. It is because the filler fills the composite very effectively. The

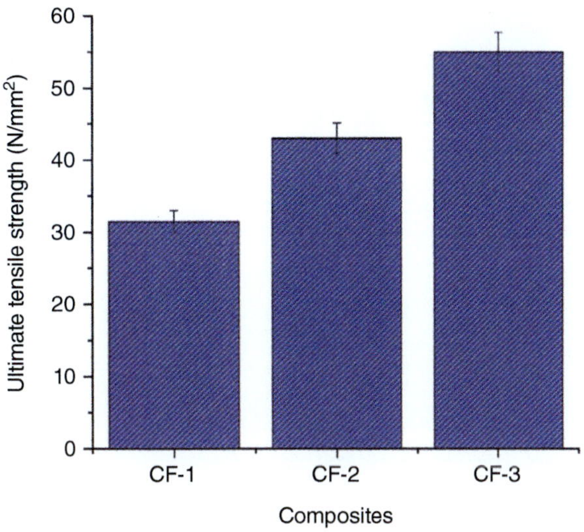


**Figure 11.4** Tensile strength of various composites. Source: From Kumaran et al. [17].

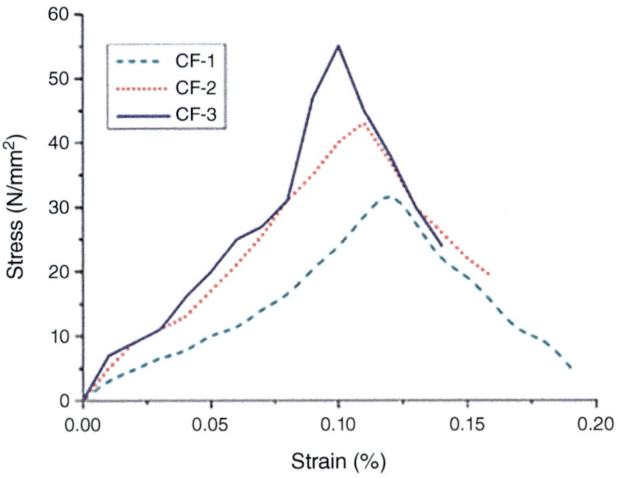


**Figure 11.5** Tensile stress–strain graph of various composites. Source: From Kumarani et al. [17].

stress–strain curve in Figure 11.5 indicates that the CF-3 composite had better firm bonding between the fillers and the matrix. Crack propagation was arrested drastically by the filler and the matrix.

The better performance of CF-3 composite continues for flexural, shear, and impact strengths. In flexural strength the highest value achieved by the CF-3 composite was 5.45 MPa. But the CF-2 and CF-1 composite specimens achieved 4.2 and 4.0 MPa respectively. This is because of the interlocks formed by the filler with the matrix. The impact strength of CF-3 composite was measured as 6.21 J, which exceeds the values of the other two combinations. This is because the filler shell powders are very rich in calcium, which makes a strong interfacial bond and reduces the easy pullout of the fiber.

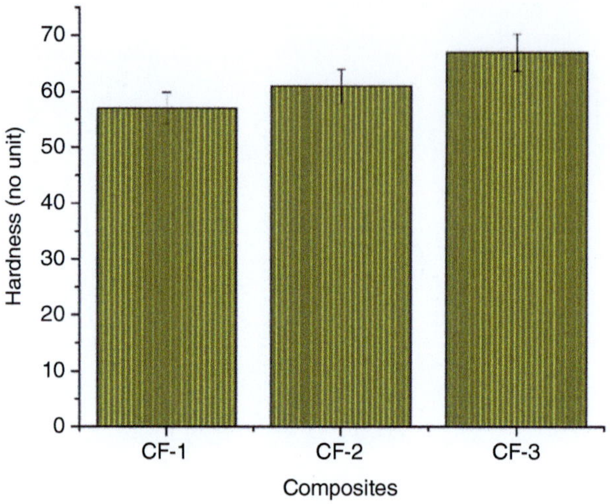


**Figure 11.6** Hardness (Shore D) of various composites. Source: From Kumaran et al. [17].

The hardness values (Shore D) of the CF-3 composite specimen from Figure 11.6 showed better performance than the other two. The highest value of 67 was achieved by CF-3 composite followed by CF-2 (61) and CF-1 (57). The reason for the peak value of hardness was because of the uniform mixing of fiber in the resin and very less voids present in the composite specimen.

## 11.8 Influence of Nano-$SiO_2$ Filler on the *Phaseolus vulgaris* Fiber–Polyester Composite

Gurukarthik Babu et al. [18] investigated the properties of *Phaseolus vulgaris* fiber with polyester resin as matrix and nano-$SiO_2$ as filler. This research evaluated the dielectric property, thermal conductivity, hardness properties, tensile properties, impact properties, water absorption test, and chemical resistance of different weight proportions (2–10 wt.% with the increment of 2 wt.%) of *Phaseolus vulgaris* fiber added with one percentage of nano-$SiO_2$.

From Figure 11.7, it can be observed that the tensile strength of the composite specimen increases until 6 wt.% of *Phaseolus vulgaris* fiber and further increment in weight percentage of the fiber results in a decreasing trend. This was due to the thermoplastic behavior of the adder filler. After 6 wt.% the filler quantity increases so the poor interfacial bonding between the filler and matrix increases the pullout of the fiber.

The water absorption property and thermal conductivity show a decreasing trend up to 6 wt.% of *Phaseolus vulgaris* fiber and after that an increasing trend occurs. Generally, the water absorption is related to the bond between the fiber and the matrix and the voids present in the composite. The increased fiber weight decreases the voids in the composite. Thus, increase in fiber content weakens the bond in one stage and leads to increased water absorption at another stage. This investigation concluded that the composite prepared with 6 wt.% of *Phaseolus vulgaris* fiber

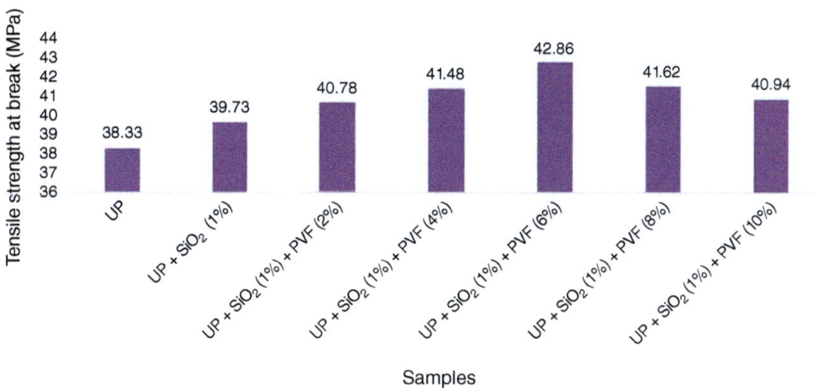


**Figure 11.7** Tensile strength comparison of various combination specimens. Source: From Gurukarthik Babu et al. [18].

and 1% $SiO_2$ as filler delivers better properties (electrical, thermal, and mechanical) among all the prepared combinations [18].

## 11.9 Influence of Aluminum Hydroxide (Al(OH)$_3$) Filler on the Vulgaris Banana Fiber–Epoxy Composite

Shivamurthy et al. 2020 [19] investigated the sliding wear, flammability, water absorption, hardness, and tensile behavior of banana fiber with $Al(OH)_3$ as fillers in epoxy composites. Composite specimens were prepared with 5 and 10 vol.% of banana fiber and varying weight percentages of $Al(OH)_3$ filler (0, 2.5, 5, and 7.5 wt.%).

The stress–strain curve was plotted for varying combinations of fibers (5 and 10 vol.% banana fiber/$Al(OH)_3$/epoxy composites). It is observed that the 10 vol.% banana fiber composite displays a comparatively higher value than 5 vol.% banana fiber composite. It may be because of the higher number of fibers in the 10 vol.% banana fiber for a unit area, and it takes much tensile stress. On tensile test, the effect of filler ($Al(OH)_3$) increases significantly up to 2.5 wt.% due to the pullout restriction of the fillers. Further increase in the filler is not beneficial to the matrix, but it increases the water uptake capacity of the composite [19].

From Figure 11.8, it is observed that the hardness value increases with increase in fiber content for both 10 vol.% banana fiber composite and 5 vol.% banana fiber composites up to 5 wt.%. Owing to the voids, further addition of fillers reduces the hardness. On comparing both the 10 and 5 vol.% banana fiber composites, it was noted that the 10 vol.% banana fiber composites always had higher hardness value. This is because of the higher content of banana fiber that improves the tensile strength and stiffness of the composite.

It is observed that the 10 vol.% banana fiber composite displays a comparatively higher value than the 5 vol.% banana fiber composite. The higher number of fibers and the bonding between the filler and the matrix in the 10 vol.% banana fiber for a

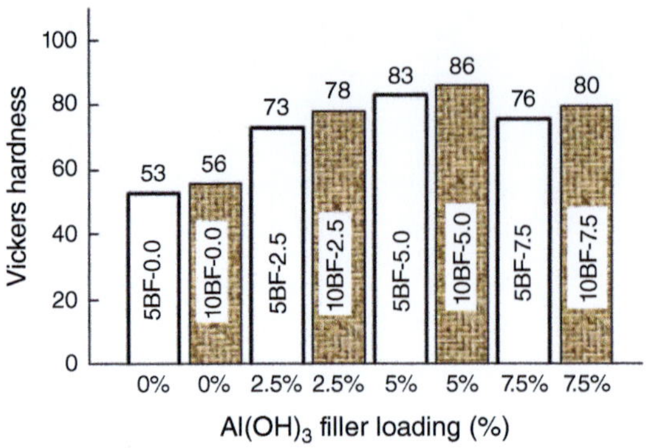


**Figure 11.8** Hardness against the filler loading. Source: From Shivamurthy et al. [19].

unit area makes a better tensile strength. On tensile test, the effect of filler ($Al(OH)_3$) increases significantly up to 2.5 wt.%. Further increase in the filler is not beneficial to the matrix [19].

## 11.10 Influence of Palm and Coconut Shell Filler on the Hemp–Kevlar Fiber–Epoxy Composite

Jani et al. [20] studied the mechanical properties of Hemp–Kevlar hybrid composites by adding coconut shell and palm filler (CP) with a size of 75–150 μm. Chemical treatment increases the adhesion property of the natural fibers. Hence, in the experiment, the fibers and fillers were treated with 5% NaOH for eight hours. Composite specimens are prepared with Hemp–Kevlar fibers and coconut shell and palm filler in 1 : 1 ratio.

The mechanical properties of the test composites are listed in Table 11.4. The flexural strength improves moderately due to filler stiffness. When the added filler exceeds the limit of 5%, then the flexural strength gets reduced. The findings of Brinell hardness number revealed that the hardness value increased by increasing the filler content. However, the rate of increment is found to be slow [20].

**Table 11.4** Mechanical properties of composites with filler contents.

| Sample | Flexural modulus ($N/mm^2$) | Flexural strength ($N/mm^2$) | Brinell hardness number | Impact strength (J) |
|---|---|---|---|---|
| Neat epoxy | 9640 | 141.47 | 188.96 | 3.1 |
| CP-2.5 | 10 400 | 154.353 | 200.12 | 3.52 |
| CP-5 | 8000 | 146.079 | 238.03 | 3.65 |
| CP-7.5 | 7200 | 92.039 | 249.03 | 2.7 |

Source: Adapted from Jani et al. [20].

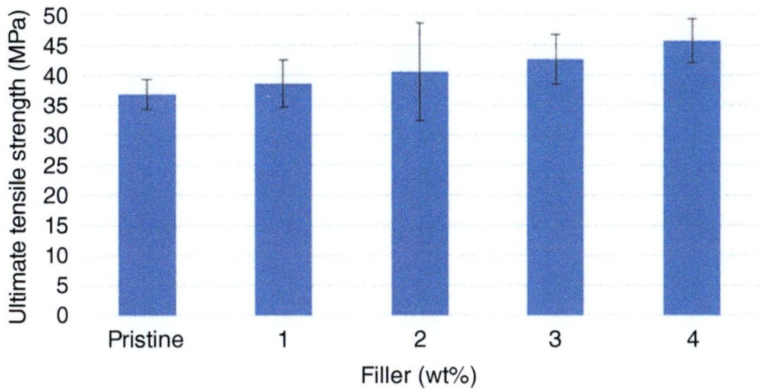


**Figure 11.9** Tensile strength of various weight percentage specimens. Source: From Karthik Babu et al. [21].

## 11.11 Influence of Coir Powder Filler on Polyester Composite

Karthik Babu et al. [21] conducted a study on the thermal and mechanical properties of polyester composites filled with coir powder filler. The specimens were prepared according to the ASTM standards, such as ASTM D638-10 for tensile testing and the ASTM D790-10 for flexural testing. Polyester resin was used as a matrix in different ranges (0–4 wt.%), and 1 g of coir filler was used as reinforcement. For the prepared specimens, the different values of force, stress, and elongation at break were found. From Figure 11.9 it was observed that the ultimate tensile strength of the composites was found to be improved by increasing the filler weight (%). It showed better adhesion of fillers with the polyester matrix and also that the bonding between the filler and the matrix is good when compared with the neat polyester [21].

## 11.12 Influence of $CaCO_3$ (Calcium Carbonate) Filler on the Luffa Fiber–Epoxy Composite

Mohana Krishnudu et al. [22] investigated the influence of $CaCO_3$ filler on the mechanical properties of coir and luffa fiber/epoxy composite. Specimens were prepared with varying filler percentages of 0, 2, and 4. For easy identification, the specimens with different filler addition were noted as SP1 for plain composite, SP2 for 2 g filled composite, and SP3 for 4 g filled composite.

It was observed that the SP3 specimen achieved a higher tensile strength of 49 MPa. It showed that the highest filler content achieves the highest tensile strength. Almost 29% of the tensile strength increased when compared to the neat specimen. A flexural strength of 360 MPa was observed for the SP3 specimen and the flexural strengths obtained for other specimens were 300 and 270 MPa for SP2 and SP1, respectively. From Figure 11.10, the impact strength of 50 kJ/$m^2$ was

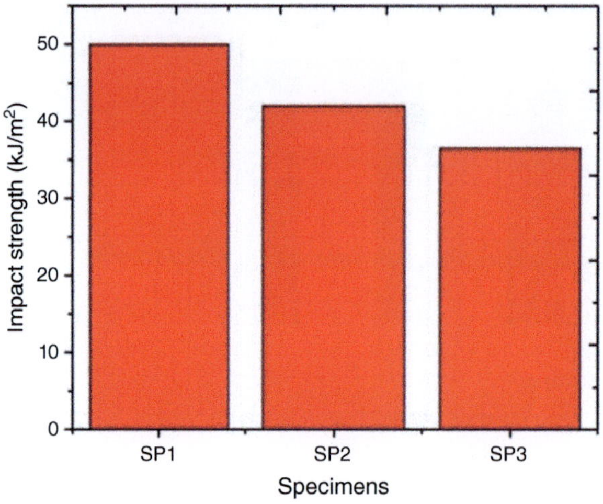


**Figure 11.10** Impact strength. Source: From Mohana Krishnudu et al. [22].

observed for without-filler content specimen. The addition of filler changes the composite sample from ductile to brittle and decreases the impact properties [22].

## 11.13 Influence of Pineapple Leaf, Napier, and Hemp Fiber Filler on Epoxy Composite

Ridzuan et al. [23] analyzed the effect of fillers on scratch resistance with epoxy as a matrix. They used PALF, Napier, and hemp fibers as fillers in particulate form. Furthermore, they analyzed the effect of fillers on horizontal loading, penetration depth, friction coefficient, scratch hardness (Figure 11.11), and fracture toughness. The fiber particles are mixed with epoxy matrix in different weight percentage such as 5%, 7.5%, and 10%.

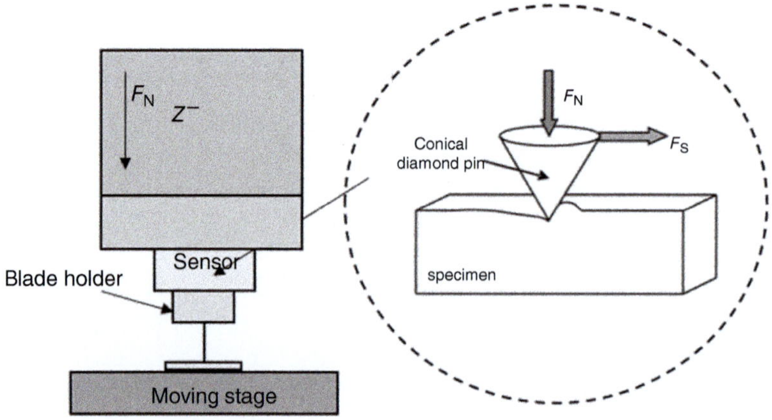


**Figure 11.11** Diagram of the scratch tester. Source: From Ridzuan et al. [23].

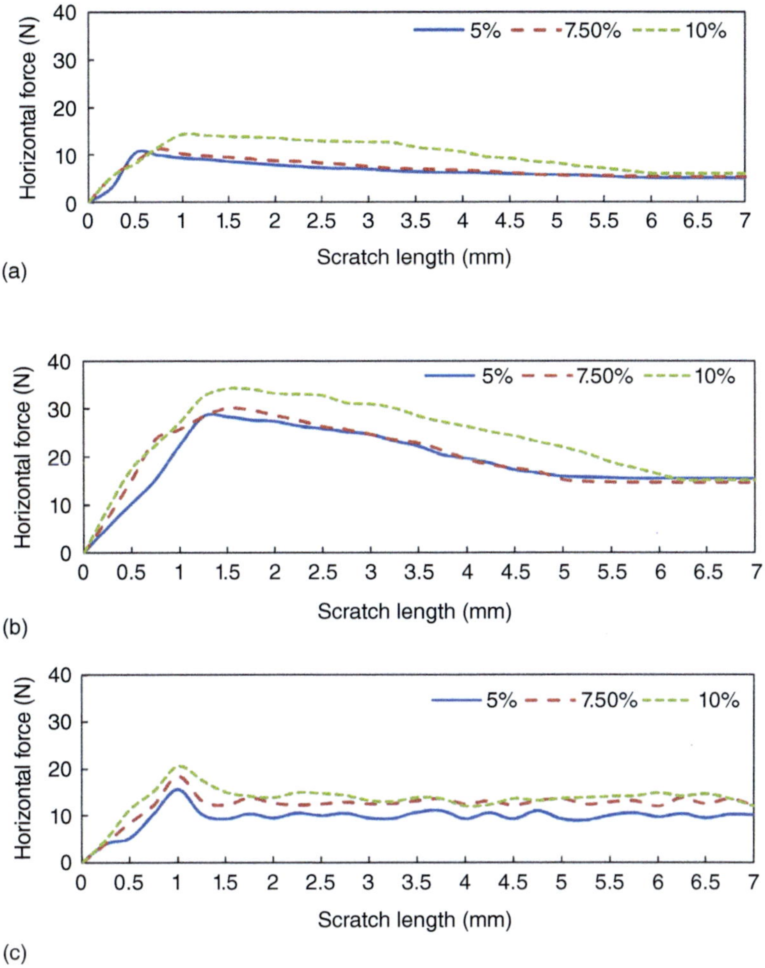


**Figure 11.12** Effect of horizontal force on natural fiber-filled epoxy composites (a) PALF fiber-filled epoxy composites, (b) Napier fiber-filled epoxy composites, and (c) Hemp fiber-filled epoxy composites. Source: From Ridzuan et al. [23].

From a comparison of the results from Figure 11.12, it was observed that the highest value of peak load was achieved in 10 wt.% PALF fibers (14.3 N) when compared with other weight percentage combinations. For hemp fiber, a 21 N peak load was observed for the 10 wt.% composite specimen. Finally, the highest value of 36.4 N peak loads was achieved with the 10 wt.% Napier fiber composite.

It was identified that peak load increases with increase in filler content. The strength of the composite specimen improved due to higher filler content because of the robust molecular movement of the molecules in the filler and the resin. The higher filler content resists the load before the brittle fractures.

The results also indicate that the Napier fiber particle-filled epoxy composite specimens had a better peak load capacity than the PALF and hemp fiber. It was ascribed to the higher lignin content of Napier fiber when compared with the hemp and PALF fibers [23].

## 11.14 Influence of Dipotassium Phosphate Filler on Wheat Straw Fiber–Natural Rubber Composite

Rybiński et al. [24] evaluated the thermal and mechanical properties of natural rubber composites with wheat straw fiber treated with alkali and Dipotassium phosphate as filler.

The chemical treatment started with the alkalization and phosphorylation of fillers. For all types of experiments, the specimens were prepared in dumbbell shape (five numbers) and the average value was reported. Filler incorporation decreases the tensile strength of the NR composite. The reason behind this is that the fillers in the rubber matrix decrease the mobility of rubber chains and act as a barrier. But the incorporation of fillers increases the hardness of the composite. According to observations, the dynamic modulus of rubber composite with fillers was found to be increased when compared with that of the virgin polymers. This was because of the filler interaction that makes the composite more strain dependent. The breakdown of the filler leads to a downward trend of the modulus value because of the increasing strain rate. The fillers modified by NaOH increase the elastic modulus of composite specimens [24].

## 11.15 Influence of Groundnut Shell, Rice Husk, and Wood Powder Fillers on the *Luffa cylindrica* Fiber–Polyester Composite

Dhanola et al. [25] analyzed the mechanical properties of *Luffa cylindrica* fiber with natural fillers (groundnut shell, rice husk, and wood powder) and polyester resin as a matrix. Samples for tensile testing were prepared using the ASTM D303. ASTM E23 was used to test samples for impact strength. For all the composite samples, *Luffa cylindrica* fiber was maintained at 30% and the fillers at 3%, 7%, and 11% with polyester resin. By the experimental analysis, it was observed that the tensile strength of rice husk and wood powder-filled composite shows a downtrend when the filler content is increased. When analyzing the groundnut shell-filled composites, the increase in filler content increases the tensile strength. Overall, the luffa fiber with 11% groundnut shell filler achieves the highest tensile strength of 31.5 MPa. Particle size in luffa fiber and high dispersion enable the composite to get better result. It was observed that the luffa fiber with 7% groundnut shell filler achieved the highest impact strength of 9 J. For rice husk filler, the impact strength increases with an increase in filler content [25].

## 11.16 Influence of Rice Husk Fillers on the *Bauhinia vahlii* – Sisal Fiber–Epoxy Composite

Kumar et al. [26] discussed the mechanical properties of bauhinia vahlii–sisal fiber with rice husk filler epoxy composites. Fillers were mixed at 0, 2, 4, and 6 wt.% with epoxies for producing composite specimens. This investigation includes two different varieties of hybrid composites, namely hybrid bauhinia-vahlii-weight (BVW-R) and bauhinia-vahlii-weight/sisal (BVWS-R) with a different weight percentage of rice husk filler. From Figure 11.13a,b, it was identified that increasing the filler content increases the tensile strength and the flexural strength of both the composites. The highest tensile strength was observed in 6 wt.% of the rice husk filler in both the BVW-R and BVWS-R composites. The results proved the better adhesion capacity of fiber–filler and epoxy matrix.

Unlike the tensile strength and flexural strength, the impact energy decreased slightly because of the deformability of the epoxy resin. The BVWS-R composite showed improved mechanical properties when compared with the BVW-R composites [26].

## 11.17 Influence of Areca Fine Fiber Fillers on the *Calotropis gigantea* Fiber Phenol Formaldehyde Composite

Venkatarajan et al. [27] determined the effect of areca fine fiber (AFF) fillers on the mechanical properties of *Calotropis gigantea* fiber (CGF) as reinforcement and phenol PF as matrix. Hybrid composites (CGF/AFF/PF) were prepared with varying weight percentage (25, 35, and 45 wt.%). From Figure 11.14, it is observed that the 35 wt.% (CGF 17.5 wt.% and AFF 17.5 wt.%) composite delivers better tensile properties than all other hybrid composites.

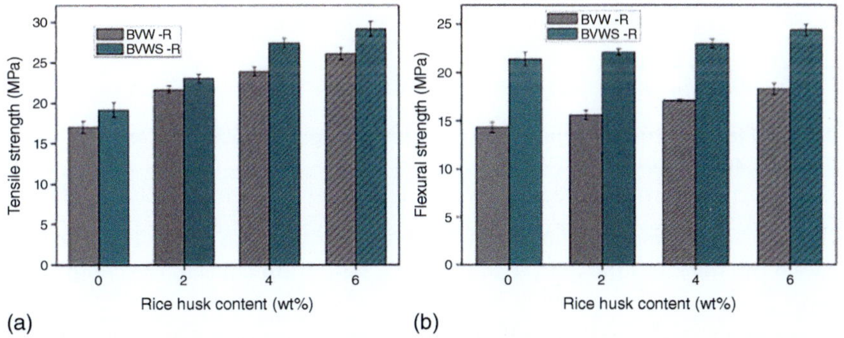


**Figure 11.13** (a) Tensile strength and (b) flexural strength of various composites. Source: From Kumar et al. [26].

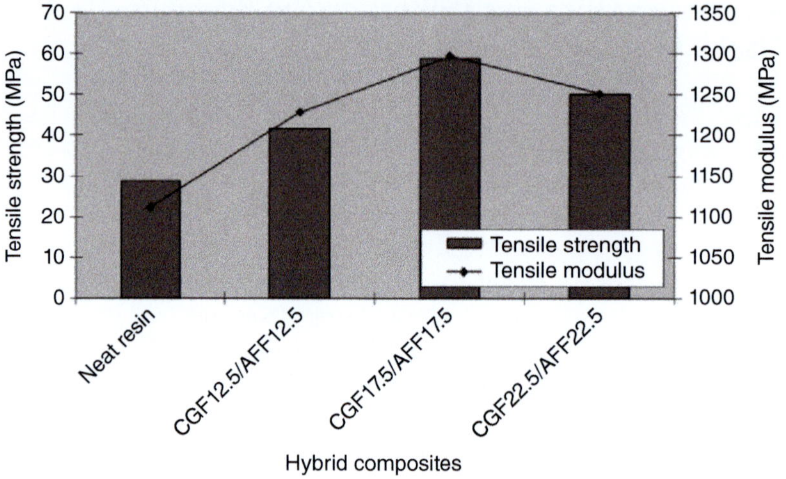


**Figure 11.14** Tensile strength comparison of various composites. Source: From Venkatarajan et al. [27].

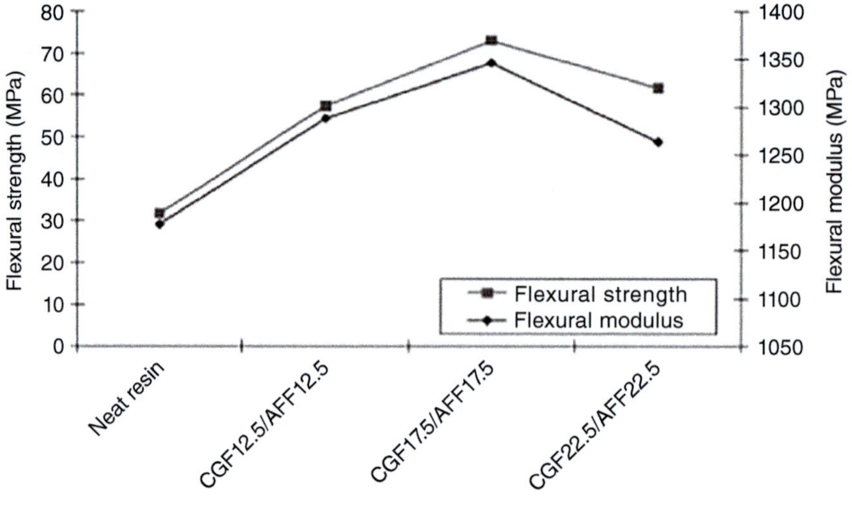


**Figure 11.15** Flexural strength comparison of various composites. Source: From Venkatarajan et al. [27].

This highest achievement among other combinations was due to the strong bonding between the fiber and the matrix and also the minimum voids in the composite specimen.

From Figure 11.15 it is observed that the flexural strength of the 35 wt.% (CGF 17.5 wt.% and the AFF 17.5 wt.%) composite attained a maximum level when compared with other composites. The highest impact strength was also achieved by the 35 wt.% composite and a 23.68% increase in impact strength was noted when compared to the neat resin. The 4.44% and 2.29% increase in impact strength was achieved when compared to 25 and 50 wt.% composites respectively. The 35 wt.% composite had better load-carrying capacity and resisted the applied load effectively.

## 11.18 Influence of Tamarind Seed Fillers on the Flax Fiber–Liquid Thermoplastic Composite

Selvaraj et al. [28] carried out an investigation on the effects of the addition of tamarind seed filler on the properties of flax fiber / liquid thermoplastic resin composite. The outcome of the study clearly showed that the added additive enhanced the mechanical properties of the composites. The tensile strength of the composite specimen increased by 6% when compared with the unfilled filler composite [28].

## 11.19 Influence of Walnut Shell, Hazelnut Shell, and Sunflower Husk Fillers on the Epoxy Composites

Barczewski et al. [29] reported the properties of epoxy resin (Epidian 624) with the addition of walnut shell (WS), hazelnut shell (HS), and sunflower husk (SH) fillers. Mechanical properties of the composites with different fillers and their content variation were studied.

The mechanical properties considering the surface area, particle size, and aspect ratio of fillers were tested and evaluated for 15–35 wt.%. From Figure 11.16 it was observed that the impact strength of HS composite proves to be better among other composite specimens. The composite specimen with HS fillers increased

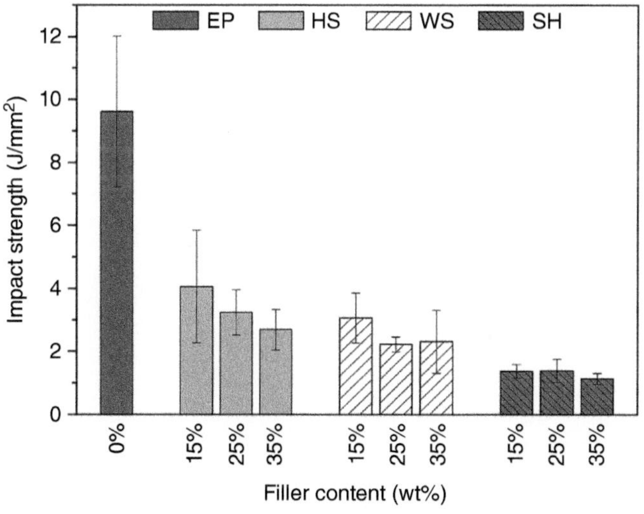


**Figure 11.16** Impact strength comparison of various composites. Source: From Barczewski et al. [29].

the stiffness by 20% and the hardness by 8%. The HS fillers improved the overall mechanical property because of the exothermic cross-linking with the matrix [29].

## 11.20 Influence of Waste Vegetable Peel Fillers on the Epoxy Composite

Patil et al. [30] reported new materials for light-duty applications. In this investigation, biocomposite materials (waste vegetable peels) from the outermost peels of onion, potato, and carrot were used as fillers with epoxy matrix. The filler weight percentage varied from 0%, 10%, 20%, and 30 vol.% with different fillers and the epoxy resin. With any type of natural fillers, particle size and volume play a major role in the mechanical properties. The lower value of tensile properties proves the above clearly. In Table 11.5, a comparison of the results of various combinations of composite specimens was tabulated. The results show that 20 and 30 vol.% filler-reinforced composites have a lower value than the 10% filler-reinforced composites.

It is observed from Table 11.5 that the onion/epoxy matrix composites showed improved results than the rest of the composites. It is attributed to the fact that the onion micro-fillers made a good bonding and correlation with the resin [30].

## 11.21 Influence of *Clusia multiflora* Saw Dust Fillers on the Rubber Composite

Delgado et al. 2020 [31] evaluated mineral fillers (*Clusia multiflora* sawdust (CMS)) as reinforcement in rubber composite (synthetic styrene butadiene rubber (SBR)).

**Table 11.5** Comparison of results of various composite specimens.

| Configurations | Composite (vol.%) | Tensile strength (MPa) | Flexural strength (MPa) | Hardness (Rockwell B hardness) |
|---|---|---|---|---|
| Onion/epoxy | 10 | 20.8 | 36.06 | 50.75 |
| | 20 | 19.13 | 28.85 | 41.54 |
| | 30 | 16.22 | 28.85 | 36.32 |
| Potato/epoxy | 10 | 20.75 | 35 | 49.75 |
| | 20 | 16.79 | 26.25 | 39.45 |
| | 30 | 9.64 | 18.85 | 32.33 |
| Carrot/epoxy | 10 | 19.33 | 34.55 | 37.75 |
| | 20 | 10.62 | 19 | 30.50 |
| | 30 | 10.09 | 17.44 | 26.74 |
| Epoxy resin | 100 | 85 | 112 | 81.50 |

Source: Based on Patil et al. [30].

From the investigation, it was identified that the mechanical properties of SBR reduced due to the addition of CMS. Addition of wood dust into the neat polymer had a negative effect because of the restriction of molecular mobility of the polymer chain. The cross-linked structure developed by the striking force between the filler and the polymer molecules limits the mobility of the polymer chain [31].

## 11.22 Influence of Wood Flour Fillers on the Red Banana Peduncle Fiber Polyester Composite

Manimaran et al. [32] reported the mechanical properties of red banana peduncle/wood flour/polyester matrix composites. In this investigation, the samples are prepared with red banana peduncle fiber/polyester composite (RBPF) and red banana peduncle wood fiber/polyester composite (RBWF) with 10 wt.% of wood flour added as filler. Figure 11.17 shows that the ultimate tensile strength achieved by the RBPF was 9.14 MPa and that achieved by RBWF was 11.61 MPa. Here the filler addition confirms the strength increment in the composite. The flexural and the impact strength also increase in performance, similar to the tensile strength [32].

## 11.23 Influence of Wood Dust Fillers (Rosewood and Padauk) on the Jute Fiber–Epoxy Composite

Dinesh et al. 2020 [33] studied the effect of wood dust (rosewood and padauk) as a filler on the mechanical properties of jute fiber/epoxy composites. This investigation analyzed the tensile (Figure 11.18), compression, flexural, and impact properties of the composites prepared as per the ASTM standards. The specimens were prepared in four different types such as W1 – rosewood/jute/epoxy, W2 – padauk/jute/epoxy, W3 – rosewood + padauk/jute/epoxy, and W4 – jute/epoxy.

The results clearly state that the tensile strength of padauk wood filler-based jute epoxy composite (W2) reached the maximum strength (43 MPa) when compared with all other composites. The findings of the mechanical property revealed that the W2 configuration showed higher strengths in flexural strength, CS, shore D

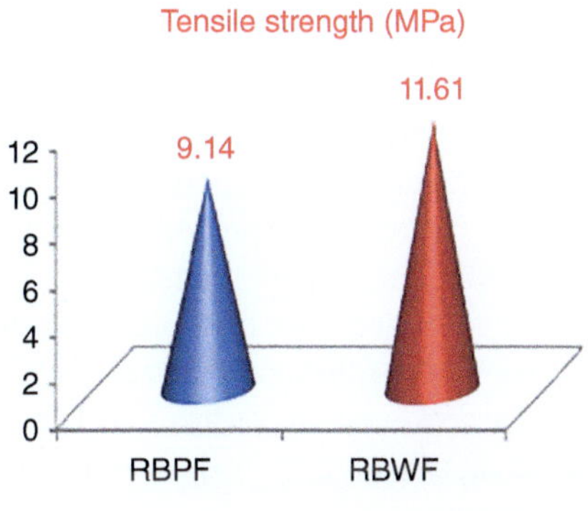


**Figure 11.17** Mean tensile strength comparison of RBPF and RBWF composites. Source: From Manimaran et al. [32].

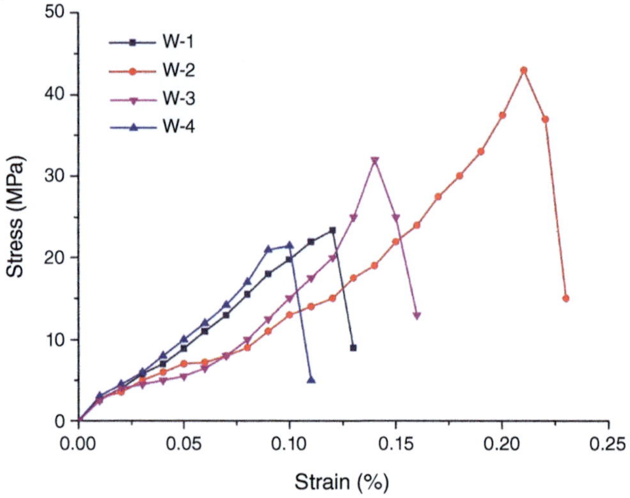


**Figure 11.18** Tensile strength stress–strain graph of prepared composites. Source: From Dinesh et al. [33].

hardness, and impact strength. The higher results of the padauk wood filler-based jute epoxy composite (W2) was due to the better adhesion between the filler and the matrix [33].

## 11.24 Summary

The above study shows that filler addition helped enhance the mechanical properties of their respective composites. From Table 11.6 it is observed that the hybrid composites made by natural fibers and natural fillers with some unique combination have got some enhanced properties compared with other combinations. As per the reviewed papers, the composites toughened by fillers showed significant improvements in the mechanical properties. Hybrid composites with fillers exhibited improved properties than the single fiber-reinforced composites.

## 11.25 Conclusions

This review concludes the following:

i. The addition of fillers is the major reason for the improvement of the physical and mechanical properties of composites.
ii. The treated fillers helped to improve the mechanical properties when compared with the untreated fillers.
iii. The unfilled filler matrix composites showed lower mechanical properties than the fillers-filled composites.
iv. This review also confirms that finding the optimum loading percentage of filler is significant as it leads to achieving the best properties in the fiber matrix composites.

**Table 11.6** Summary of the effect of nano-fillers on the physical and mechanical properties.

| Matrix | Fiber | Filler | Loading percentage | Improvement | Reason for improvement | References |
|---|---|---|---|---|---|---|
| Epoxy | Flax/pineapple | Cellulose | 30 and 35 wt.% | The inclusion of 3% CMF with 35 wt.% of PF improved the overall mechanical properties | 35 wt.% composite specimen had good fiber–matrix adhesion properties | [12] |
| Cardanol Polymer | — | Sugarcane bagasse | 0 (neat), 5, 10, 15, and 20 wt.% | Increasing trend has occurred in mechanical property up to 15 wt.% composite | At high weight percentage of reinforcement, it was very tough for polymers to entirely impregnate the fibers. So it leads to poor interfacial bonding | [13] |
| Natural rubber | — | Sugarcane bagasse | 0% (Neat), 5 phr, 10 phr, 15 phr, and 20 phr | When adding sugarcane bagasse in the composite specimen, the tensile strength of both (NR/$SCB_U$ and NR/$SCB_T$) specimens get reduced | This was because of less filler adhesiveness to the matrix | [14] |
| Geopolymer | Wood fiber | Fly ash | 5–35 wt.% with an increment of 5 wt.% | 100% fly ash composites showed a better compressive strength | Good particle size distribution makes the 100% fly ash better in compressive strength | [15] |
| Polyester | Jute fiber | Egg shell powder/nanoclay | (NJM/UJM)<br>(NJM/UJM)/NC<br>(NJM/UJM)//ESP<br>(NJM/UJM)/ESP/NC | Flexural strength in the hybrid filler NJM composite showed a considerable increment when compared with the UJM composites | The filler composition resists crack formation and propagation during the bending load | [16] |

*(continued)*

**Table 11.6** (Continued)

| Matrix | Fiber | Filler | Loading percentage | Improvement | Reason for improvement | References |
|---|---|---|---|---|---|---|
| Epoxy | Jute fiber | *Portunus sanguinolentus* shell powder | CF-1<br>CF-2<br>CF-3 | Chemically treated 10 wt.% *Portunus sanguinolentus* sheel filler (CF-3) improves the overall mechanical property | The filler shell powders are very rich in calcium, which makes a strong interfacial bond and reduces easy pullout of the fiber | [17] |
| Polyester | *Phaseolus vulgaris* fiber | Fly ash waste (nano-$SiO_2$) filler | 2, 4, 6, and 10 wt.% | The composite prepared with 6 wt.% of *Phaseolus vulgaris* fiber and 1% $SiO_2$ as filler delivers better properties | The thermoplastic behavior of the adder filler improves the properties | [18] |
| Epoxy | Banana fiber | $Al(OH)_3$ fillers | 0, 2.5, 5, and 7.5 wt.% | The tensile property increases significantly up to 2.5% | Because of voids, further addition of fillers reduces the tensile property | [19] |
| Araldite epoxy LY536 | Hemp –kevlar | Palm and coconut shell filler | 5%, 7.5% and 10 wt.% | The addition of fillers improves the tensile strength and ductility of the composites | Flexural strength improves moderately due to filler stiffness | [20] |
| Polyester | — | Coir powder filler | 0, 1, 2, 3, 4 wt.% | The filler content and the tensile strength are in direct proportion for all the test specimens | It showed better adhesion of fillers with the polyester matrix | [21] |
| Epoxy | Coir and luffa fiber | $CaCO_3$ filler | SP1 – plain composite, SP2 – 2 g filled and SP3 – 4 g filled | Addition of filler changes the composite sample from ductile to brittle and increases the flexural properties | Highest filler content achieves the highest flexural strength | [22] |

| | | | | | | |
|---|---|---|---|---|---|---|
| Epoxy | — | Pineapple leaf, napier, and hemp fiber fillers | 5%, 7.5% and 10 wt.% | The Napier fiber particle-filled epoxy composite specimens have a good peak load capacity | Robust molecular movement of the molecules in the filler and the resin | [23] |
| Natural rubber | — | Wheat straw filler | 10 wt.% | The fillers modified by NaOH increase the elastic modulus of the composite specimens | The filler interaction makes the composite more strain dependent | [24] |
| Polyester | *Luffa cylindrica* fiber | Groundnut shell, rice husk, and wood powder fillers | 3%, 7% and 11 wt.% | 7% groundnut shell filler content achieves the highest strength | Particle size in groundnut shell fiber and high dispersion make the composite to get better result | [25] |
| Epoxy | *Bauhinia vahlii* – sisal fiber | Rice-husk filler | 0, 2, 4 and 6 wt.% | The highest tensile strength was observed in 6 wt.% of rise husk filler | The better adhesion capacity of fiber – filler and matrix | [26] |
| Phenol formaldehyde | *Calotropis gigantea* fiber | Areca fine fiber filers | 25, 35 and 45 wt.% | Flexural strength and the impact strength proved the 35 wt.% (CG 17.5 wt.% and the AF 17.5 wt.%) composite obtained a maximum level | The strong bonding between the fiber and the matrix and also the minimum voids in the composite specimen | [27] |
| Liquid thermoplastic resin | Flax fiber | Tamarind seed filler | 5–15% of the weight | The added additive enhanced the mechanical properties of the composites | The added additive has more influence on the composite | [28] |
| Epoxy resin (Epidian624) | — | Walnut shell, hazelnut shell (HS) and sunflower husk fillers | 15–35 wt.% | HS fillers increased the stiffness by 20% and the hardness by 8% | The exothermic cross-linking of HS fillers with the matrix | [29] |

*(continued)*

**Table 11.6** (Continued)

| Matrix | Fiber | Filler | Loading percentage | Improvement | Reason for improvement | References |
|---|---|---|---|---|---|---|
| Epoxy | — | Bio-composite fillers (peels of onion, potato and carrot) | 0%, 10%, 20% and 30 wt.% | Higher results in all aspects are achieved by the onion/epoxy composite | The onion peel easily blends with all types of polymer matrix | [30] |
| Synthetic styrene butadiene rubber | — | Mineral fillers (Clusia multiflora sawdust) | Different varieties | Addition of wood dust into the neat polymer had a negative effect | The cross-linked structure developed by the filler and polymer molecules limits the mobility of the polymer chain | [31] |
| Polyester | Banana peduncle fiber | Wood flour fillers | 10 wt.% | The filler addition confirms the strength increment in the composite | The better adhesion capacity of fiber – filler and matrix | [32] |
| Epoxy | Jute fiber | Wood dust fillers (rosewood and padauk) | W1 – Rose wood/jute/epoxy<br>W2 – Padauk/jute/epoxy<br>W3 – Rose wood + Padauk/jute/epoxy<br>W4 – Jute/epoxy | Higher results were achieved for padauk wood filler-based jute–epoxy composite (W2) | The thermoplastic behavior improves the properties | [33] |

## References


1 Saravanakumaar, A., Senthilkumar, A., Sornakumar, T. et al. (2019). Physico-chemical properties of new cellulosic fiber extracted from *Carica papaya* bark. *J. Nat. Fibers* 16 https://doi.org/10.1080/15440478.2017.1410514.

2 Karabulut, N., Aktaş, M., and Balcíoğlu, H.E. (2018). Surface modification effects on the mechanical properties of woven jute fabric reinforced laminated composites. *J. Nat. Fiber* https://doi.org/10.1080/15440478.2018.1431995.

3 Chavhan, G.R. and Wankhade, L.N. (2019). Improvement of the mechanical properties of hybrid composites prepared by fibers, fiber-metals, and nano-filler particles – a review. *Mater. Today: Proc.* https://doi.org/10.1016/j.matpr.2019.08.240.

4 Devnani, G.L. and Sinha, S. (2019). Effect of nanofillers on the properties of natural fiber reinforced polymer composites. *Mater. Today: Proc.* 18: 647–654. https://doi.org/10.1016/j.matpr.2019.06.460.

5 Singh, H. and Singh, T. (2019). Effect of fillers of various sizes on mechanical characterization of natural fiber polymer hybrid composites: a review. *Mater. Today: Proc.* 18: 5345–5350. https://doi.org/10.1016/j.matpr.2019.07.560.

6 Reddy, B.M., Mohana Reddy, Y.V., Mohan Reddy, B.C., and Reddy, R.M. (2018). Mechanical, morphological, and thermogravimetric analysis of alkali-treated Cordia-Dichotoma natural fiber composites. *J. Nat. Fibers* 17 (5): 759–768. https://doi.org/10.1080/15440478.2018.1534183.

7 Dong, C. (2019). Mechanical properties of natural fibre-reinforced hybrid composites. *J. Reinf. Plast. Compos.* 38: 910–922. https://doi.org/10.1177/0731684419856686.

8 Saravanakumaar, A., Senthilkumar, A., Saravanakumar, S.S. et al. (2018). Impact of alkali treatment on physico-chemical, thermal, structural and tensile properties of *Carica papaya* bark fibers. *Int. J. Polym. Anal. Charact.* 23: 529–536. https://doi.org/10.1080/1023666X.2018.1501931.

9 Saravanakumaar, A., Senthilkumar, A., Saravanakumar, S.S. et al. (2020). Mechanical properties of alkali-treated carica papaya fiber-reinforced epoxy composites. *J. Nat. Fibers*: 1–11. https://doi.org/10.1080/15440478.2020.1739590.

10 Shanmugasundaram, N., Rajendran, I., and Ramkumar, T. (2018). Static, dynamic mechanical and thermal properties of untreated and alkali treated mulberry fiber reinforced polyester composites. *Polym. Compos.* 39: E1908–E1919. https://doi.org/10.1002/pc.24890.

11 Adesina, O.T., Jamiru, T., Sadiku, E.R. et al. (2019). Mechanical evaluation of hybrid natural fibre-reinforced polymeric composites for automotive bumper beam: a review. *Int. J. Adv. Manuf. Technol.* 103: 1781–1797. https://doi.org/10.1007/s00170-019-03638-w.

12 Sumesh, K.R., Kanthavel, K., and Kavimani, V. (2020). Machinability of hybrid natural fiber reinforced composites with cellulose micro filler incorporation. *J. Compos. Mater.* https://doi.org/10.1177/0021998320918020.

13 Balaji, A., Karthikeyan, B., Swaminathan, J., and Sundar Raj, C. (2019). Effect of filler content of chemically treated short bagasse fiber-reinforced cardanol

polymer composites. *J. Nat. Fibers* 16: 613–627. https://doi.org/10.1080/15440478.2018.1431829.

**14** De Paiva, G., Friol, F., Peixoto, V. et al. (2018). Sugarcane bagasse fiber as semi-reinforcement filler in natural rubber composite sandals. *J. Mater. Cycles Waste Manag.* https://doi.org/10.1007/s10163-018-0801-y.

**15** Furtos, G., Silaghi-Dumitrescu, L., Pascuta, P. et al. (2019). Mechanical properties of wood fiber reinforced geopolymer composites with sand addition. *J. Nat. Fibers*: 1–12. https://doi.org/10.1080/15440478.2019.1621792.

**16** Ganesan, K., Kailasanathan, C., Sanjay, M.R. et al. (2018). A new assessment on mechanical properties of jute fiber mat with egg shell powder/nanoclay-reinforced polyester matrix composites. *J. Nat. Fibers* 17: 482–490. https://doi.org/10.1080/15440478.2018.1500340.

**17** Kumaran, P., Mohanamurugan, S., Madhu, S. et al. (2019). Suchart Siengchin, Investigation on thermo-mechanical characteristics of treated/untreated Portunus sanguinolentus shell powder-based jute fabrics reinforced epoxy composites. *J. Ind. Text.*: 1–12. https://doi.org/10.1177/1528083719832851.

**18** Gurukarthik Babu, B., Prince Winston, D., Aravind Bhaskar, P.V. et al. (2020). Exploration of electrical, thermal, and mechanical properties of phaseolus vulgaris fiber/unsaturated polyester resin composite filled with nano-$SiO_2$. *J. Nat. Fibers*: 1–17. https://doi.org/10.1080/15440478.2020.1724231.

**19** Shivamurthy, B., Thimmappa, B.H.S., and Monteiro, J. (2020). Sliding wear, mechanical, flammability, and water intake properties of banana short fiber/$Al(OH)_3$/epoxy composites. *J. Nat. Fibers* 17: 337–345. https://doi.org/10.1080/15440478.2018.1492489.

**20** Jani, S.P., Kumar, A.S., Khan, M.A. et al. (2019). Influence of natural filler on mechanical properties of hemp/kevlar hybrid green composite and analysis of change in material behavior using acoustic emission. *J. Nat. Fibers*: 1–12. https://doi.org/10.1080/15440478.2019.1692321.

**21** Karthik Babu, N.B., Muthukumaran, S., Arokiasamy, S., and Ramesh, T. (2018). Thermal and mechanical behavior of the coir powder filled polyester micro-composites. *J. Nat. Fibers*: 1–11. https://doi.org/10.1080/15440478.2018.1555503.

**22** Mohana Krishnudu, D., Sreeramulu, D., Reddy, P.V., and Rajendra Prasad, P. (2020). Influence of filler on mechanical and di-electric properties of coir and luffa cylindrica fiber reinforced epoxy hybrid composites. *J. Nat. Fibers*: 1–10. https://doi.org/10.1080/15440478.2020.1745115.

**23** Ridzuan, M.J.M., Abdul Majid, M.S., Khasri, A. et al. (2019). Effect of pineapple leaf (PALF), napier, and hemp fibres as filler on the scratch resistance of epoxy composites. *J. Mater. Res. Technol.* 8: 5384–5395. https://doi.org/10.1016/j.jmrt.2019.09.005.

**24** Rybiński, P., Syrek, B., Masłowski, M. et al. (2018). Influence of lignocellulose fillers on properties natural rubber composites. *J. Polym. Environ.* 26: 2489–2501. https://doi.org/10.1007/s10924-017-1144-9.

**25** Dhanola, A., Bisht, A.S., Kumar, A., and Kumar, A. (2018). Influence of natural fillers on physico-mechanical properties of luffa cylindrica/ polyester composites. *Mater. Today: Proc.* 5: 17021–17029. https://doi.org/10.1016/j.matpr.2018.04.107.

**26** Kumar, S., Mer, K.K.S., Gangil, B., and Patel, V.K. (2019). Synergy of rice-husk filler on physico-mechanical and tribological properties of hybrid Bauhinia-vahlii/sisal fiber reinforced epoxy composites. *J. Mater. Res. Technol.* 8: 2070–2082. https://doi.org/10.1016/j.jmrt.2018.12.021.

**27** Venkatarajan, S., Bhuvaneswari, B.V., Athijayamani, A., and Sekar, S. (2019). Effect of addition of areca fine fibers on the mechanical properties of *Calotropis Gigantea* fiber/phenol formaldehyde biocomposites. *Vacuum* 166: 6–10. https://doi.org/10.1016/j.vacuum.2019.04.022.

**28** Selvaraj, D.K., Silva, F.J.G., Campilho, R.D.S.G. et al. (2019). Influence of the natural additive on natural fiber reinforced thermoplastic composite. *Procedia Manuf.* 38: 1121–1129. https://doi.org/10.1016/j.promfg.2020.01.200.

**29** Barczewski, M., Sałasińska, K., and Szulc, J. (2019). Application of sunflower husk, hazelnut shell and walnut shell as waste agricultural fillers for epoxy-based composites: a study into mechanical behavior related to structural and rheological properties. *Polym. Test.* 75: 1–11. https://doi.org/10.1016/j.polymertesting.2019.01.017.

**30** Patil, A.Y., Banapurmath, N.R., Yaradoddi, J.S. et al. (2019). Experimental and simulation studies on waste vegetable peels as bio-composite fillers for light duty applications. *Arab. J. Sci. Eng.* 44: 7895–7907. https://doi.org/10.1007/s13369-019-03951-2.

**31** Delgado, E., Espitia, A., and Aperador, W. (2020). Comparative evaluation of *Clusia multiflora* wood flour, against mineral fillers, as reinforcement in SBR rubber composites. *Iran. Polym. J.* 29: 13–23. https://doi.org/10.1007/s13726-019-00768-6.

**32** Manimaran, P., Jeyasekaran, A.S., Purohit, R., and Pillai, G.P. (2018). https://doi.org/10.1080/15440478.2018.1558148). An experimental and numerical investigation on the mechanical properties of addition of wood flour fillers in red banana peduncle fiber reinforced polyester composites. *J. Nat. Fibers.*

**33** Dinesh, S., Kumaran, P., Mohanamurugan, S. et al. (2020). Influence of wood dust fillers on the mechanical, thermal, water absorption and biodegradation characteristics of jute fiber epoxy composites. *J. Polym. Res.* 27 (1): 9.

# 12

# Temperature-Dependent Dynamic Mechanical Properties and Static Mechanical Properties of *Sansevieria cylindrica* Reinforced Biochar-Tailored Vinyl Ester Composite

*Rajendran Deepak Joel Johnson[1], Veerasimman Arumugaprabu[2], Rajini Nagarajan[2], Fernando G. Souza[3], and Vigneswaran Shanmugam[1,2]*

[1] *Saveetha School of Engineering, Saveetha Institute of Medical and Technical Sciences, Department of Mechanical Engineering, Chennai, Tamilnadu, India*
[2] *Department of Mechanical Engineering, School of Automotive and Mechanical Engineering, Kalasalingam Academy of Research and Education, Srivilliputhur 626128, Tamil Nadu, India*
[3] *IMA/UFRJ, Laboratório de Biopolímeros e Sensores/LaBioS Centro de Tecnologia – Cidade Universitária, Rio de Janeiro, 21941-909, Brazil*

## 12.1 Introduction

In contemporary times it has become necessary for the researchers to craft composite materials that are highly effective and environmentally friendly, which can be utilized in almost every industry such as aerospace, automotive, biomedicine, and navy. In India, the annual maize production is up to 24.17 million tonnes and is expected to increase in the coming years [1]. Owing to the high production rate of maize (*Zea mays* L.), *Z. mays* residues are abundant in India, and these abundant *Z. mays* cobs are burned for disposal, which causes severe hazards to the environment, such as air pollution and soil degradation. Alternative methods must be instituted to protect the environment [2]. Maize residues can be used as substratum material for biochar preparation through the slow pyrolysis process [3, 4]. Biochar is a carbon-rich material derived from the natural waste by a slow pyrolysis process with limited presence of oxygen [5–8]. Biochar produced from maize cob has the highest carbon content of 72.58–86.92% compared to the other residues of *Z. mays*. Also, the carbon content can be increased using a higher pyrolysis temperature. The optimal temperature for producing biochar from the *Zea maize* cob is 600–700 °C [9]. Polymer strengthening was done using natural fillers due to their indicative mechanical properties and cost-effective aspects. Also, bio-fillers constitute an increased application due to their accessibility, viability, and eco-friendly nature [10–12]. Dynamic properties of the polymeric composite are notably significant while establishing the temperature range and frequency of the material and are essential in determining the properties of the material [13]. *Arundo donax* was used as a natural filler to reinforce the epoxy resin. The effects of filler content and size on the static and dynamic mechanical properties

*Mechanical and Dynamic Properties of Biocomposites,* First Edition. Edited by Senthilkumar Krishnasamy, Rajini Nagarajan, Senthil Muthu Kumar Thiagamani, and Suchart Siengchin.

were studied and a high modulus and low strength values compared to the neat resin were inferred [14]. Static and dynamic mechanical properties for untreated, 5%, and 10% alkali-treated *Pennisetum purpureum*/glass fiber epoxy composite were evaluated. It was found that maximum mechanical properties and the glass transition temperature were attained for 5% alkaline-treated *P. purpureum*/glass fiber epoxy composite [15]. In polymer composite, filler content and the type of filler play a vital role in the mechanical as well as viscoelastic properties [16–20]. Filler content should be optimized because increased filler content, to a certain extent, causes failure of material due to agglomeration and void formation [21]. A perfect alternative to utilize the maize residue is to produce biochar from the *Z. mays* cob (maize residue) and to use this biochar as filler material. To check the potentiality of biochar to enhance the mechanical property, commercial accessibility, and technical feasibility of the composite, investigating its static and dynamic mechanical properties is necessary. This research focuses on preparing the biochar and utilizing it as filler for the fabrication of biochar-filled *Sansevieria cylindrica*-reinforced vinyl ester composite (SCVEC). The tensile property, flexural property, impact strength, and dynamic mechanical properties were evaluated to showcase the effectiveness of biochar as potential low-cost filler material in composite fabrication.

## 12.2 Materials and Method

### 12.2.1 Materials

Vinyl ester resin is used as the matrix material, and purchased from Vasavibala Pvt. Ltd., Chennai, Tamilnadu, India. VBR 4508 is the trade name of vinyl ester resin that is used in our research work, which has a viscosity and specific gravity of 300 cps and 1.04 measured at 25 °C room temperature. *Sansevieria cylindrica* fiber (SCF) of an optimum length of 40 mm (fiber length), which is treated with a 3% NaOH solution for two hours, was used as the reinforcing material in this research work. SCF, which is a naturally available material, was purchased from a private vendor in Coimbatore, India. Biochar was produced from the maize cob or corn cob (*Z. mays* cob) using a slow pyrolysis process at 800 °C in the absence of oxygen. *Z. mays* cob was collected from the nearby villages of Virudhunagar District, Tamilnadu, India.

Biochar produced from the pyrolysis process was ball milled at 200 rpm for four hours. Figure 12.1 shows the ball-milled biochar particle. Horiba SZ-100 particle analyzer was used to measure the biochar particle size distribution (Figure 12.2). The average particle size of the ball-milled *Z. maize* cob biochar particle was measured to be in the range of 413 nm. Scanning electron microscopy combined with energy dispersive analysis of X-rays (SEM-EDAX) of the *Z. mays* cob biochar was taken to evaluate the chemical composition of the biochar, which is shown in Figure 12.3. SEM image of the ball-milled biochar is shown in Figure 12.4.

**Figure 12.1** Ball-milled biochar particle.

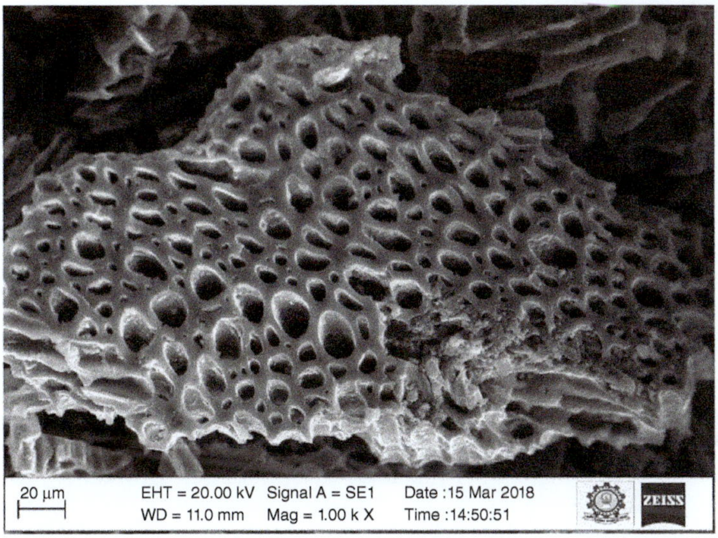


**Figure 12.2** SEM micrograph of biochar particle.

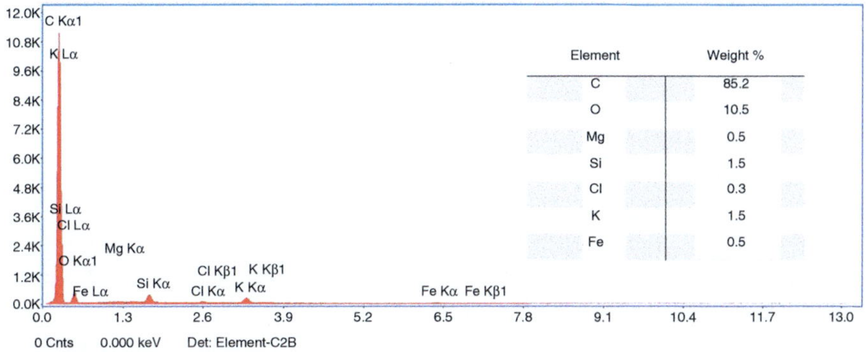


| Element | Weight % |
|---|---|
| C | 85.2 |
| O | 10.5 |
| Mg | 0.5 |
| Si | 1.5 |
| Cl | 0.3 |
| K | 1.5 |
| Fe | 0.5 |

**Figure 12.3** SEM-EDAX of *Zea mays* cob biochar.

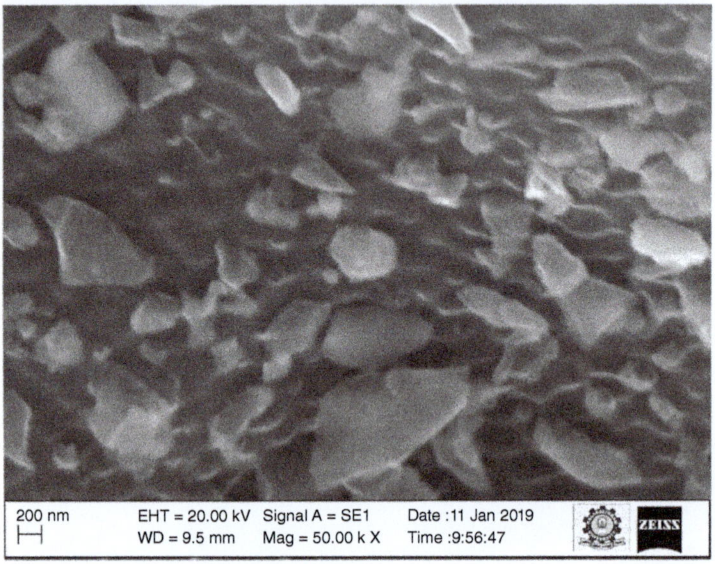


**Figure 12.4** SEM image of biochar after ball milling.

## 12.2.2 Biochar Characterization

### 12.2.2.1 Particle Size Analyzer

Horiba SZ-100 particle analyzer with standard distribution was used to determine the *Z. mays* cob particle size distribution. The particle size of the biochar will be represented by scattered light intensity. The holder temperature was maintained at 25 °C with dispersion medium viscosity as 0.896 mPa s. The scattering angle for biochar particle was measured as 173°. The particle size distribution was measured using the angle of light scattering; higher the angle of light scattering, smaller the particle size.

### 12.2.2.2 X-ray Diffraction

X-ray diffraction patterns of the biochar particle prepared from *Z. mays* cob were recorded using Bruker D8 Advance ECO X-ray diffractometer using coupled two theta/theta scan type. Room temperature (25 °C) was maintained while recording the X-ray diffraction pattern with a step size of 0.02, and Cu anode was used for effective diffraction patterning of powdered samples.

### 12.2.2.3 FTIR Spectroscopy

Fourier Transform Infrared Spectroscopy (FTIR) was used to characterize the *Z. mays* cob biochar particle using SHIMADZU IR Tracer 100 FTIR spectrophotometer. Infrared light was passed through the *Z. mays* cob biochar powder; the infrared frequency was checked with the vibrational frequency. When these two frequencies become equal, absorption takes place. FTIR spectrometer gives an interferogram, and with Fourier transform, this interferogram was analyzed to obtain the FTIR spectrum. With this spectrum obtained using Fourier transform, the presence of the functional group in the *Z. mays* cob biochar particle was observed.

### 12.2.3 Composite Fabrication

*Zea mays* cob biochar-filled SCVEC was fabricated using a compression molding method. *Sansevieria cylindrica* of fiber length 40 mm treated with 3% NaOH solution was used as reinforcing material.

The rectangular mold of dimensions 300 mm × 130 mm × 3 mm made of steel was used for the fabrication of composite plates. Composites plates were fabricated for different biochar weight percentages of 0, 2, 4, 6, 8, and 10 wt.%, respectively. SICK C 2000 compression molding machine was used to fabricate the composite plate. Biochar-filled SCVEC was allowed to cure under a compression pressure of 200 psi with the temperature maintained at 30 °C (room temperature) for six hours. The composite fabricated with varying biochar weight percentage was designated for testing purposes, as shown in Table 12.1.

### 12.2.4 Dynamic Mechanical Analysis (DMA)

Dynamic mechanical analysis (DMA) of biochar-reinforced SCVEC was performed on a Hitachi DMA 7100 thermal analyzer system. The experiments were performed under three-point bending mode at a frequency of 1 Hz. The experimental tests were conducted from 25 to 220 °C with a heating rate of 5 °C/min maintained under a nitrogen atmosphere to protect the sample from oxidation.

### 12.2.5 Tensile Testing

Testing of the tensile properties of the biochar-filled SCVEC was performed according to ASTM D3039 standard using the universal testing machine. The fabricated

**Table 12.1** Composite designation.

| Fabricated composite description | Designation |
|---|---|
| Unfilled + 40 wt.% *Sansevieria cylindrica* fiber + vinyl ester | 0 wt.% BC |
| 2 wt.% *Zea mays* cob biochar + 40 wt.% *Sansevieria cylindrica* fiber + vinyl ester | 2 wt.% BC |
| 4 wt.% *Zea mays* cob biochar + 40 wt.% *Sansevieria cylindrica* fiber + vinyl ester | 4 wt.% BC |
| 6 wt.% *Zea mays* cob biochar + 40 wt.% *Sansevieria cylindrica* fiber + vinyl ester | 6 wt.% BC |
| 8 wt.% *Zea mays* cob biochar + 40 wt.% *Sansevieria cylindrica* fiber + vinyl ester | 8 wt.% BC |
| 10 wt.% *Zea mays* cob biochar + 40 wt.% *Sansevieria cylindrica* fiber + vinyl ester | 10 wt.% BC |

composites were cut to 200 mm × 20 mm × 3 mm dimensions. Five samples of each composite were taken for the tensile test, and the average values were reported.

### 12.2.6 Flexural Testing

A three-point bending test was performed using a universal testing machine for determining the flexural properties of the biochar-filled SCVEC. ASTM D790 standard was followed, and the sample was cut to a dimension of 125 mm × 13 mm × 3 mm. Five samples of each composite were tested, and the average was taken as the flexural property.

### 12.2.7 Impact Testing

The impact strength of the biochar-filled SCVEC was measured according to ASTM D256 with sample dimensions of 65 mm × 3 mm × 3 mm using Izod Impact Tester. Similar to other mechanical testing, impact tests were also conducted for five samples of each composite, and the average was taken as the impact strength of each composite.

### 12.2.8 Scanning Electron Microscopy

Mechanical-tested biochar-filled SCVEC was investigated for its failure mechanism using Zeiss EVO 18 research SEM. The SEM micrograph of the powdered biochar particle was investigated for its microstructure.

## 12.3 Results and Discussion

### 12.3.1 Biochar Characterization

#### 12.3.1.1 Particle Analyzer

The particle distribution curve for the *Z. mays* cob biochar particle is shown in Figure 12.5. It can be observed that the size distribution of the particles was in the range of 300–450 nm. The mean particle size measured using the particle size analyzer for the prepared ball-milled *Z. mays* cob biochar was found to be 413 nm. Effective mechanical properties can be achieved for biochar having the particle size in nanoscale due to the proper and easy dispersion of biochar particles into the matrix material [22].

#### 12.3.1.2 Fourier Transform (InfraRed) Spectroscopy

FTIR absorbance spectrum for the *Z. mays* cob biochar is shown in Figure 12.6. The peak at 875.68 $cm^{-1}$ represents the aromatic C–H strong atom, which creates out of plane deformation. The peak at 1386.82 $cm^{-1}$ corresponds to the C–H bending vibration of the aldehyde compound class with a methylene group. *Z. mays* cob biochar exhibits distinct peaks at 2883.58 and 2945.3 $cm^{-1}$, which are related to the medium

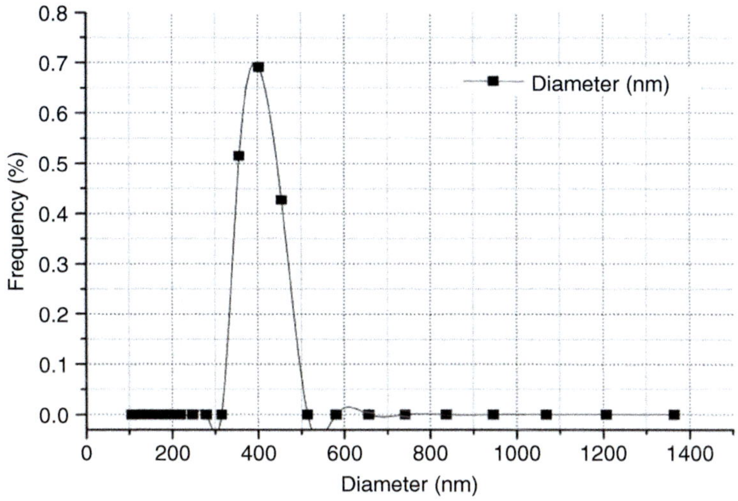


**Figure 12.5** Particle size distribution of *Zea mays* cob biochar.

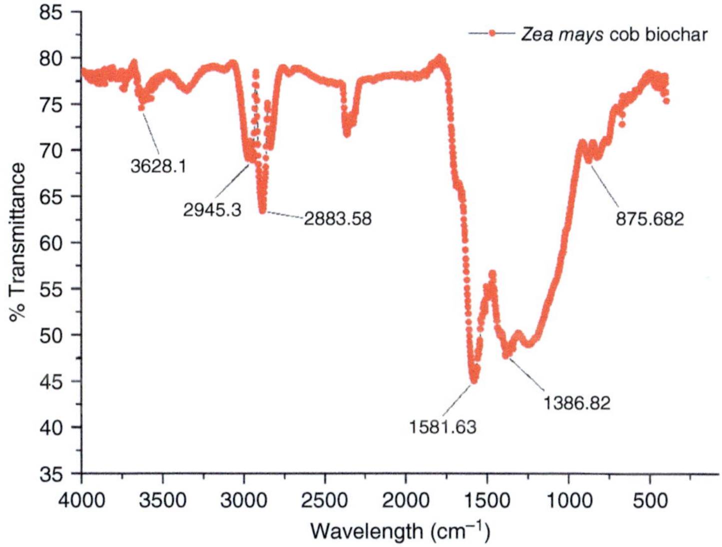


**Figure 12.6** FTIR Spectrum of biochar particle.

C–H stretching vibration and deformation vibration. These peaks confirm the presence of alcohol and phenol structures. Generally, these types of peaks formed by carbonyl groups of the carboxyl group in the *Z. mays* cob biochar make it more compatible with the polymer matrix. The peak at 1581.63 cm$^{-1}$ represents medium N–H bending vibration with amine presence, which may be due to the slow pyrolysis process. Heating till 800 °C converts the carboxyl group to ammonium, and a further rise in temperature transforms ammonium to amides. Mostly agricultural residues show a medium sharp peak that corresponds to 3628.1 cm$^{-1}$, which represents an

O–H stretching alcoholic compound. This stretching occurs due to intramolecular hydrogen bonds between the cellulose chains. A similar display of peaks for various biochars under FTIR spectroscopy was studied by a few researchers [23, 24].

#### 12.3.1.3 X-ray Diffraction

X-ray diffraction analysis of *Z. mays* cob biochar displayed two distinct peaks at $2\theta = 28°$ and $2\theta = 40°$ whose miller indices were (1 1 0) and (1 1 3), which correspond to a layer-to-layer distance, i.e. d-spacing of 3.24 and 2.2487 nm. These d-spacing values of the *Z. mays* cob biochar only slightly vary from the experimental values of graphite of 3.18 and 2.2482 nm, which is evident from Figure 12.7. These slight variations may be due to the presence of a small amount of oxygen holding functional groups formed during the slow pyrolysis process [25]. Figure 12.7 shows the X-ray diffraction pattern of *Z. mays* cob biochar plotted against the experimental X-ray diffraction pattern of graphite (carbon). It is evident that the biochar prepared from the *Z. mays* cob (residues of maize) has a similar peak formation and d-spacing, which matches the peaks formed from graphite, a carbon-rich material. Thus, it was proved from the X-ray powder diffraction (XRD) pattern of the *Z. mays* cob biochar that it can be a better replacement for carbon-rich graphite material, which is costlier anyway. Curve fitting was performed to determine the position of diffraction peak using Fityk 1.3.1 software. The Levenberg–Marquardt algorithm was used to fit the

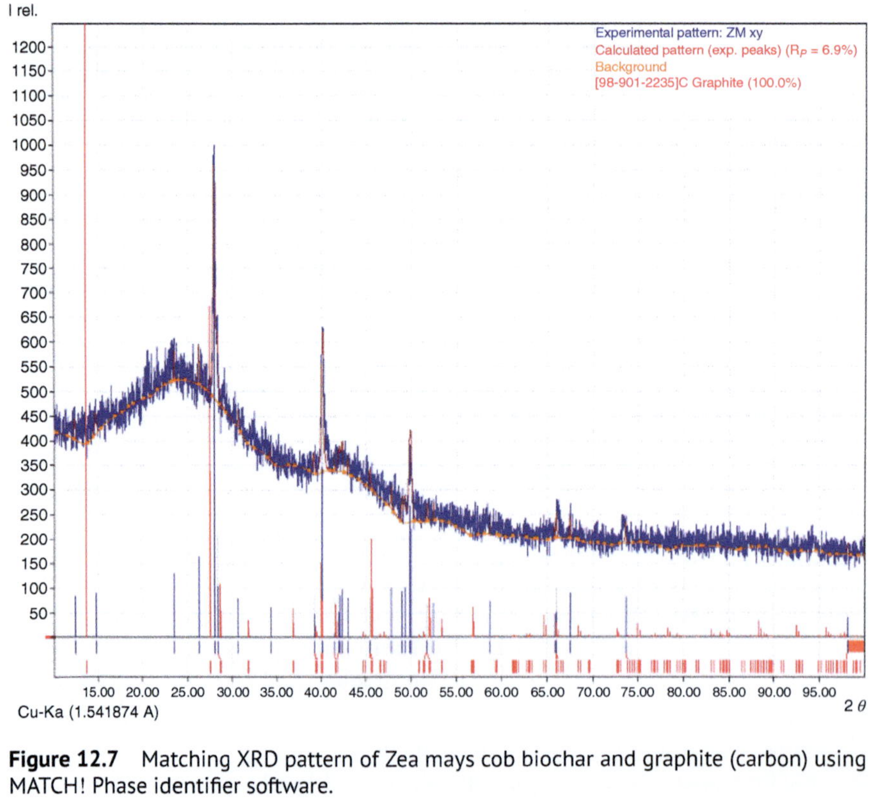


**Figure 12.7** Matching XRD pattern of Zea mays cob biochar and graphite (carbon) using MATCH! Phase identifier software.

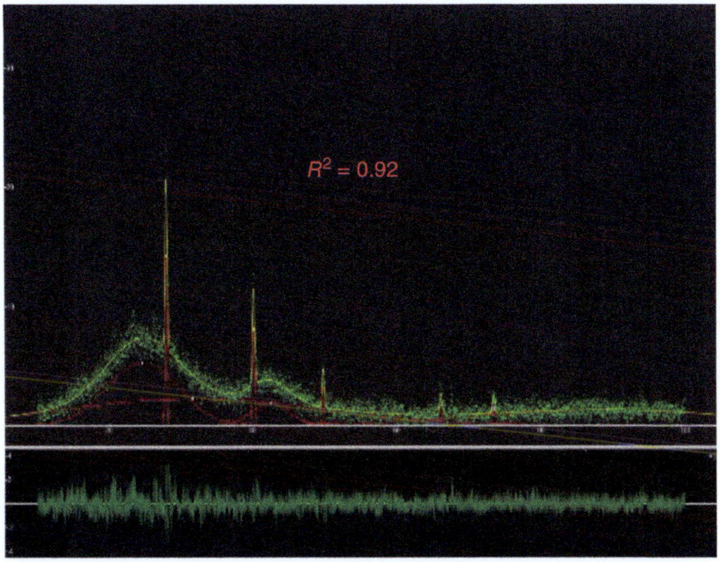


**Figure 12.8** Gaussian fit for the XRD pattern of *Zea mays* cob biochar using Fityk 1.3.1.

XRD spectrum to a Gaussian fit curve, and the goodness of the fit was represented by the residual value. Figure 12.8 shows the Gaussian fit curve for the *Z. mays* cob biochar XRD spectrum with a residual value ($R^2 = 0.92$).

### 12.3.2 Dynamic Mechanical Analysis

The complete characterization of a material's rheology needs elasticity information along with viscosity information. A DMA is a unique technique providing both information about a material. The viscoelastic properties of the biochar-filled SCVEC was examined by DMA, and the dynamic mechanical responses were represented by storage modulus (the ability of the polymer material to store energy elastically), loss modulus (viscous property of the material energy lost due to heat dissipation), and damping factor or tan $\delta$ (damping factor is the ratio between storage modulus and loss modulus). Usually, with the measure of damping factor, a material's ability to absorb energy can be determined. Figure 12.9 shows the variation in the storage modulus with respect to the increase in temperature for different loading of *Z. mays* cob biochar particle.

It was observed from Figure 12.9 that the shape of storage modulus curves for varying biochar loading was mostly similar to the decrease in storage modulus values as the temperature increases from 20 to 200 °C. It is due to the transformation of the material under study from glassy to rubbery state along with the temperature range. Further, it was noted that the storage modulus increases with an increase in biochar loading, which proves the supercilious property of the biochar prepared from *Z. mays* cob. Intermolecular forces will be good in the glassy region since the components present in the material will be closer, tightly packed, and immobile.

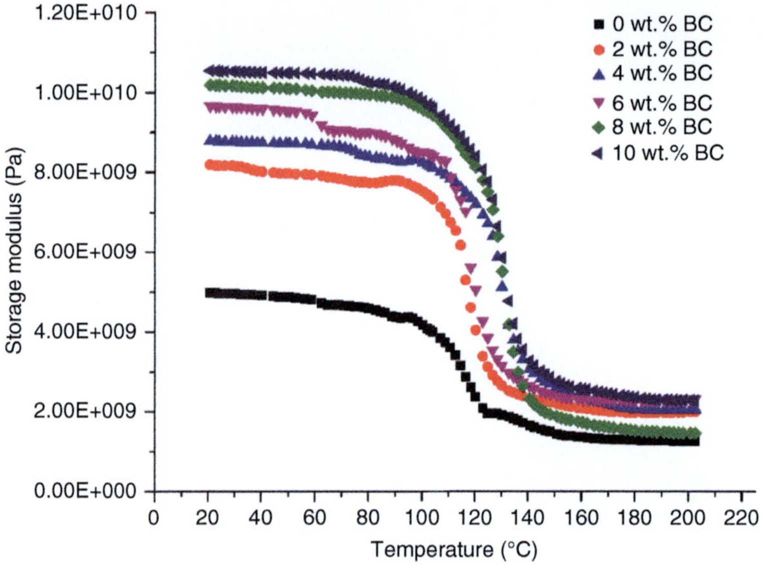


**Figure 12.9** Storage modulus plotted with respect to temperature for varying biochar wt.%. Source: Based on Adak et al. [26].

As the temperature increases, intermolecular forces become weakened and cause the components to move easily, and alteration in the packaging arrangement occurs. This results in the lowering of storage modulus due to decreased stiffness [26]. Figure 12.9 reveals that composite designation of 0 wt.% BC has 4.986 GPa storage modulus measured at 20.62 °C. Also, it is observed that as the biochar wt.% increases, the storage modulus improves. Increase in the storage modulus for biochar-filled SCVEC was noted as 64%, 75%, 93%, 99%, and 101% for 2, 4, 6, 8, and 10 wt.% of BC, respectively. As the temperature increases to 200 °C, a similar trend was observed. From the results obtained, it was inferred that biochar loading to the SCVEC increases the storage modulus as a result of increased stiffness for the entire temperature ranges, i.e. in both glassy and rubbery regions. Such findings from the storage modulus curves of the composite are attributed to the fact that addition of biochar to the SCVEC improves the surface of the composite, resulting in good interfacial adhesion between the reinforcement and the matrix material. These phenomena limit the mobility of the polymer chain and improve the stress transfer at the interface.

Figure 12.10a,b shows the loss modulus and damping factor curves as a function of temperature for different biochar particle loading. Loss modulus represents the viscous part or energy dissipation for the heat in the sample under study. It was observed in all the samples of biochar-filled SCVEC that loss modulus increased and then decreased as the temperature increases. Heat dissipation will be higher at a temperature where the loss modulus is the maximum, which denotes the glass transition temperature of the material [28]. The free movement of the polymer chain increases when a rapid increase in the loss modulus occurs at a

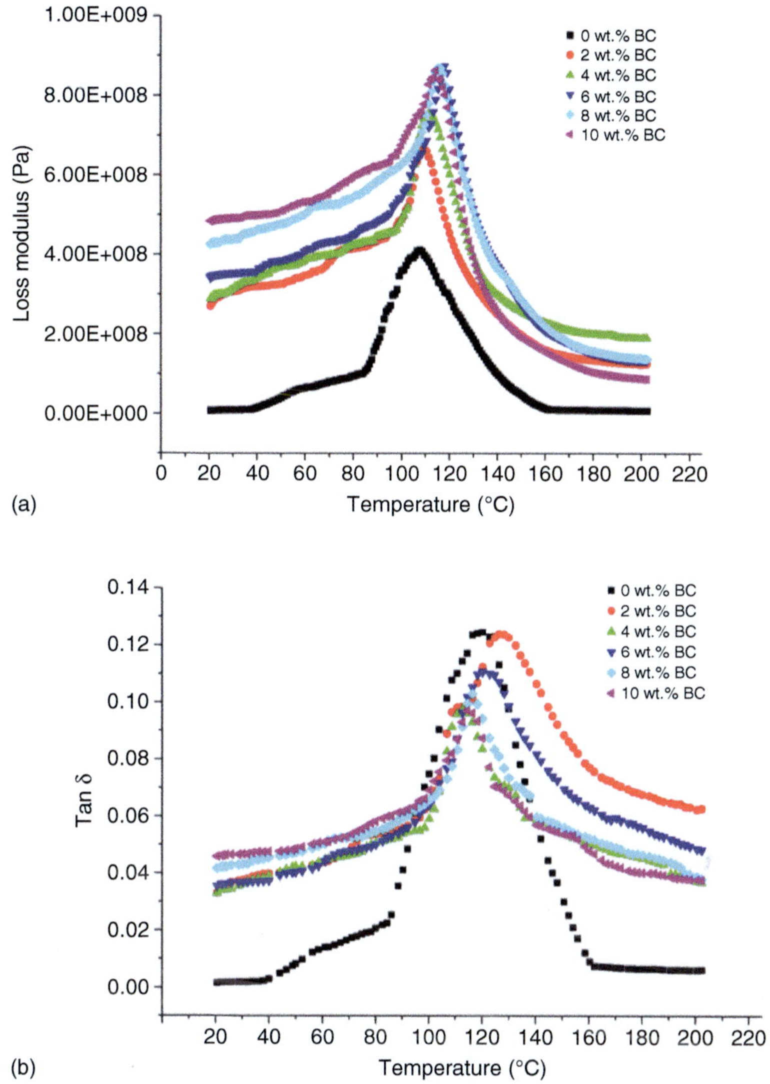


**Figure 12.10** (a) Loss modulus versus temperature for varying biochar wt.% and (b) damping factor versus temperature for varying biochar wt.%. Source: Based on Zeltmann et al. [27].

higher temperature. This enables a more significant amount of motion along the chain, which is impossible below the glass transition temperature [27]. It was also observed in Figure 12.10a that loss modulus increases as the biochar particle loading increases, which denotes a higher amount of heat dissipation linked with internal friction between the molecules. Glass transition temperature can be obtained from the curves of loss modulus as well as the damping factor. The glass transition temperature determined from loss modulus will be more substantial compared to the glass transition temperature found from the damping factor [29].

There is a general perception that the incorporation of fiber in the composite system leads to damping failure. This happens mainly due to shear stress concentration of fiber material at its end region, accompanied by the inclusion of viscoelastic energy indulgence in the matrix material [30]. This difficulty is overcome by introducing proper secondary filler to the composite system to achieve all-around mechanical properties. From Figure 12.10b it can be observed that an increase in biochar particle loading affects the lowering of $\tan\delta$ peak height subsequent to restricted movement of polymer molecules in the composite system. The maximum tan delta peak height of 0.123 was observed for 0 wt.% BC, and the minimum peak height observed was 0.096 for 10 wt.% BC, which is a 28.12% decrease in energy dissipation for increased biochar particle loading.

Figure 12.11 shows the Cole–Cole plot of biochar-filled SCVEC. Cole–Cole plot is considered to be the highly recommended method for evaluating the relation between loss modulus and storage modulus [31]. Figure 12.11 shows the Cole–Cole plot for loss modulus as a function of storage modulus of the composite with varying biochar loading. It is observed from the figure that semicircular arc-like curves were formed. It is known that when the curves are smooth, regular, and parabolic, it represents the polymeric system. It was observed that for 0 wt.% biochar-filled composite the curves were partially smooth projecting the weak interfacial bonding between the matrix and the fiber. However, for biochar-filled composite, the curves were irregular in shape resembling good bonding between the matrix and the fiber material. As the biochar content increases, the irregularity in the curve increases,

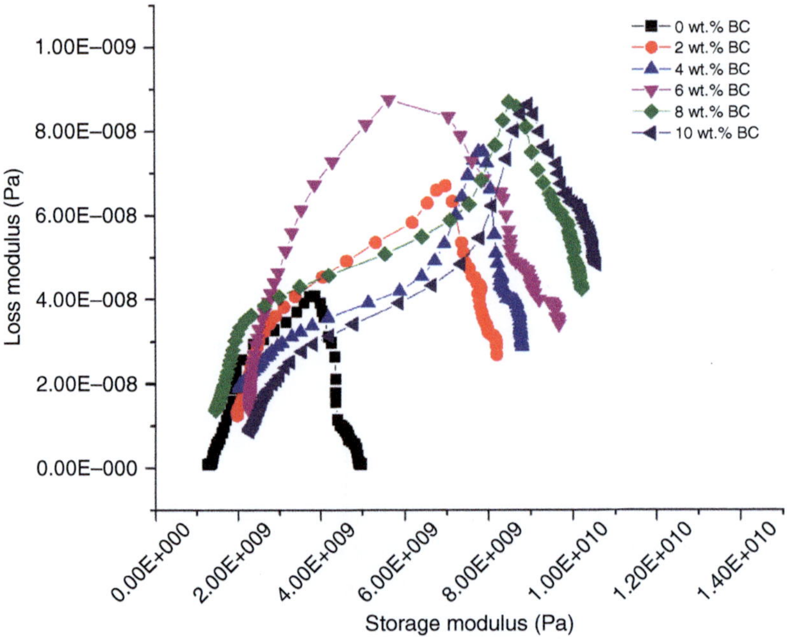


**Figure 12.11** Cole–Cole plot for the biochar-filled *Sansevieria cylindrica*-reinforced vinyl ester composite for varying biochar wt.%. Source: Based on Lavoratti et al. [28].

demonstrating the effectiveness of the biochar filler. As the shape of the Cole–Cole plot depends on the biochar loading, the inclusion of biochar influences the dynamic mechanical properties of the SCVEC.

### 12.3.3 Tensile Tests

Figure 12.12 shows the tensile strength and tensile modulus in its double *y*-axis plotted against the biochar particle loading in terms of weight percentage for the SCVEC. It was observed that tensile strength increases as the biochar loading increases until it reaches 6 wt.%. Further increase in the biochar weight percentage of the composite system shows a fall in its tensile strength. Unfilled (i.e. 0 wt.%) biochar composite showed more brittle nature compared to the biochar-filled composite because the incorporation of biochar filler improves the adhesion property of the matrix, resulting in enhanced tensile property of the composite [32]. The tensile strength of the 6 wt.% biochar loading was 72.71 MPa, which was 64.65% more compared to the unfilled SCVEC.

Further increase in biochar loading decreases the tensile strength; it is due to the agglomeration of biochar into the composite system. The tensile modulus of the composite system increases from 0.74 to 1.22 GPa for 0 and 6 wt.% biochar loading, which is 64.86% enhancement of the tensile modulus; it is because the inclusion of the filler to the matrix improves the stress-transferring ability of the matrix material. Further increase in biochar loading creates an adverse effect on the composite system because of insufficient wetting. Young's modulus of the filler material will be higher compared to that of the matrix material, causing an increase in tensile modulus as the filler loading occurs [20].

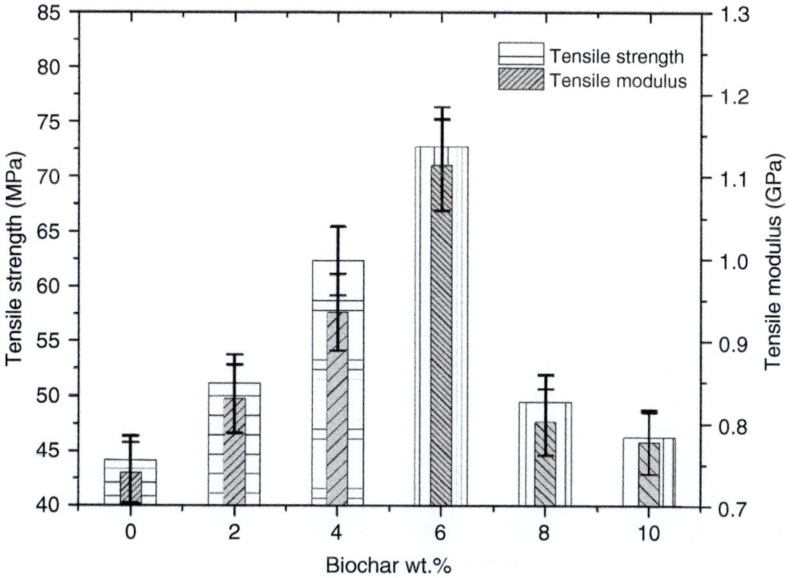


**Figure 12.12** Tensile strength and tensile modulus for the biochar-filled SCVEC for varying biochar wt.%.

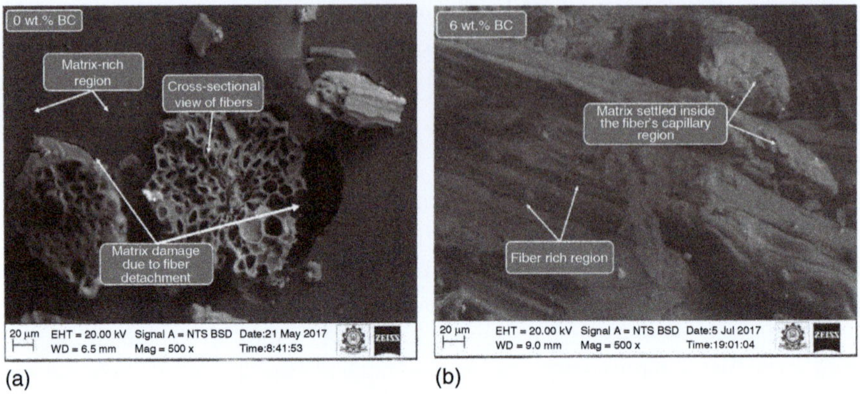


**Figure 12.13** SEM micrograph of tensile-tested (a) 0 wt.% biochar-filled SCVEC. (b) 6 wt.% biochar-filled SCVEC.

Figure 12.13 shows the SEM micrograph of tensile-tested fabricated composites. It can be observed that fibers in the 0 wt.% biochar-filled composite were clear without any matrix settled on it, whereas in the case of 6 wt.% biochar-filled composite the fibers were entirely covered with a matrix material, proving the statement that fillers naturally enhance the bonding nature of the matrix material. Thus, improving the bonding strength between the matrix and reinforcement results in improved tensile properties. It was also observed that the unfilled composite possesses matrix damage caused by fiber detachment due to weak adhesive property. This proves the superior nature of the *Z. mays* cob biochar as filler material for the SCVEC.

### 12.3.4 Flexural Tests

Flexural properties of biochar-filled SCVEC specimens in three-point bending against the biochar loading in weight percentage are shown in Figure 12.14. It is obvious from the graph that increases in biochar wt.% result in increased flexural strength of the composite till 6 wt.% biochar loading. These tendencies can be enlightened on the basis of improved interfacial strength with respect to the incorporation of biochar filler. The highly porous nature of the biochar particle plays an essential role in achieving good mechanical strength by allowing infiltration of the matrix into its pores resulting in enhanced mechanical strength [33]. Drastic improvement in the flexural strength was observed, 115.52 and 347.63 MPa for 0 and 6 wt.% biochar loading. Moreover, flexural properties also have similar trends as observed in the tensile property of the biochar-filled SCVEC, i.e. decrease in flexural strength beyond 6 wt.% biochar loading. Further increase in its loading has agglomeration effect, causing a decrease in the flexural property as the loading goes to 8 wt.% biochar.

SEM micrograph of the flexural-tested specimen shown in Figure 12.15 explains more clearly the effect of adding biochar filler to the composite. It can be observed that 0 wt.% biochar content flexural-tested sample has more cracks in its matrix

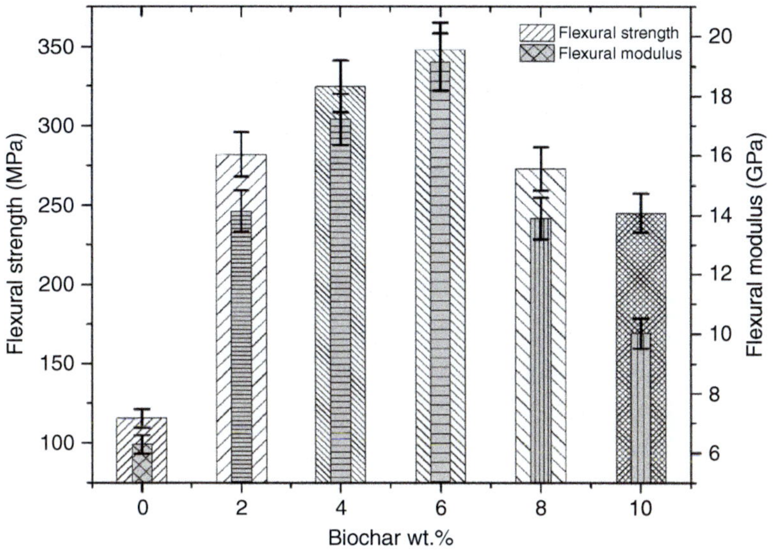


**Figure 12.14** Flexural strength and flexural modulus for the biochar-filled SCVEC for varying biochar wt.%.

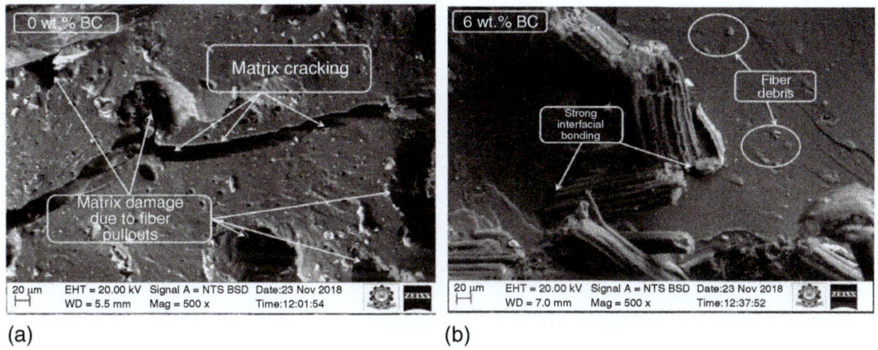


**Figure 12.15** SEM Micrograph representing the three-point bending on the (a) 0 wt.% biochar-filled SCVEC and (b) 6 wt.% biochar-filled SCVEC.

region due to fiber pullouts. But 6 wt.% biochar-loaded composite has strong interfacial bonding and smooth matrix surface. This proves that the natural ability of biochar particle enhances the flexural strength of the composite under a three-point bending test.

### 12.3.5 Impact Tests

Impact response of the biochar-filled SCVEC was measured by varying the biochar weight percentage, and the results are shown in Figure 12.16. It was observed that the impact strength of the composite follows a similar trend as the tensile

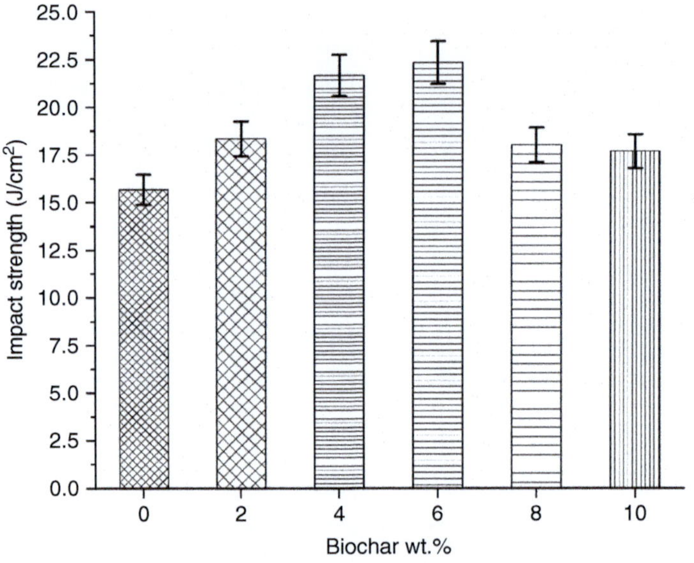


**Figure 12.16** Impact response of the biochar-filled SCVEC for different biochar loading.

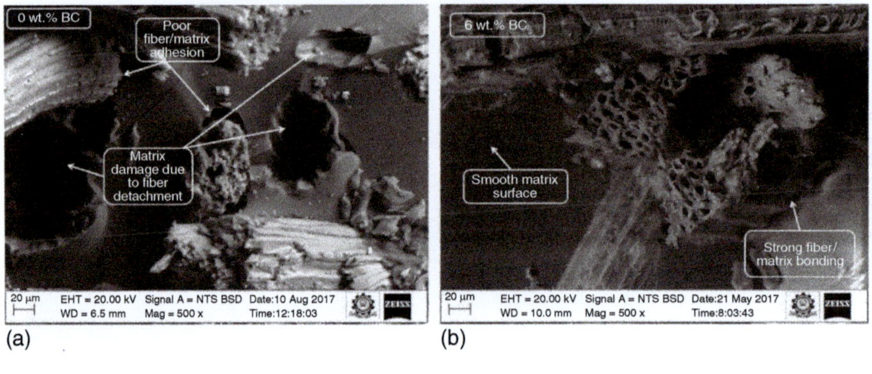


**Figure 12.17** SEM images for impact tested specimens (a) 0 wt.% biochar-filled SCVEC. (b) 6 wt.% biochar-filled SCVEC.

and flexural properties of the composite. However, only a marginal improvement in impact strength of 42.22% was observed when adding 6 wt.% of biochar to the composite system; still, it can contribute a good result for the composite in terms of tribological aspects.

The SEM micrograph in Figure 12.17 represents the biochar-filled and unfilled composite system under the Izod impact test. It is evident from the micrograph that adding biochar to the composite also increases the adhesive strength of the matrix, resulting in increased impact strength. In general, a particle will have more rigidity compared to the polymer material, so adding biochar filler to the composite decreases the amount of matrix material in the composite, thus improving the rigidity of the composite [34].

## 12.4 Conclusions

The following conclusions were drawn from the research work carried out on the biochar-filled SCVEC.

- Biochar was successfully derived from the *Z. mays* cob (maize residues) using a slow pyrolysis process. Characterization of the prepared biochar with X-ray diffraction patterns points to the enhancement of the SCVEC when used as a filler material.
- Dynamic mechanical response of the SCVEC with varying biochar wt.% was expressed in terms of storage modulus, loss modulus, and damping factor. It was inferred from the results that the incorporation of biochar filler material to the composite system, i.e. SCVEC, showed an increase in storage and loss modulus as 97% and 104% respectively. Also, it was observed that the peak broadening occurs in the curves plotted for the damping factor as a function of temperature as increase in the filler wt.% results in improved interfacial strength. Thus, biochar inclusion plays a significant role in enhancing the dynamic mechanical properties of the SCVEC. Irregular curves in the Cole–Cole plot display more heterogeneity owing to the hardness and stiffness of the incorporated biochar to the composite system with enhanced dynamic mechanical property.
- The increase in tensile strength, tensile modulus, flexural strength, flexural modulus, and impact strength of 64.52%, 64.86%, 200%, 204%, and 42.22% showed a similar trend of increase in strength and modulus as increasing the filler addition until 6 wt.% of biochar loading due to the improved interfacial adhesion and stress transferring capability. However, adverse effect was observed with biochar content beyond 6 wt.% due to insufficient wetting and agglomeration of biochar particles.
- SEM micrograph for the unfilled and 6 wt.% biochar-filled composite was investigated and it was inferred that biochar-filled SCVEC has a better interface strength to hold the fibers rigidly to the matrix material compared to the unfilled, i.e. 0 wt.% biochar-filled SCVEC.

## References


1 Jadav, V.M., Patel, P.M., Chaudhari, J.B. et al. (2018). Effect of integrated nutrient management on growth and yield of rabi forage maize (*Zea mays* L.). *Int. J. Chem. Stud.* 6 (1): 2160–2163.

2 Shareef, T.M.E., Zhao, B., and Filonchyk, M. (2018). Characterization of biochars derived from maize straw and corn cob and effects of their amendment on maize growth and loess soil properties. *Fresenius Environ. Bull.* 27 (5A): 3678–3686.

3 Peterson, S.C. and Jackson, M.A. (2014). Simplifying pyrolysis: using gasification to produce corn stover and wheat straw biochar for sorptive and horticultural media. *Ind. Crops Prod.* 53: 228–235.

4 Mullen, C.A., Boateng, A.A., Goldberg, N.M. et al. (2010). Bio-oil and biochar production from corn cobs and stover by fast pyrolysis. *Biomass Bioenergy* 34 (1): 67–74.

5 Lehmann, J., Gaunt, J., and Rondon, M. (2006). Bio-char sequestration in terrestrial ecosystems – a review. *Mitig. Adapt. Strateg. Glob. Chang.* 11 (2): 403–427.

6 Das, O., Bhattacharyya, D., Hui, D., and Lau, K.T. (2016). Mechanical and flammability characterizations of biochar/polypropylene biocomposites. *Composites Part B* 106: 120–128.

7 Srinivasan, P., Sarmah, A.K., Smernik, R. et al. (2015). A feasibility study of agricultural and sewage biomass as biochar, bioenergy and biocomposite feedstock: production, characterization and potential applications. *Sci. Total Environ.* 512: 495–505.

8 Zhang, M., Gao, B., Yao, Y. et al. (2012). Synthesis, characterization, and environmental implications of graphene-coated biochar. *Sci. Total Environ.* 435: 567–572.

9 Intani, K., Latif, S., Cao, Z., and Müller, J. (2018). Characterisation of biochar from maize residues produced in a self-purging pyrolysis reactor. *Bioresour. Technol.* 265: 224–235.

10 Thakur, V.K., Thakur, M.K., and Gupta, R.K. (2014). raw natural fiber–based polymer composites. *Int. J. Polym. Anal. Charact.* 19 (3): 256–271.

11 Qaiss, A.E.K., Bouhfid, R., and Essabir, H. (2014). Natural fibers reinforced polymeric matrix: thermal, mechanical and interfacial properties. In: *Biomass and Bioenergy* (eds. K. Hakeem, M. Jawaid and U. Rashid), 225–245. Cham: Springer.

12 Qaiss, A., Bouhfid, R., and Essabir, H. (2015). Characterization and use of coir, almond, apricot, argan, shells, and wood as reinforcement in the polymeric matrix in order to valorize these products. In: *Agricultural Biomass Based Potential Materials* (eds. K. Hakeem, M. Jawaid and Y.O. Alothman), 305–339. Cham: Springer.

13 Pothan, L.A., Oommen, Z., and Thomas, S. (2003). Dynamic mechanical analysis of banana fiber reinforced polyester composites. *Compos. Sci. Technol.* 63 (2): 283–293.

14 Fiore, V., Scalici, T., Vitale, G., and Valenza, A. (2014). Static and dynamic mechanical properties of *Arundo Donax* fillers-epoxy composites. *Mater. Des.* 57: 456–464.

15 Ridzuan, M.J., Majid, M.A., Afendi, M. et al. (2016). Thermal behaviour and dynamic mechanical analysis of *Pennisetum purpureum*/glass-reinforced epoxy hybrid composites. *Compos. Struct.* 152: 850–859.

16 Chen, R.Y., Zou, W., Zhang, H.C. et al. (2015). Thermal behavior, dynamic mechanical properties and rheological properties of poly(butylene succinate) composites filled with nanometer calcium carbonate. *Polym. Test.* 42: 160–167.

17 Kumar, R., Kumar, K., and Bhowmik, S. (2018). Mechanical characterization and quantification of tensile, fracture and viscoelastic characteristics of wood filler reinforced epoxy composite. *Wood Sci. Technol.* 52 (3): 677–699.

18 Lin, Y., Chen, Y., Zeng, Z. et al. (2015). Effect of ZnO nanoparticles doped graphene on static and dynamic mechanical properties of natural rubber composites. *Composites Part A* 70: 35–44.

19 Wang, F., Drzal, L.T., Qin, Y., and Huang, Z. (2015). Mechanical properties and thermal conductivity of graphene nanoplatelet/epoxy composites. *J. Mater. Sci.* 50 (3): 1082–1093.

**20** Kaya, N., Atagur, M., Akyuz, O. et al. (2018). Fabrication and characterization of olive pomace filled PP composites. *Composites Part B* 150: 277–283.

**21** Bleach, N.C., Nazhat, S.N., Tanner, K.E. et al. (2002). Effect of filler content on mechanical and dynamic mechanical properties of particulate biphasic calcium phosphate—polylactide composites. *Biomaterials* 23 (7): 1579–1585.

**22** Richard, S., Rajadurai, J.S., and Manikandan, V. (2017). Effects of particle loading and particle size on tribological properties of biochar particulate reinforced polymer composites. *J. Tribol.* 139 (1): 012202.

**23** Biswas, B., Pandey, N., Bisht, Y. et al. (2017). Pyrolysis of agricultural biomass residues: comparative study of corn cob, wheat straw, rice straw and rice husk. *Bioresour. Technol.* 237: 57–63.

**24** Shaaban, A., Se, S.M., Dimin, M.F. et al. (2014). Influence of heating temperature and holding time on biochars derived from rubber wood sawdust via slow pyrolysis. *J. Anal. Appl. Pyrolysis* 107: 31–39.

**25** Yu, J., Zhao, Y., and Li, Y. (2014). Utilization of corn cob biochar in a direct carbon fuel cell. *J. Power Sources* 270: 312–317.

**26** Adak, N.C., Chhetri, S., Kim, N.H. et al. (2018). Static and dynamic mechanical properties of graphene oxide-incorporated woven carbon fiber/epoxy composite. *J. Mater. Eng. Perform.* 27 (3): 1138–1147.

**27** Zeltmann, S.E., Prakash, K.A., Doddamani, M., and Gupta, N. (2017). Prediction of modulus at various strain rates from dynamic mechanical analysis data for polymer matrix composites. *Composites Part B* 120: 27–34.

**28** Lavoratti, A., Scienza, L.C., and Zattera, A.J. (2016). Dynamic-mechanical and thermomechanical properties of cellulose nanofiber/polyester resin composites. *Carbohydr. Polym.* 136: 955–963.

**29** Jabbar, A., Militký, J., Wiener, J. et al. (2017). Nanocellulose coated woven jute/green epoxy composites: characterization of mechanical and dynamic mechanical behavior. *Compos. Struct.* 161: 340–349.

**30** Etaati, A., Pather, S., Fang, Z., and Wang, H. (2014). The study of fibre/matrix bond strength in short hemp polypropylene composites from dynamic mechanical analysis. *Composites Part B* 62: 19–28.

**31** Rajesh, M. and Pitchaimani, J. (2016). Dynamic mechanical analysis and free vibration behavior of intra-ply woven natural fiber hybrid polymer composite. *J. Reinf. Plast. Compos.* 35 (3): 228–242.

**32** DeVallance, D.B., Oporto, G.S., and Quigley, P. (2016). Investigation of hardwood biochar as a replacement for wood flour in wood–polypropylene composites. *J. Elastomers Plast.* 48 (6): 510–522.

**33** Ikram, S., Das, O., and Bhattacharyya, D. (2016). A parametric study of mechanical and flammability properties of biochar reinforced polypropylene composites. *Composites Part A* 91: 177–188.

**34** Richard, S., Rajadurai, J.S., and Manikandan, V. (2016). Influence of particle size and particle loading on mechanical and dielectric properties of biochar particulate-reinforced polymer nanocomposites. *Int. J. Polym. Anal. Charact.* 21 (6): 462–477.

# 13

# Development and Sustainability of Biochar Derived from Cashew Nutshell-Reinforced Polymer Matrix Composite

*Rajendren Sundarakannan[1], Vigneswaran Shanmugam[1,4], Veerasimman Arumugaprabu[1], Vairavan Manikandan[2], and Paramasivan Sivaranjana[3]*

[1] *Kalasalingam Academy of Research and Education, Department of Mechanical Engineering, Anand Nagar, Krishnankovil, Tamilnadu 626126, India*
[2] *PSN College of Engineering and Technology, Department of Mechanical and Automation, PSN College Road, Melathediyoor, Tirunelveli, Tamilnadu 627152, India*
[3] *Kalasalingam Academy of Research and Education, Department of Chemistry, Krishnankovil, Anand Nagar, Tamilnadu 626126, India*
[4] *Saveetha School of Engineering, Saveetha Institute of Medical and Technical Sciences, Department of Mechanical Engineering, Chennai, Tamilnadu, India*

## 13.1 Introduction

Polymer composites have been used in numerous applications such as defense, automobile, aerospace, and domestic applications due to their advantages such as low density, biodegradability, low price, and good mechanical properties. The filler is the particle reinforcement added to the polymer composites to tailor the properties. In filler-reinforced polymer composites, the properties mainly depend on the dispersion of the particle in the matrix. The addition of filler in the polymer matrix exhibits improved strength; thus, it helps to utilize the filled composites in various applications. Various forms of filler have been found and reinforced in polymer composites. Arumugaprabu and coworkers investigated banana fiber-reinforced and red mud-filled polyester composites. ANOVA analysis was conducted at different parameters such as fiber length, treatment, weight fraction, and weight composition. They formed an $L_9$ orthogonal array to find the optimum values of the reinforced polyester composites. Optimum mechanical properties were observed in untreated fibers of 5 mm length, 15 wt.% of red mud-filled composites [1]. Vigneshwaran et al. observed an improvement in the mechanical strength of polyester composites owing to red mud reinforcement [2]. Sreekanth et al. investigated the polyester composites with fly ash reinforcement at different particle sizes and weight percentages. An increase in the fly ash weight percentage increased the composite flexural strength and reduced the tensile strength [3]. Rama et al. fabricated fly ash-reinforced epoxy composite through the solution blending method. This study found that the tensile and flexural properties of composites filled with fly ash increased up to an average of 1% compared to those of unfilled epoxy [4]. Vigneshwaran et al. analyzed filler-reinforced polymer

*Mechanical and Dynamic Properties of Biocomposites*, First Edition. Edited by Senthilkumar Krishnasamy, Rajini Nagarajan, Senthil Muthu Kumar Thiagamani, and Suchart Siengchin.

composites. The erosion behavior was determined from various parameters, and the optimum conditions were also analyzed [5]. Carbon is one of the fillers used as reinforcement in the polymer composite.

In recent years carbon-rich materials have been paid great attention owing to their ability to enhance the composite properties. Much research has been reported in recent years on carbon-rich filler reinforcement. Notably, carbon nanotubes are the most common carbon-rich filler utilized so far, but the major disadvantage is the cost. Since carbon nanotubes are costlier, researchers have initiated their search to find an alternative to solve the issue. As a result, a new area of biomasses has come into focus, and much interest has been given for the production of the carbon-rich material called biochar.

Biochar is a carbonaceous material and is derived from solid waste by the slow pyrolysis process. In biochar production, temperature plays a vital role; a controlled temperature is more important in biochar production. Two methods have been followed to prepare biochar. One is the pyrolysis process at low temperature, and the other one is the pyrolysis process at high temperature. Both methods have their advantages and disadvantages. The biochar derived from theses process is proved to be rich in carbon content. This would be a possible replacement for the high-cost carbon nanotubes. Also, various studies have demonstrated the properties of biochar composites. In recent years biochar has been produced from agricultural residues. Biochar from the agricultural residue as reinforcement in polymer composites creates interest due to their low cost, low density, biodegradability, and renewability [6]. It can be extracted from various agricultural wastes such as rice husk [7], wood [8], tea [9], and maize [10]. It has been observed that the value of this agricultural residue can be upgraded by bonding with the resin to produce composites suitable for building materials. The advantages of incorporating biochar as reinforcement in plastic composites enhance their mechanical and thermal properties at a reasonable cost [11]. Besides these, various researchers studied the effect of biochar reinforcement in the polymer matrix. Ikram et al. investigated the mechanical properties of biochar- and wood particles-reinforced polypropylene (PP) and maleic anhydride-grafted polypropylene (MAPP) composites. It has been observed that the biochar reinforcement increased the tensile modulus of the composites [12]. Biochar derived from corn starch, corn flour, and corn stover was investigated by Peterson [13]. Maximum tensile strength was noted in the corn starch biochar-filled composites. Poulose et al. investigated the date palm agricultural residue-derived biochar prepared at varying temperatures incorporated in polypropylene matrix. The maximum tensile strength value of 1.36 GPa is recorded on the biochar processed at 900 °C [14]. Ehsan Behazin et al. prepared the biochar at two different sizes and studied their reinforcement effect in the polypropylene matrix. The results conclude that at smaller size the particles present in the polypropylene matrix revealed better flexural properties [15]. Lucas Bowlby et al. synthesized the biochar from biomass feedstock by microwave pyrolysis method. The biochar particles were reinforced with glass fiber. Maximum flexural strength was noted on 10 wt.% composites [16]. Zhang et al. derived biochar from rice husk and studied their reinforcement effect in the high density poly ethylene (HDPE) plastic composites.

The biochar reinforcement is varied from 30 to 70 wt.%. Biochar reinforcement showed good dispersion with the HDPE matrix. The maximum impact strength was achieved by a 30% HDPE matrix [17]. Das et al. studied the biochar reinforcement in polypropylene composites. The results reported improvement in the mechanical characteristics, but at high biochar reinforcement, the mechanical strength was reduced due to improper dispersion [8]. Behazin et al. [15] studied the impact strength of low- and high-temperature reinforced PP composites. The biochar was prepared from miscanthus gross by slow pyrolysis process at 500 and 900 °C. The impact strength of the composite with 20% of high-temperature biochar is 39% higher than that of the low-temperature biochar composite [18].

Although biochar has been derived from various sources, no study has reported on the biochar derived from cashew nutshell waste. Cashew nutshell waste is one of the natural wastes available from agricultural residue, which has been found abundantly. Cashew nutshell waste is used as bio-oil extracted by using the pyrolysis process [19]. Currently, cashew nutshell waste, coming from the extract of cashew nuts from the food industry, is mainly burned as a fuel. Conversion of these wastes into biochar could increase the composite properties; more specifically, these wastes can be effectively used.

Based on the literature, the present work has been undertaken to develop a polymer matrix composite (polyester resin) using cashew biochar as reinforcement with volume fractions 5, 10, and 15 wt.% and subjected to a study of the mechanical properties such as tensile, flexural, and impact properties.

## 13.2 Materials and Methods

Cashew nutshell waste is collected from the local market at Krishnankovil, Tamilnadu, India. The unsaturated polyester resin, initiator, and accelerator were purchased from Vasavibala Resin Private Limited, Chennai, India.

### 13.2.1 Biochar Preparation

In the present work, biochar is prepared using slow pyrolysis at controlled temperature. The cashew nutshell waste collected was cut down into smaller pieces and dried well for the removal of moisture. After that, the well-dried cashew waste pieces are tightly packed in the airtight steel container and kept inside the oven at 500 °C for one hour. After one hour of slow pyrolysis process, the container is taken out carefully. Further, they are processed by ball milling for converting them to uniformly sized particles.

### 13.2.2 Composite Preparation

Cashew nutshell biochar polyester composites are fabricated using a solution dispersion method. Glass mold is used for fabricating the composite samples. The mold is separated by an intermediate leather liner of 3 mm thickness with 20 mm width on

**Figure 13.1** Cashew nutshell waste extracted biochar-reinforced polyester composite.

three sides to get the proper thickness. Wax is applied on the glass plates for easy removal of the final composite plate. Then the matrix material of 300 g is taken in a beaker, and the processed biochar particle is added at three varying weight percentages i.e. 5, 10, and 15 wt.%. The mixture is then mixed well manually for 10 minutes to achieve uniform dispersion. After that, the accelerator and catalyst mixture are added to the homogeneous mixture and stirred well to activate the curing. The mixture is then poured gently into the glass mold. It is left to cure for three hours to attain the final sample (Figure 13.1) from the glass mold.

### 13.2.3 Mechanical Testing

Tensile and flexural tests have been carried out on a universal tensile machine. The tensile test was conducted as per ASTM D3039. The specimen size used for the tensile test was 200 mm × 20 mm × 3 mm. The specimen size used for the flexural test was 125 mm × 13 mm × 3 mm. ASTM D790 was followed for flexural test. Impact test was conducted as per ASTM D256 for the dimension 65 mm × 13 mm × 3 mm. The five specimens were tested in all three mechanical tests and the average value was reported.

## 13.3 Results and Discussion

### 13.3.1 Tensile Strength

The tensile strength of the fabricated cashew biochar polyester composites is shown in Figure 13.2. It is observed from the figure that the tensile strength of filled biochar composites (5%) increased by 7% when compared with the virgin matrix. Maximum tensile strength is reported for 10 wt.% biochar composite, which is 25% higher than that of the pure polyester matrix. The addition of 15 wt.% biochar on polyester decreases the tensile strength drastically by 8%. It could be due to better bonding

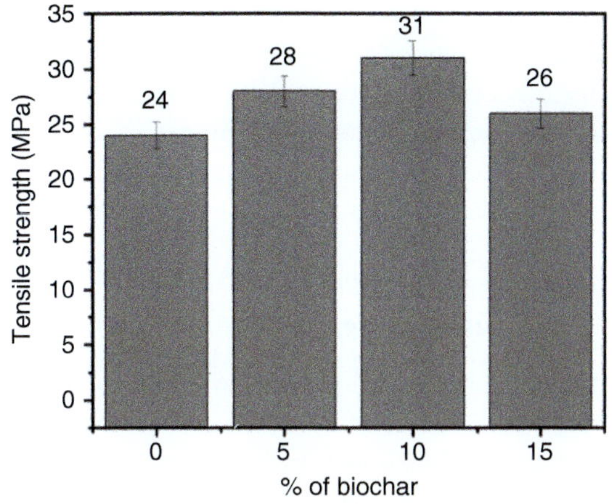


**Figure 13.2** Tensile strength values of cashew nutshell waste extracted biochar-reinforced polyester composites.

and fine dispersion of biochar particles in the matrix. Furthermore, it is noticed from the figure that the increase in biochar weight percentage of 15 wt.%. leads to a slight decrease in the tensile strength, which is due to nonuniform distribution leading to the clustering of biochar powder.

### 13.3.2 Flexural Strength

The flexural test analysis has been done to know the bending strength of the polymer composites prepared. The biochar-reinforced polyester composites and their flexural strength values are shown in Figure 13.3. The addition of biochar into the polymer matrix leads to the enhancement of the flexural properties. In this case, when compared to pure polyester, the addition of biochar increases the flexural strength by 14%, 24%, and 27%, respectively. It is noted that for 15 wt.% biochar addition the flexural strength value is 113 MPa, which is high among all composite combinations. This shows the potential of biochar addition as reinforcement along with the pure polyester, which creates more resistance against the bending load. The flexible resistance of the polyester matrix enhances considerably by means of reinforcing with

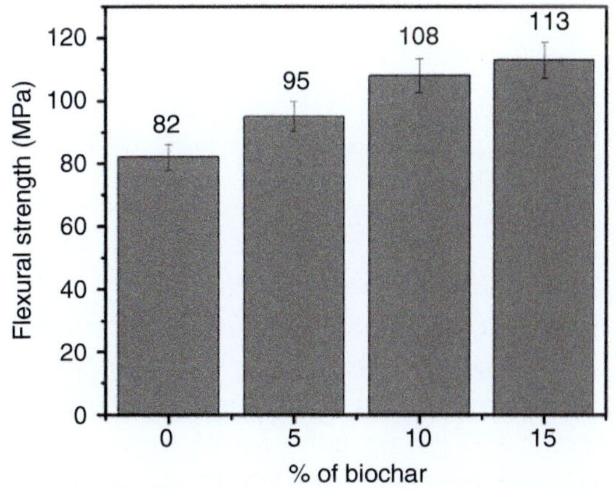


**Figure 13.3** Flexural strength values of cashew nutshell waste extracted biochar-reinforced polyester composites.

biochar, which contributes to the rigidity that gives high resistance to withstand the load. The composites become stiffer with more addition of biochar, yielding good flexural strength. Interestingly, in this case, an increment in the flexural strength is noted on increasing the biochar reinforcement weight percentage. The polyester matrix exhibited strong bonding with the biochar particles, which could be the reason for this continuous increase in strength concerning the weight percentage.

### 13.3.3 Impact Strength

The impact strength of cashewnut biochar composites is shown in Figure 13.4. The incorporation of cashew biochar improved the impact properties of the polyester composites. Maximum impact strength is observed for 10 wt.% filler composite. Incorporation of various biochar-filled composites increases the impact strength due to the absorption of more energy, which ultimately yields high resistance. The addition of biochar along with the polyester increases the impact strength by 11%, 20%, and 11%, respectively. The energy-absorbing capability has been increased by the addition of biochar. Addition of biochar provides good adhesion with the matrix, which is evident from their improved strengths.

### 13.3.4 Hardness

The hardness test is performed on biochar-reinforced composites, and their values are shown in Figure 13.5. The addition of filler leads to the improvement of the hardness values. It is observed that addition of biochar gives an enhancement in all the loaded composites: 5%, 10%, and 15%. Biochar creates a good inter-bonding with the polyester matrix to produce a crystalline structure of the composites. The crystalline structure improves the toughness behavior of the composites. Better bonding between the biochar reinforcement and polyester matrix yields improvement in the hardness of the biochar-reinforced composite.

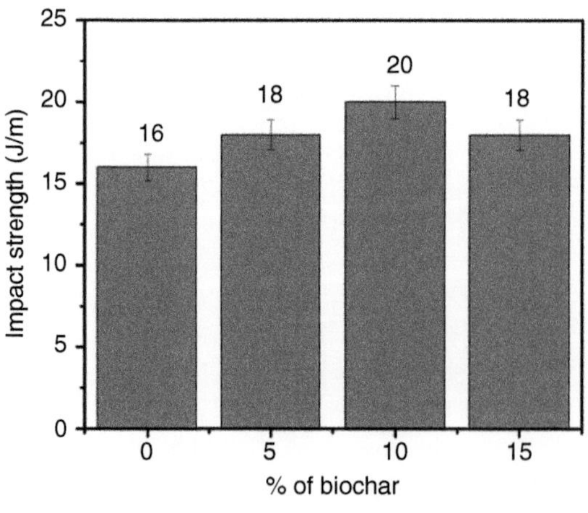


**Figure 13.4** Impact strength values of cashew nutshell waste biochar-reinforced polyester composites.

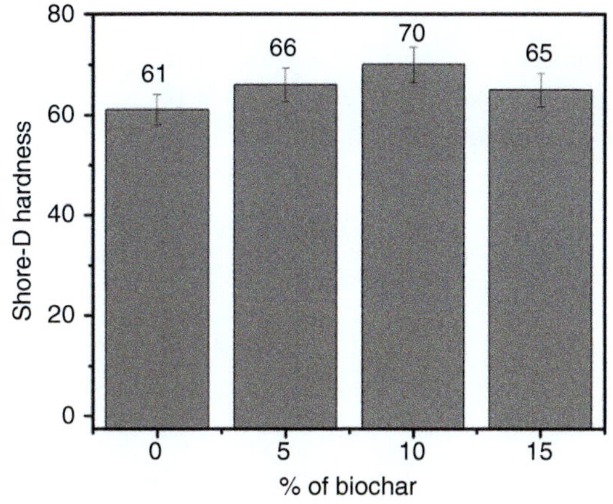


**Figure 13.5** Hardness values of cashew nutshell waste extracted biochar-reinforced polyester composites.

### 13.3.5 Failure Analysis of Cashew Nutshell Waste Extracted Biochar-Reinforced Polymer Composites

#### 13.3.5.1 Tensile Strength Failure Analysis

The fractured specimen's tensile test was analyzed using scanning electron microscope (SEM) images to understand the failure mechanism. The SEM image of 10 wt.% cashew nutshell waste extracted biochar tensile-tested specimen is shown in Figure 13.6a,b where the cashew nutshell waste extracted char and the polyester matrix hold together strongly. The fine dispersion of cashew nutshell waste extracted biochar in the polymer matrix results in better bonding. This in turn gives high strength. In the case of 15 wt.% cashew nutshell waste extracted biochar, from the SEM image shown in Figure 13.6b it is clearly evident that there is more formation of blow holes due to the entrapment of air particles, ultimately leading to fracture. Also, the flow of the matrix as it gets detached from the biochar is evident from the image, leading to weak strength.

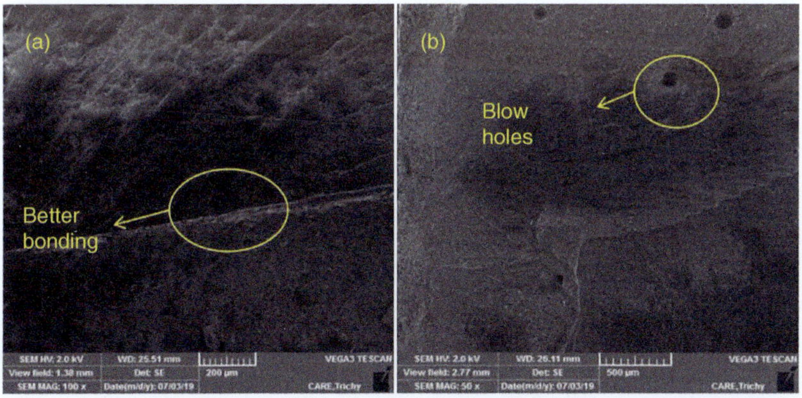


**Figure 13.6** Scanning electron microscopy images of tensile tested fractured cashew nutshell waste extracted biochar-reinforced polymer composites.

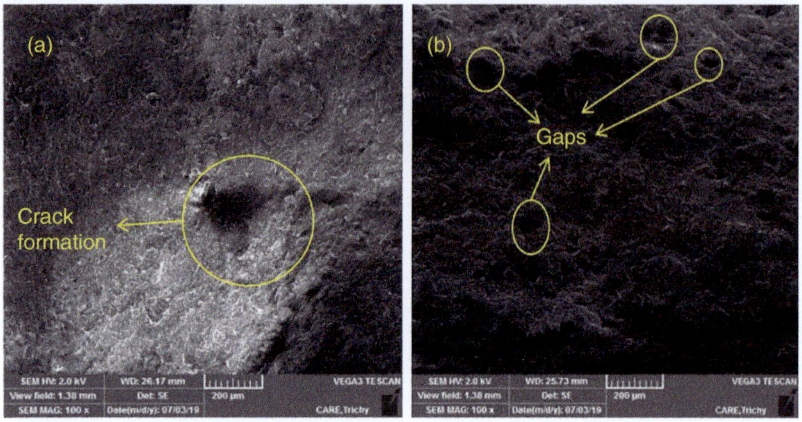


**Figure 13.7** Scanning electron microscopy images of impact tested fractured cashew nutshell waste extracted biochar-reinforced polymer composites.

#### 13.3.5.2 Flexural Strength Failure Analysis

The cashew nutshell biochar-reinforced polymer composite flexural test result was subjected to SEM studies, as shown in Figure 13.7. In all cases, the flexural resistance improves relative to the pure polyester matrix irrespective of biochar weight percentage. As shown in Figure 13.7a for 10 wt.%, the bending resistance of the composite is evident from the crack formation that takes place. Figure 13.7b shows the 15 wt.% filler-loaded composites; it shows no sign of particle pullout. It is ascribed to good adhesion between the particle and matrix, resulting in increased strength.

#### 13.3.5.3 Impact Strength Failure Analysis

The impact-tested specimens were subjected to SEM studies as shown in Figure 13.8. The addition of cashew nutshell waste extracted biochar to the polyester matrix increases the impact strength due to its higher energy-absorbing capability. Owing to impact, crack was developed on the matrix surface; however, crack propagation

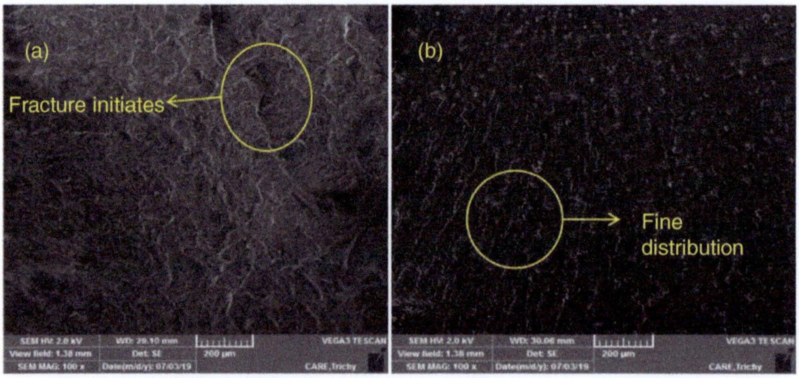


**Figure 13.8** (a,b) SEM images of impact tested cashew nutshell waste extracted biochar-reinforced polymer composites.

was controlled by the biochar particle (Figure 13.8a). The strong bonding between the particle and matrix is the reason for the resistance of the crack during impact. The uniform dispersion of the biochar particle is also the reason for the improved strength as shown in Figure 13.8b.

## 13.4 Conclusion

Elimination of waste is one of the important aspects for the promotion of environmental sustainability. Thus, biochar was extracted from the major agricultural waste cashew nutshell, which was utilized as a potential reinforcement in the polymer composite. The utilization of biochar waste in the polymer composite exemplifies an efficient way for waste management, thereby reducing the environmental hazard. From this experimental work, the following conclusions have been drawn.

- Biochar from cashew nutshell waste-reinforced polyester composites were fabricated successfully using the solution dispersion method.
- The mechanical performance of this composite was found to be enhanced by the incorporation of biochar.
- Among all the composite combinations, biochar of 10 wt.% addition possesses better tensile, flexural, and impact strength. Tensile, flexural, and impact strength increase by 25%, 27%, and 20% respectively by the addition of biochar compared to the virgin matrix. Hardness results also revealed that the addition of biochar increased the surface properties of the composites.
- The newly developed composites are suggested for applications in construction sector, automotive sector, and sports items.

## References


1 Prabu, V.A., Manikandan, V., and Uthayakumar, M. (2013). Effect of red mud on the mechanical properties of banana/polyester composites using design of experiments. *Proc. Inst. Mech. Eng., Part L* 227 (2): 143–155.

2 Vigneshwaran, S., Uthayakumar, M., and Arumugaprabu, V. (2019). Development and sustainability of industrial waste-based red mud hybrid composites. *J. Cleaner Prod.* 230: 862–868.

3 Sreekanth, M.S., Bambole, V.A., Mhaske, S.T., and Mahanwar, P.A. (2009). Effect of particle size and concentration of flyash on properties of polyester thermoplastic elastomer composites. *J. Miner. Mater. Charact. Eng.* 8 (03): 237.

4 Rama Shetty Ravindra, S.R. and Rai, S.K. (2010). Studies on physicomechanical properties of fly ash-filled hydroxyl-terminated polyurethane-toughened epoxy composites. *J. Reinf. Plast. Compos.* 29 (14): 2099–2104.

5 Vigneshwaran, S., Uthayakumar, M., Arumugaprabu, V., and Deepak Joel Johnson, R. (2018). Influence of filler on erosion behavior of polymer composites: a comprehensive review. *J. Reinf. Plast. Compos.* 37 (15): 1011–1019.

**6** Das, O., Sarmah, A.K., and Bhattacharyya, D. (2015). A novel approach in organic waste utilization through biochar addition in wood/polypropylene composites. *Waste Manage. (Oxford)* 38: 132–140.

**7** Masulili, A., Utomo, W.H., and Syechfani, M.S. (2010). Rice husk biochar for rice based cropping system in acid soil 1. The characteristics of rice husk biochar and its influence on the properties of acid sulfate soils and rice growth in West Kalimantan, Indonesia. *J. Agric. Sci.* 2 (1): 39.

**8** Das, O., Sarmah, A.K., and Bhattacharyya, D. (2016). Biocomposites from waste derived biochars: mechanical, thermal, chemical, and morphological properties. *Waste Manage. (Oxford)* 49: 560–570.

**9** Vithanage, M., Mayakaduwa, S.S., Herath, I. et al. (2016). Kinetics, thermodynamics and mechanistic studies of carbofuran removal using biochars from tea waste and rice husks. *Chemosphere* 150: 781–789.

**10** Głąb, T., Żabiński, A., Sadowska, U. et al. (2018). Effects of co-composted maize, sewage sludge, and biochar mixtures on hydrological and physical qualities of sandy soil. *Geoderma* 315: 27–35.

**11** Das, O., Bhattacharyya, D., and Sarmah, A.K. (2016). Sustainable eco–composites obtained from waste derived biochar: a consideration in performance properties, production costs, and environmental impact. *J. Cleaner Prod.* 129: 159–168.

**12** Ikram, S., Das, O., and Bhattacharyya, D. (2016). A parametric study of mechanical and flammability properties of biochar reinforced polypropylene composites. *Composites Part A* 91: 177–188.

**13** Peterson, S.C. (2012). Evaluating corn starch and corn stover biochar as renewable filler in carboxylated styrene–butadiene rubber composites. *J. Elastomers Plast.* 44 (1): 43–54.

**14** Poulose, A.M., Elnour, A.Y., Anis, A. et al. (2018). Date palm biochar-polymer composites: an investigation of electrical, mechanical, thermal and rheological characteristics. *Sci. Total Environ.* 619: 311–318.

**15** Behazin, E., Misra, M., and Mohanty, A.K. (2017). Sustainable biocomposites from pyrolyzed grass and toughened polypropylene: structure–property relationships. *ACS Omega* 2 (5): 2191–2199.

**16** Bowlby, L.K., Saha, G.C., and Afzal, M.T. (2018). Flexural strength behavior in pultruded GFRP composites reinforced with high specific-surface-area biochar particles synthesized via microwave pyrolysis. *Composites Part A* 110: 190–196.

**17** Zhang, Q., Yi, W., Li, Z. et al. (2018). Mechanical properties of rice husk biochar reinforced high density polyethylene composites. *Polymers* 10 (3): 286.

**18** Li, X., Lei, B., Lin, Z. et al. (2014). The utilization of bamboo charcoal enhances wood plastic composites with excellent mechanical and thermal properties. *Mater. Des.* 53: 419–424.

**19** Das, P. and Ganesh, A. (2003). Bio-oil from pyrolysis of cashew nut shell—a near fuel. *Biomass Bioenergy* 25 (1): 113–117.

# 14

# Influence of Fiber Loading on the Mechanical Properties and Moisture Absorption of the Sisal Fiber-Reinforced Epoxy Composites

*Banisetti Manoj, Chandrasekar Muthukumar, Chennuri Phani Durga Prasad, Swathi Manickam, and Titus I. Benjamin*

*Hindustan Institute of Technology & Science, School of Aeronautical Sciences, Rajiv Gandhi Salai, Padur, Kelambakkam, Chennai, Tamilnadu 603103, India*

## 14.1 Introduction

The rising environmental concern due to the landfills caused by the disposal of synthetic fiber-reinforced composites has laid the emphasis on research on environment-friendly materials [1]. Moreover, the increasing demand for human-made fibers such as carbon, kevlar, and glass derived from fossil fuel-based resources, depletion of the fossil fuel, and the stringent government regulation on the disposal and recycling of synthetic fiber-based composites have pushed the need for sustainable materials [2]. In lieu of this, natural fibers have gained significant attention in the recent decades as a potential candidate for reinforcement in composite materials. Other benefits of natural fibers such as low density (which makes them suitable for lightweight structures), non-abrasiveness to tooling, low cost, abundant availability, and biodegradability make their use as reinforcement inevitable in composites [3–6]. Although biocomposites cannot match the performance of synthetic fiber-based composites in structural applications such as the aircraft and aerospace sector, it can be used in nonstructural applications such as interior furnishing and seats. It can also be used in structural applications in land transport such as automotive parts, construction industries, household appliances, and sports equipment.

### 14.1.1 Sisal Fibers

Among the various natural fiber, sisal fiber is of particular interest to this study. It is usually obtained from the leaf of plants by the retting and scrapping process. Sisal fiber has good specific strength and stiffness. Sisal plant grows quickly in a few months and requires shorter cultivation time [7].

*Mechanical and Dynamic Properties of Biocomposites*, First Edition. Edited by Senthilkumar Krishnasamy, Rajini Nagarajan, Senthil Muthu Kumar Thiagamani, and Suchart Siengchin.

### 14.1.2 Fiber Parameters Affecting Mechanical Properties of the Composite

It is well known that thermal, physical, and mechanical properties of the composites are dependent on the fiber parameters and the type of resin. Various properties of the composite can be tailored by varying the fiber parameters such as fiber architecture and fiber loading. It was found from the literature that the sisal fiber has been extensively used as reinforcement in the form of chopped fibers [8, 9], long fibers, mat form [10], etc. Fiber loading represents the weight percentage or volume fraction of the fiber in the composite. In general, various studies in the literature indicate that fiber loading as low as 10–50 wt.% has been used as reinforcement in biocomposites [11–14]. Since the fibers are the primary load carriers in the composite, increase in the fiber weight percentage improves their strength, stiffness, and rigidity.

Thus, in this study fiber loading has been varied from 15–25 wt.% and the influence on mechanical and moisture diffusion properties has been explored.

## 14.2 Materials and Methods

### 14.2.1 Materials

Sisal fabric was purchased from a local supplier in Andhra Pradesh, India. Araldite LY 556 epoxy resin and Aradur HY 951 hardener were purchased from Go Green Products Pvt. Ltd., Chennai. Table 14.1 presents the technical specifications of the fiber and resin.

### 14.2.2 Fabrication Method

Sisal fabric and the resin mixture were laid into the mold size of 300 mm × 300 mm × 4 mm. Three composite samples were fabricated with a fiber weight of 15, 20, and 25 wt.% respectively. A resin and hardener mixture in the ratio of 10 : 1 was poured on each fabric layer and was spread using a roller until the fabric layers were completely wetted by the resin mixture. Then the mold with the fabrics and resin mixture at room temperature was cured for three hours under 5 bar pressure. After 24 hours, the composite samples were removed from the mold and cut as per the specimen dimension requirements of ASTM testing standards.

### 14.2.3 Characterization

Mechanical properties and moisture diffusion characteristic of the sisal/epoxy composite were investigated.

#### 14.2.3.1 Tensile Test

Sisal/epoxy composite specimens were subjected to the uniaxial tensile test as per the ASTM D3039 standard. Five specimens of dimension 200 mm × 20 mm × 4 mm were tested and a cross-head displacement rate of 1 mm/min was maintained during the test.

**Table 14.1** Material specifications of the sisal fiber and epoxy resin.

| Specification | Sisal fiber | Resin |
|---|---|---|
| Density (g/cm$^3$) | 1.5 | 1.15–1.20 |
| Tensile strength (MPa) | 511–635 | 35–38 |
| Young's modulus (GPa) | 9.4–22 | — |
| Elongation at break (%) | 2–2.5 | — |

#### 14.2.3.2 Flexural Test

Three-point bending test on the composite specimens was performed as per the ASTM D790 standard. Five specimens of dimension 130 mm × 15 mm × 4 mm were tested and a cross-head displacement rate of 1 mm/min was maintained during the test.

#### 14.2.3.3 Moisture Diffusion

Moisture diffusion characteristics of the composites were studied according to the ASTM D570 standard. The specimen of dimension 2 mm × 2 mm × 4 mm was immersed in distilled water until 888 hours. Dry weight of the specimen before immersion was measured followed by measurement of weight after immersion at specific intervals. Weight gain is calculated as shown in the Eq. (14.1):

$$\text{Weight gain (\%)} = \frac{W_{\mathrm{W}} - W_{\mathrm{D}}}{W_{\mathrm{D}}} \times 100\% \tag{14.1}$$

## 14.3 Results and Discussion

In this section, mechanical properties and moisture diffusion characteristic of the composite against the fiber loading are discussed.

### 14.3.1 Tensile Properties

Figure 14.1a–c presents the load–elongation plot from the tensile test. The peak load increased with the increase in fiber loading, which indicates the improvement in load-bearing capability of the composite. Another observation was that load increased almost linearly until certain point followed by a sudden drop and further increase in load until failure. The load drop within the final failure load was more frequent for the composite with higher fiber loading (Figure 14.1c). The possible explanation for this behavior is the delamination between the interplies in the composite. The number of layers increased from five layers at 15 wt.% to eight layers at 20 wt.%. Hence, delamination between the interplies was more prominent as evident from the frequent load drop before the final failure. Similar observations have been reported in the earlier work on the natural fiber-reinforced composite laminates [2].

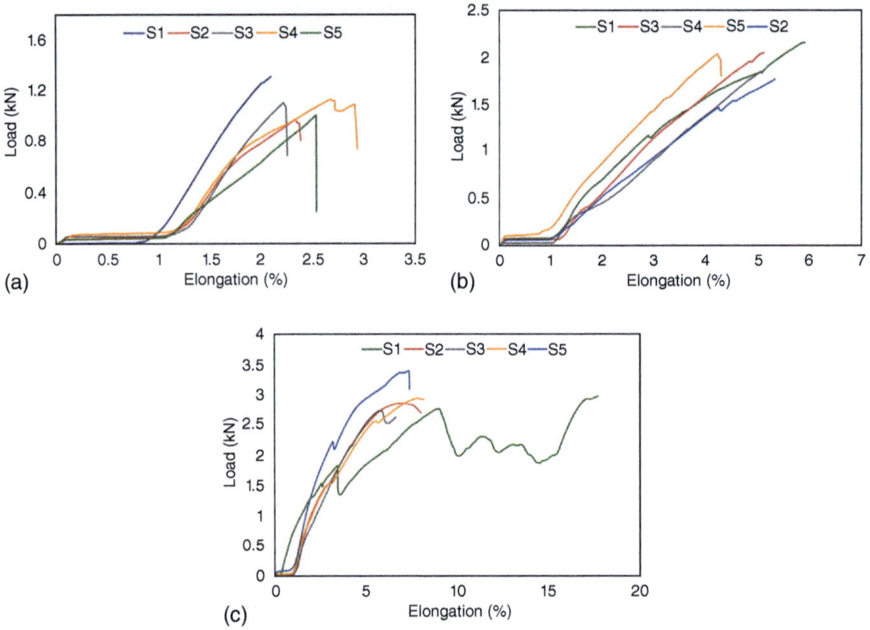


**Figure 14.1** Tensile load–elongation plot of the composites: (a) 15 wt.%, (b) 20 wt.%, and (c) 25 wt.%.

**Table 14.2** Tensile properties of the composites against the fiber loading.

| Fiber loading | Peak load (kN) | Tensile strength (MPa) | Young's modulus (GPa) | Elongation at break (%) |
|---|---|---|---|---|
| 15 wt.% | 1.11 ± 0.13 | 18.09 ± 2.52 | 1.70 ± 0.23 | 2.44 ± 0.32 |
| 20 wt.% | 1.97 ± 0.16 | 31.05 ± 2.28 | 3.13 ± 0.37 | 5.14 ± 0.59 |
| 25 wt.% | 2.98 ± 0.25 | 39.74 ± 3.21 | 5.66 (0.55) | 9.56 (4.59) |

Table 14.2 shows the tensile properties of the composites at different fiber loading. It can be noticed that the composite with 25 wt.% displayed the maximum strength of 40 MPa, stiffness of 7 GPa, and 9.56% value for the elongation at break. The superior tensile properties of the composite with the increase in fiber loading indicate the effectiveness of sisal in taking up the applied load. The trend of increasing the percentage elongation at break signifies the improvement in ductility of the composite. Since fibers are the primary load carrier, enhancement in the tensile properties with the increment in fiber loading was anticipated. In general, tensile strength and Young's modulus reach optimum values until a certain fiber loading where the fiber reinforcement is effective in taking up the load and efficient in stress transfer within the matrix.

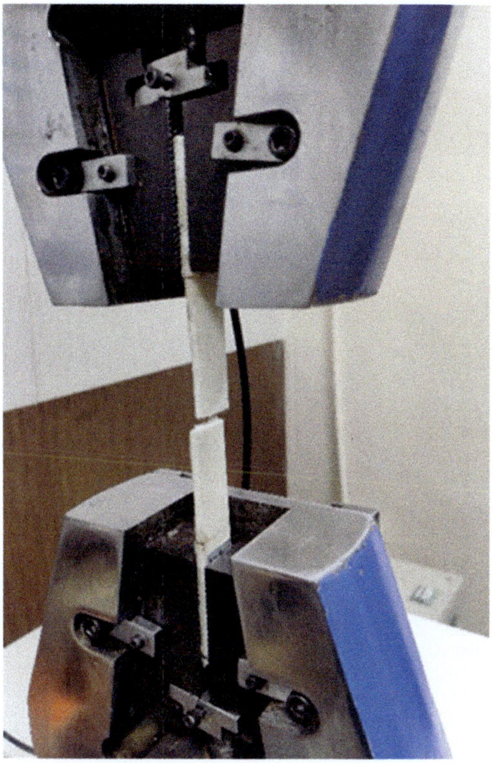

**Figure 14.2** Tensile failure in the composite.

The composite specimens subjected to tensile load displayed typical failures such as fiber breakage and matrix cracking as shown in Figure 14.2. The fracture mechanism due to the applied tensile load was visually observed during the test. As per the visual observation, failure initiates in the form of matrix crack in the gauge length followed by the propagation of cracks until the entire width, and fiber breakage before the specimen endures the final breakage.

### 14.3.2 Flexural Properties

The load–deflection plot obtained from the three-point bending test is presented in the Figure 14.3a–c. It can be seen that the peak load at which the failure occurs increased with the fiber loading. Also, the load–deflection plot showed two stages before final failure. Initially, the relationship between load and deflection was linear. The applied load became constant for a short duration in which the crack initiated at the bottom surface of the composite below the loading nose. As the applied load was further increased, crack propagation continued to increase until it reached the full width of the specimen. Beyond this point, the final load drop occurred as shown in Figure 14.3.

Parameters such as the peak load, flexural strength, modulus, and peak deflection obtained from the three-point bending test are tabulated in Table 14.3. The composite specimens showed increasing trend with the fiber loading similar to the trend

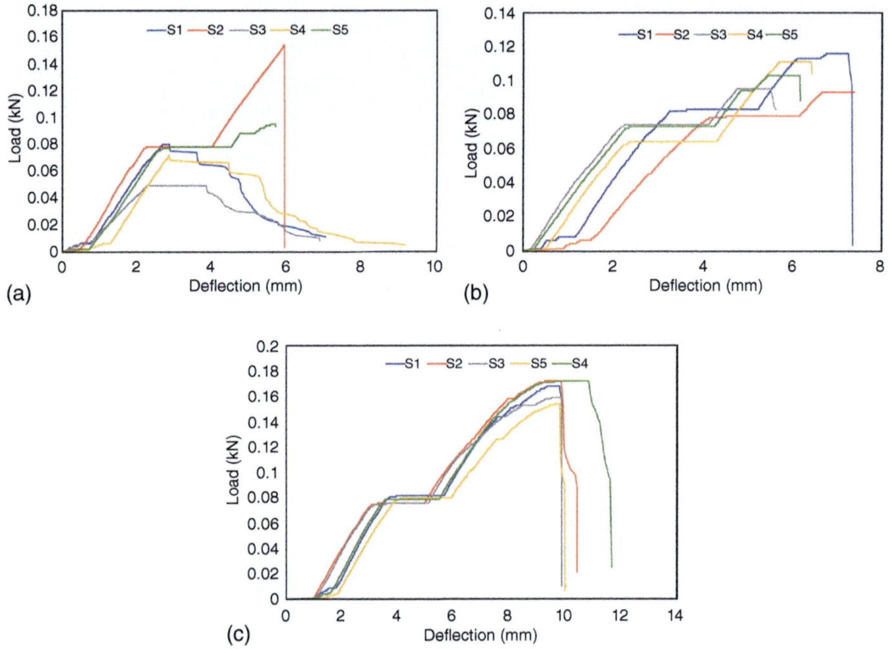


**Figure 14.3** Flexural load–deflection graph plot of the composites: (a) 15 wt.%, (b) 20 wt.%, and (c) 25 wt.%.

**Table 14.3** Flexural properties of the composites against the fiber loading.

| Fiber loading | Peak load (kN) | Flexural strength (MPa) | Flexural modulus (GPa) | Deflection at peak load (mm) |
|---|---|---|---|---|
| 15 wt.% | 0.09 ± 0.04 | 45.54 ± 18 | 1.48 ± 0.40 | 4.25 ± 1.49 |
| 20 wt.% | 0.10 ± 0.01 | 58.28 ± 6.20 | 1.09 ± 0.14 | 6.53 ± 0.78 |
| 25 wt.% | 0.165 ± 0.01 | 73.15 ± 5.72 | 1.12 ± 0.05 | 10.05 ± 0.45 |

observed in the tensile properties. The composite reinforced with the 25 wt.% fiber loading displayed the best performance under bending.

Figure 14.4 shows the failed specimen under three-point bending. It can be noticed that the composite specimen failed by crack initiation and propagation in the bottom surface. The crack formation at the bottom face due to the applied bending is explained as follows: it can be observed from Figure 14.4 that the bending load is applied in three specific points on the specimen. As the loading nose moves down, the top surface of the composite experiences compression whereas the bottom surface experiences tension. Owing to the applied point load in the middle of the gauge length and tension on the bottom surface, crack initiates at the bottom face of the composite and keeps propagating until full width. At this point, the load-bearing capacity drops and the specimen fails.

**Figure 14.4** Failure of the composite under flexural load.

### 14.3.3 Water Absorption

The weight gain–time plot obtained from the water absorption study is presented in Figure 14.5. Moisture diffusion into the composite is evident from the weight gain over the immersion time for all the composite specimens immersed in water. Weight gain increased with the immersion time and reached the maximum at 888 hours of exposure.

Moisture diffusion into the natural fiber-reinforced composite occurs due to the following reasons: (i) presence of manufacturing defects such as voids, microcracks, and resin starvation exposing the natural fiber and (b) hydrophilic nature of the bio-fiber, which absorbs moisture [15]. Owing to the above reasons, the composites with 15, 20, and 25 wt.% showed a maximum weight gain of approximately 3%, 4%, and 6% respectively until 888 hours. The increase in weight gain against the increase

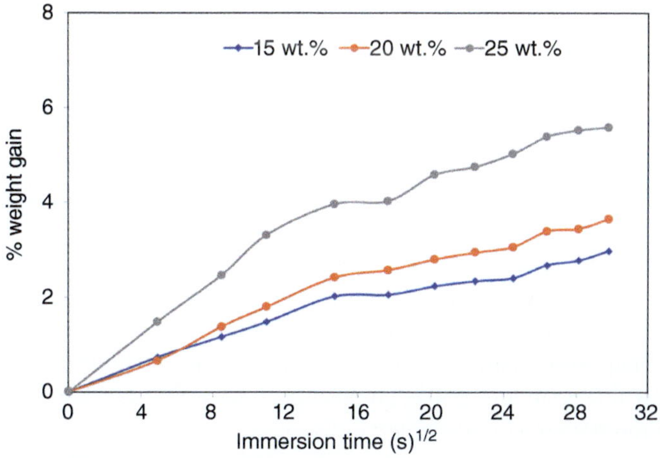


**Figure 14.5** Moisture absorption versus time graph as a function of fiber loading.

in fiber weight was in agreement with the previous work on the sisal/polyester composite [16]. The only difference is in the percentage weight gain, which is dependent on the fiber architecture and the type of fiber used as reinforcement. The maximum weight gain could also be lower for synthetic fiber-based composites, since the synthetic fibers are hydrophobic and resistant to moisture absorption.

Moisture diffusion into the composite can be classified into Fickian, Non-Fickian, and anomalous behavior based on the shape of the weight gain–time plot [17]. A composite that shows linear increase in weight over time until the saturation follows Fick's law. A composite that shows a two-stage absorption curve and s-shaped curve before reaching the saturation state falls under the category of the Non-Fickian and anomalous diffusion behavior. The sisal/epoxy composite employed in this study showed two-stage absorption with an initial linear increase until 216 hours followed by a slight inflection and further increase until 888 hours as could be seen in Figure 14.5. Thus, it is clear that sisal/epoxy composite with various fiber loadings followed the non-Fickian diffusion behavior.

## 14.4 Conclusion

The experimental characterization of the sisal/epoxy composite against fiber loading (15, 20, and 25 wt.%) led to the following conclusions:

- Composites with higher fiber loading showed better performance under the tensile and flexural load.
- Mechanical properties of the composite followed the rule of mixtures. Among the fiber loadings, 25 wt.% had higher load-bearing capacity and possessed the maximum strength, Young's modulus, and percentage elongation at break. A similar trend was observed for the composites subjected to the flexural loads.
- Typical failure behavior such as the fiber breakage and matrix cracking were observed from the tensile fractured specimens. On the other hand, composites subjected to bending load failed by crack initiation and crack propagation until full width in the bottom face of the composite.
- Weight gain due to the moisture diffusion into the composite increased with the immersion time. Also, the composite with higher fiber loading showed substantially higher weight gain. Weight gain of the composite immersed in water is due to the fact that sisal fibers are hydrophilic in nature and have natural affinity to water.

## References

1 Chandrasekar, M., Ishak, M.R., Sapuan, S.M. et al. (2016). Tensile and flexural properties of the hybrid flax/carbon based fibre metal laminate. *Proceedings of the 5th Postgraduate Seminar on Natural Fiber Composites*, pp. 5–10.

2 Chandrasekar, M. et al. (2019). Flax and sugar palm reinforced epoxy composites: effect of hybridization on physical, mechanical, morphological and dynamic mechanical properties. *Mater. Res. Express* 6 (10): 105331.

**3** Atiqah, A., Chandrasekar, M., Senthil Muthu Kumar, T. et al. (2019). *Characterization and Interface of Natural and Synthetic Hybrid Composites*. Elsevier Ltd.

**4** Shahroze, R.M. et al. (2019). Sugar palm fiber/polyester nanocomposites: influence of adding nanoclay fillers on thermal, dynamic mechanical, and physical properties. *J. Vinyl Add. Technol.* 26 (3): 236–243.

**5** Sapuan, S.M., Ishak, M.R., Chandrasekar, M. et al. (2018). Preparation and characterization of sugar palm fibers. In: *Sugar Palm Biofibers, Biopolymers, and Biocomposites*, 71–88. CRC Press.

**6** Chandrasekar, M., Ishak, M.R., Sapuan, S.M. et al. (2017). A review on the characterisation of natural fibres and their composites after alkali treatment and water absorption. *Plast. Rubber Compos.* 46 (3): 119–136.

**7** Senthilkumar, K. et al. (2018). Mechanical properties evaluation of sisal fibre reinforced polymer composites: a review. *Constr. Build. Mater.* 174: 713–729.

**8** Krishnasamy, S. et al. (2019). Effect of fibre loading and $Ca(OH)_2$ treatment on thermal, mechanical, and physical properties of pineapple leaf fibre/polyester reinforced composites. *Mater. Res. Express* 6 (8): 085545.

**9** Senthilkumar, K. et al. (2017). Static and dynamic properties of sisal fiber polyester composites - effect of interlaminar fiber orientation. *BioResources* 12 (4): 7819–7833.

**10** Thiagamani, S.M.K. et al. (2019). Investigation into mechanical, absorption and swelling behaviour of hemp/sisal fibre reinforced bioepoxy hybrid composites: effects of stacking sequences. *Int. J. Biol. Macromol.* 140: 637–646.

**11** Senthilkumar, K. et al. (2019). Evaluation of mechanical and free vibration properties of the pineapple leaf fibre reinforced polyester composites. *Constr. Build. Mater.* 195: 423–431.

**12** Kumar, K.S., Siva, I., Rajini, N. et al. (2014). Tensile, impact, and vibration properties of coconut sheath/sisal hybrid composites: effect of stacking sequence. *J. Reinf. Plast. Compos.* 33 (19): 1802–1812.

**13** Kumar, K.S., Siva, I., Jeyaraj, P. et al. (2014). Synergy of fiber length and content on free vibration and damping behavior of natural fiber reinforced polyester composite beams. *Mater. Des.* 56: 379–386.

**14** Krishnasamy, S. et al. (2019). Effects of stacking sequences on static, dynamic mechanical and thermal properties of completely biodegradable green epoxy hybrid composites. *Mater. Res. Express* 6 (10): 105351.

**15** Shahroze, R.M., Ishak, M.R., Sapuan, S.M. et al. (2019). Effect of silica aerogel additive on mechanical properties of the sugar palm fiber-reinforced polyester composites. *Int. J. Polym. Sci.* 2019: 1–4.

**16** Sreekumar, P.A., Joseph, K., Unnikrishnan, G., and Thomas, S. (2007). A comparative study on mechanical properties of sisal-leaf fibre-reinforced polyester composites prepared by resin transfer and compression moulding techniques. *Compos. Sci. Technol.* 67 (3–4): 453–461.

**17** Chandrasekar, M., Ishak, M.R., Jawaid, M. et al. (2017). An experimental review on the mechanical properties and hygrothermal behaviour of fibre metal laminates. *J. Reinf. Plast. Compos.* 36 (1).

# 15

# Mechanical and Dynamic Properties of Ramie Fiber-Reinforced Composites

*Manickam Ramesh[1], Lakshminarasimhan Rajeshkumar[2], and Devarajan Balaji[2]*

[1] *KIT-Kalaignarkarunanidhi Institute of Technology, Department of Mechanical Engineering, Coimbatore, Tamil Nadu 641402, India*
[2] *KPR Institute of Engineering and Technology, Department of Mechanical Engineering, Coimbatore, Tamil Nadu 641407, India*

## 15.1 Introduction

Composite materials find majority of applications in marketable, engineering and industrial areas such as vehicles, aircrafts, and ship building. Such materials possess better properties such as high stiffness-to-weight ratio and strength-to-weight ratio [1]. Polymer matrix composites with natural and artificial fibers evolved very rapidly for usage in many engineering fields such as automobile and aerospace fields [2]. Preparation of composite materials is a broad area of research with a wide scope since these materials possess numerous density-related, properties-related, and weight-related advantages when compared with the monolithic materials. Such advantages find applications for the composite materials broadly in marine, automobile, sports, and aerospace industries [3]. Research is under way to find an optimized processing route for manufacturing the composite materials, which renders the same physical and mechanical properties as obtained from the traditionally made materials.

Ramie fiber (*Boehmerianivea*) comes under the Urticacease family, which is extracted from the bast and is considered to be the fiber with highest strength among its other counterparts. This fiber is extracted from cane barks and can be a potential ingredient to be used for hybridization with other natural or human-made fibers. Much research has been carried out to explore the physical properties of ramie fibers and it was found that the fibers possessed good resistance toward acid, alkali, heat, and light, high glossiness and brightness, and high tenacity [4]. A few experiments compared the ramie fiber–epoxy composites with ramie fiber–polyester composites and glass fiber–polyester composites and concluded that epoxy composites behaved better than polyester and glass fiber composites in terms of mechanical characteristics of the composites [5]. Since the epoxy composites are very light in weight, possess more comfort, are available easily, and are readily acceptable in

*Mechanical and Dynamic Properties of Biocomposites*, First Edition. Edited by Senthilkumar Krishnasamy, Rajini Nagarajan, Senthil Muthu Kumar Thiagamani, and Suchart Siengchin.

biomechanical and psychosocial terms, they can be prospective substitutes for glass fiber composites in a few medical applications such as socket prosthesis.

Surface-treated ramie fabrics reinforced in epoxy composites were prepared by hot compaction method by using vacuum-assisted resin infusion molding and tested for compressibility [6]. It was found that surface-treated ramie fiber–epoxy composites possessed less compressibility when compared with the untreated ramie composites. It was also noticed that the mechanical properties of ramie fabric composites enhanced due to the increase in ramie fabric content in the ramie/epoxy composites, which was influenced by hot compaction method. A few experiments were conducted on untreated ramie fiber composites reinforced in polypropylene (PP), prepared by using melting hybrid technology, to assess their physical properties by changing the fiber volume fraction of ramie fibers [7]. From the results, it could be seen that the thermal stability was high and degradation temperature was low for the composites when it encompassed higher volume fraction of ramie fibers. Moreover, ramie fibers speeded up the rate of crystallization of polypropylene matrix. It was also inferred that due to the employment of ramie fibers with higher length while manufacturing the composites, the dispersibility of the fibers within the matrix could not be maintained properly, which ultimately resulted in weak interfacial bonding between the matrix and the fiber, thus reducing the tensile strength of the composites.

Ramie fiber paves the way for a better rural society economization by allowing itself to be used in its raw form rather than processing it. As the fiber processing is eliminated, toxic waste generated during the processing is also eliminated. Ramie fibers were also proved to be compatible with almost all matrix materials commonly used and the RF composite renders better mechanical properties in all respects. Many experimenters tried to prepare ramie fiber-reinforced polyester composites hybridized with cotton fabric by compression moulding to assess their mechanical properties. It was observed from the results that RF composites possess 338% more tensile strength than the pure matrix, revealing that ramie fibers were better candidate materials to form composites with any thermosetting matrix material [8]. Ramie fibers hybridized with eco-flex and cellulose nanofibers were separately manufactured as composites with corn starch as the matrix in order to evaluate the degradation rate and incubation time of the composites by exposing them to natural environments such as compost and sea water [9]. It was noticed from the experiments that ramie/eco-flex nanofiber composites degraded more rapidly than ramie–cellulose nanofiber in both natural environments while both the ramie fiber hybrid composites underwent higher degradation in compost environment when compared with sea water due to the action of various microbes.

The mechanical properties of ramie fiber-reinforced polypropylene composites were assessed by some authors [10]. Treatment of ramie fibers with 10–15% of NaOH rendered better mechanical properties of the RF polypropylene composites. A simultaneous raise in compressive, tensile, and flexural characteristics of the composites along with the decrease of percentage elongation and impact strength of the composites was observed in the composites when the fiber volume fraction was increased beyond 20% and the fiber length was maintained between 3 and

**Figure 15.1** Ramie fibers in different forms. Source: Giridharan [12].

8 mm. Ramie fiber-reinforced polypropylene composites find better application in automobile industry due to meritorious characteristics such as eco-friendliness, enhanced mechanical properties, and reduced weight. A few authors investigated the ramie fiber–phenolic composites for their moisture absorption and mechanical behavior by measuring the immersion time of the composites [11]. The results showcased that the coefficient of diffusion and moisture absorption were high for the composites due to the high hydrophilicity of ramie fibers, and a sharp decline in short beam shear and flexural properties was also observed due to the same reason. The different forms of ramie fibers are presented in Figure 15.1 [12].

## 15.2 Mechanical Strength of Ramie Fiber Composites

Many authors determined that hybridization of ramie fiber with E-glass fiber with the matrix being epoxy to make composites by means of hand-lay-up technique is an effective method of saving the cost of composites by many means. Composites were prepared with two different weight percentages (wt.%) of both the fibers and the tensile, impact, and flexural characteristics of the composites were evaluated alongside the observation of fracture surface morphology using scanning electron microscopy (SEM). Composites with hybrid fibers showcased enhanced mechanical characteristics when compared with the individual fiber-reinforced composites. By reinforcing a hybrid fiber weight fraction of 20% and 30% with epoxy, the tensile strength was 1.44% and 2.45% higher than the individual glass fiber-reinforced composites and 70% and 65% higher than individual ramie fiber-reinforced composites respectively. Hybridization of 20% by weight of glass fiber with 10% by weight of ramie fiber in interleaving arrangement of fibers within the matrix rendered maximum tensile strength for the hybrid composites. Flexural strength of the same

20 wt.% of hybrid fibers was 7.6% and 25.7% greater than that of the individual glass fiber-reinforced and ramie fiber-reinforced composites while 30 wt.% of fiber ended up with 5.8% and 22.5% increase in flexural strength than individual fiber-reinforced composites. Meanwhile, when the glass fibers and ramie fibers were hybridized in the ratio of 1 : 2, impact strength of the hybrid composites was found to be 4.9% greater than that of the individual glass fiber-reinforced composites and 89.5% greater than that of the individual ramie fiber-reinforced composites. Hence from all the above points, it was concluded that hybridization of ramie fibers with other natural or synthetic fibers resulted in enhanced mechanical properties of the fiber-reinforced hybrid composites.

A combination of even a small proportion of glass fibers with ramie was proved to be a better composition for improving the physical properties of the composites. Apart from the improvement in physical properties, hybridization can render environment-friendly and cost-effective composite materials. Hence the application of ramie fibers was majorly witnessed in lightweight and green areas of agricultural industries. Research on ramie fibers could be extended by hybridizing them with carbon fibers for the improvement of hardness and impact characteristics and even those composites could be subjected to heat or electric treatment. When the interaction between the fiber and matrix is increased by some means, the strength of ramie fiber composites can also be increased by all means. Lightweight applications of such composites are wall cladding, room partitions, and door panels. Yet, they can be applied in many other areas such as building and construction, electrical housing, automotive parts, home appliances, and food packaging [13].

Few experimental studies were performed on ramie fiber hybridized with kevlar and glass fibers-reinforced epoxy composites where the composites were fabricated using hand lay-up technique. Those composites were subjected to tensile, impact, and flexural assessments along with the measurement of their hardness [14]. BS EN ISO 527 and 14 125 standards were adopted for performing the tensile and flexural tests respectively and the tensile and three-point bending tests were carried out in a Bent tram testing machine, Denmark. Specimen dimensions for flexural tests were maintained as 80 mm × 20 mm and the gauge length of the specimen was 64 mm as per the standard. A load cell of 50 kN and a cross-head spindle speed of 1 mm/min were maintained during the tests and the flexural modulus and flexural strength were obtained from the stress–displacement plots from the experimental setup. In the same way, the dimensions for the tensile tests were maintained as 150 mm × 20 mm and the tests were performed over a gauge length of 100 mm while the end caps were fixed at 25 mm at both the ends. Assessment of tensile characteristics was carried out to evaluate the tensile modulus, elongation at break, and tensile strength. Similar to the flexural tests, 50 kN load and a cross-head spindle velocity of 1 mm/min were adopted for tensile investigations also [15].

A few experiments focused on evaluating the properties of ramie fibers reinforced with PP matrix with organic or inorganic fillers. It was revealed from the characterization of ramie/PP composites that the range of application of these composites faces constraints due to the poor interfacial interaction between the nonpolar group PP matrix and polar group ramie fibers coupled with the high shrinkage of the PP matrix. Adhesion at the interface of the fiber and matrix was enhanced by treating the ramie fibers with silane coupling agents. Aluminum powder was mixed with the PP matrix as colorant by means of mechanical blending to render the PP matrix as a spray-free matrix. Results of tensile tests indicated that the ramie fibers increased the modulus of elasticity and tensile behavior of PP composites by 158% and 15.7% respectively. Moreover, the reinforced ramie fiber composites minimized the PP shrinkage from 1.95% to 0.26%, thus rendering stability in the final dimensions of the composites, and the application of the ramie fiber PP composites becomes more reliable due to the enhancement in quality assurance. This also gives a systematic approach to analyzing the microstructure of the ramie fiber-reinforced PP composites. Ramie was also considered to be a solid recycled fossil fuel rendered fiber material, and when it is reinforced with a colorant added PP matrix, spray-free composites result. Ramie fibers were treated with 3-aminopropyltriethoxysilane as coupling agent, which improved the adhesion between the ramie and PP matrix at its interface. When a weight fraction of 30% and 35% of ramie fibers were reinforced with the PP matrix, the Young's modulus and tensile strength were found to be increased by 160% and 16% respectively. Shrinkage of PP matrix was reduced in this case also due to the limitation posed toward the crystallization of PP matrix, which was the primary cause of the PP shrinkage. As the dimensional stability of the silane-treated ramie fiber–PP matrix was enhanced, the application warranting high dimensional accuracy also increases [16].

Many experimenters focused on developing a bio-based composite material using flax and ramie fibers reinforced in bio-epoxy matrix material with various weight percentages such as 10%, 20%, and 30% adopting compression molding method as the manufacturing technique. These composites were mainly developed for specific medical applications in orthopedic implants and bone grafting since the above materials could be considered as potential substitutes for the currently used materials. Experimental results exhibited that when the ramie fiber and flax fiber were reinforced at 15 wt.% each, the tensile, compressive, and flexural strength were 102, 130, and 138 MPa while the respective modulus values were 5.63, 8.87, and 12.41 GPa. Since the results were more alike the properties of the tibia and the human femur bone, they can be an effective replacement for orthopedic implants. Hybrid composites were also subjected to dynamic mechanical analysis (DMA) in order to determine the viscoelastic characteristics of the biocomposites. As per the results, composites with 30 wt.% of the fibers rendered better properties such as loss modulus of 1.45 GPa, glass transition temperature of 110 °C, and storage modulus of 9.03 GPa. SEM was used to analyze the fracture surface morphology of the ramie/flax/epoxy composites, which proved that the above evaluated properties were on par with the existing characteristics of bone material. Ramie/flax hybrid composites possessed less axial stiffness when compared with metal and ceramic orthopedic implant materials, which in turn renders lower stress bearing of the new material. Besides, these composites are considered to be advantageous due to characteristics such as biostability, biocompatibility and eco-friendliness in comparison with the existing materials used for orthopedic implants [17].

Few experimenters tried to prepare ramie fiber-reinforced polypropylene composites treated with graphene oxide (GO) of different particle sizes, and the interlaminar shear strength (ILSS) of the composites were evaluated by suitable methods. Composites were fabricated by hot pressing techniques and the influence of particle size and content of GO on the ILSS was evaluated. Results of the experiments revealed that the ILSS of the composites increased with increase in size and content of GO and decreased beyond certain limit of GO content and size. When ramie fibers were treated with 5 μm sized GO and reinforced in PP matrix, the ILSS was found to be increased by 40%. This could be probably due to dominant formation of C—O—C and C=O formation at this size of GO. It was also observed from the experiments that GO was a potential surface modifier that could be readily used to treat the ramie fiber surfaces, which also enhanced the fiber–matrix adhesion at the interfaces. This was the main reason behind the enhancement of ILSS of GO treated ramie/PP composites.

A few researchers worked on biocomposites with polybutylene succinate (PBS) reinforced with ramie fabric fibers treated with amino functionalized materials manufactured through thermal-compressive method. As the nonwoven form of PBS was used as matrix, the infiltration of fiber within the matrix was expected to be much better due to the three-dimensional structural network of PBF, which holds the ramie fabrics in position. Results revealed that amino functionalized ramie fabrics in PBF matrix exhibited better rheological behavior, mechanical characteristics, thermal performance, crystallization behavior, biodegradability, and wettability when compared with untreated ramie fabric PBF composites. When the mechanical performance of other composite materials were compared with the ramie fabric/PBS, they outsmarted their counterparts. Experimental results showed that when 50 wt.% of ramie fabrics were reinforced in PBS, the ILSS, tensile strength, flexural strength, flexural modulus, elongation at break, and impact strength were 3.3 MPa, 72.4 MPa, 100.5 MPa, 4.9 GPa, 19.4%, and 67.5 $kJ/m^2$ respectively. The influential factors for such good and enhanced results were the classical plastic deformation of PBS, good bonding at the fiber–matrix interfacial region, and riveting behavior of ramie fabrics. Over and above all the tests, these composites were also subjected to compost test and they behaved positively in that too. This biodegradability of ramie fabric/PBS composites rendered the usage of these composites as potential substitutes in place of nonbiodegradable and non-eco-friendly composites [18].

Biocomposites resulting from nonwoven form of ramie fabrics reinforced in PBS matrix, when prepared by thermo-compression method, have many advantages such as enhanced mechanical characteristics and ILSS due to the influence of the manufacturing method when compared with other methods such as hand lay-up. Modification of ramie fabric surface by amino functional compounds maximized the PBS crystallinity and had a significant effect over the biocomposite characteristics. Rheological characterization of ramie fabric/PBS composites revealed that the composites had a stable structural network between the fiber and the matrix built primarily by the ramie fabric fibers and this had induced a strong interfacial bond between the fiber and the matrix, thus increasing the mechanical properties and

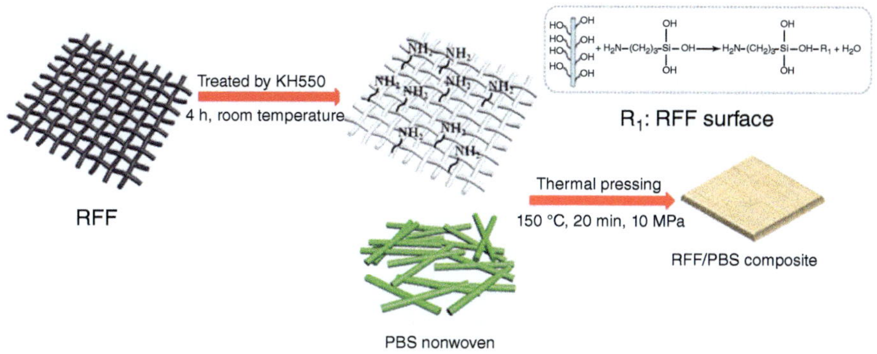


**Figure 15.2** Ramie fabric fiber surface modification with a coupling agent. Source: Based on Santhi et al. [18].

the biodegradability of the composites as stated above. This is considered to be the most novel method of manufacturing ramie fabric/PBS composite, which renders a cheap biodegradable composite with improved performance in all respects [18]. Figure 15.2 illustrates the process of surface modification of ramie fibers with amino functionalized compounds.

Some research focused on manufacturing a hybrid composite material with ramie and jute as reinforcements in polyester and epoxy matrix with the weight fraction ratio of fiber and matrix as 18 : 82 using the conventional hand lay-up technique. The lengths of both the fibers were maintained between 5 and 6 mm. Mechanical characteristics such as hardness, impact, flexural, and tensile strengths were measured for the jute–polyester, jute–epoxy, ramie–polyester, ramie–epoxy, and hybrid composites by suitable tests. Results revealed that the mechanical characteristics of hybrid fiber–epoxy composites was at least 30% higher than that of jute–polyester or ramie–polyester composites. Stress and strain induced during the loading influenced the properties of all the fabricated composite materials. It was noted from the results that the predominant reason for higher tensile strength of hybrid composites when compared with the individual fiber composites was the orientation of fibers within the matrix. Similar observations were made in case of specimens having higher impact and flexural and compressive strength [19]. Figure 15.3 depicts the fracture morphology of ramie fiber–epoxy composites with various wt.% of ramie fibers subjected to various mechanical tests.

## 15.3 Dynamic Properties of Ramie Fiber Composites

Damping behavior (tan $\delta$) and dynamic modulus ($E'$) of the major characteristics of the composites were obtained from the DMA [20]. DMA can also be used for the analysis of various other primary characteristics of a composite material such as non-Arrhenius variation of time with respect to temperature, dynamic fragility, glass transition temperature ($T_g$), and cross-linking density [21–24]. Usually, DMA tests were carried out using Rheometrics solid analyzer (RSA) in three-point bending

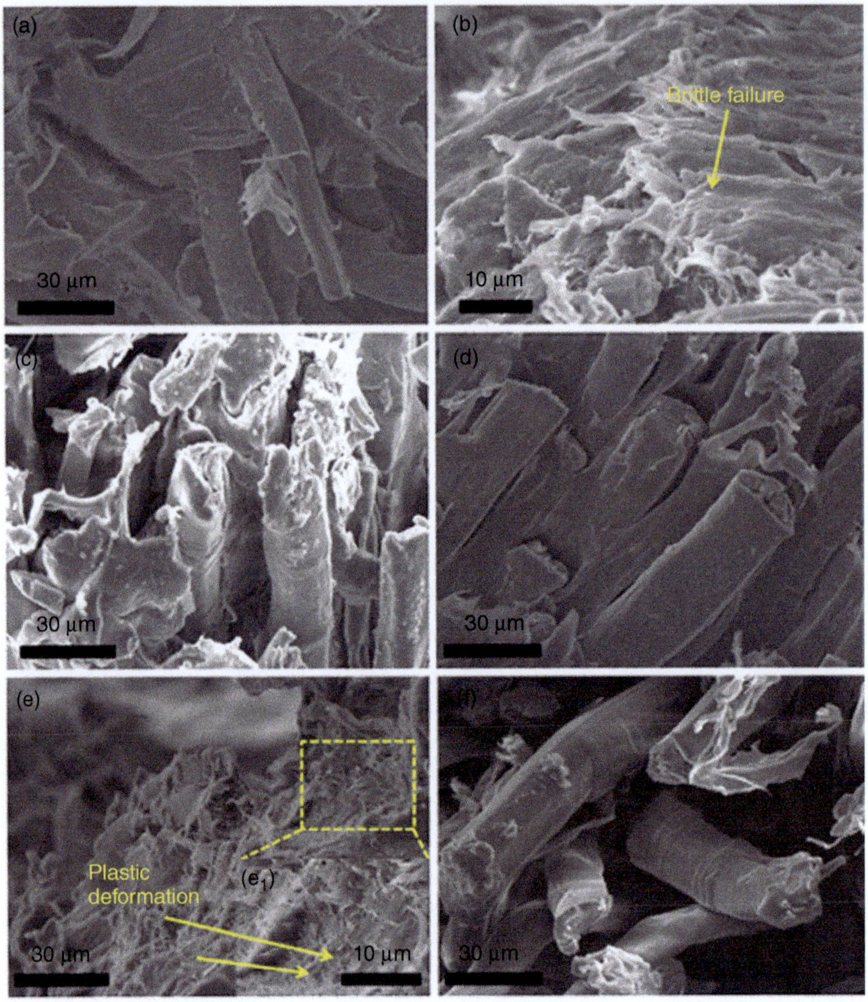


**Figure 15.3** (a) and (b) Ramie fabric fiber with 10 wt.%, (c) with 30 wt.%, (d) and (e) with 50 wt.%, and (f) with 60 wt.%.

mode tests. Specimens that are to be subjected to DMA were inured initially at a humidity of 50% and a temperature of 23 °C while the dimensions of the specimens were kept as 50 mm × 10 mm. Tests were carried out at a frequency of 1 Hz and the rate of heating would preferably be 5 °C/min. At a temperature range of 30–170 °C, the DMA values such as $E'$ and $\tan\delta$ were evaluated [15]. Dynamic mechanical characteristics including viscous modulus ($G'$), $E,'$ and $\tan\delta$ of flax/ramie hybrid composites with bio-epoxy matrix were assessed by means of three-point bending tests and the results were obtained as temperature-dependent parameters within a frequency limit of 1 Hz. The rate of stress transference within the flax and ramie reinforcements was observed owing to a better interfacial adhesion between the epoxy and the fibers, which in turn governed the viscoelastic properties of the composites.

It was determined from the results that the viscoelastic properties of the composites were mainly governed by the three factors temperature, frequency, and time.

### 15.3.1 Temperature Influence

Ramie/flax-reinforced bio-epoxy composites were investigated for their dynamic mechanical behavior along with the influence of temperature on those properties, and three different temperature regions were obtained while assessing the damping factor, loss and storage modulus. The first region was a glassy state region in which the storage modulus of the composites was high and the loss modulus and damping factor were low. Results of the experiments suggested that as the temperature increased, loss modulus and damping factor increased while the storage modulus gradually decreased and both the happenings were up to a limit. At the terminal stage of the glassy region, the storage modulus decreased steeply. The second region is the glass transition region where the loss modulus and damping factor values increased while the storage modulus value decreased with the rise in temperature. The temperature of transition, $T_g$, was measured in this region and the maximum values of loss modulus and damping factor could be attributed to the change of state of material from glassy to rubbery within this range of temperatures [25]. The third and last region was called the entropy elastic region or rubbery region in which the dynamic mechanical properties would stay as low as possible and these characteristics linearly change according to the change of temperature.

### 15.3.2 Storage Modulus

Storage modulus (E′) is the measure of stiffness of a material that specifies the quantity of energy that a composite material absorbs per oscillation of a loading cycle [26]. The influence of the ramie/flax fiber proportions on the storage modulus of the hybrid composite material was assessed at a frequency level of 1 Hz and by varying the temperature between 20 and 200 °C. The results revealed that the storage modulus of all the composite specimens, either individual fiber or hybrid composite, was found to be maximum at the glassy region due to the fact that the matrix and fiber molecules were in close proximity and packed tightly, which constrained the molecular chain movement. When the temperature crosses the glassy region, $E'$ decreases progressively owing to the fiber stiffness reduction. At the glass transition region, specifically between 50 and 90 °C, the molecular chain mobility of the matrix molecules increases suddenly in all the composite materials, which leads to a drastic fall of storage modulus, thus rendering a major change in viscoelastic properties of the ramie/flax/bio-epoxy composites [27]. As the molecular mobility further increases in the rubbery region because the closely packed arrangement of the molecules loosens its grip and the molecules are let free to move around, the storage modulus value experiences no major change and remains the same as in the glass transition region [28].

It was found from the results that the fiber with 30 wt.% rendered a maximum storage modulus value of 9.05 GPa due to enhanced fiber stiffness and improved the strength of the hybrid composites. This in turn established a rigid interlocking between the reinforcement and epoxy, which also facilitated an effective transfer medium of stresses at the fiber–matrix interfacial region [27, 29]. When compared with the storage modulus of pure epoxy (1.84 GPa), ramie and flax fiber-reinforced epoxy composites (5.43 and 4.15 GPa respectively), the hybrid composites outperformed them. It was concluded that the hybrid composites exhibited better stiffness and strength at equal weight proportions and performed dynamically far better than the ramie and flax individual fiber-reinforced epoxy composites.

### 15.3.3 Viscous Modulus

Viscous modulus ($G'$) is the degree of energy loss through dissipation in the form of heat at the time of deformation under each load cycle, which happens due to the movement of molecules of the composite material resulting in viscous response [25]. Similar to the above experiment of measuring storage modulus, loss modulus variation was recorded and measured at a frequency range of 1 Hz in the temperature range of 20 and 200 °C. Variation of loss modulus was reported in such a way that the values of loss modulus increased when the temperature moved from the glassy region to glass transition region and it was almost the same for all the specimens. When the temperature reaches the glass transition region, the loss modulus of all the composites increased rapidly and in the final rubbery region all the values dropped steeply to hit the lowest of all values for all the composites. Among all the composites, in spite of nearly equal values, hybrid composites exhibited the highest loss modulus followed by ramie, flax-reinforced composites, and then by pure epoxy composites. This variation could be attributed to the increased weight fraction of fibers, which were naturally elastic, in hybrid composites than individual fiber-reinforced composites but epoxy is viscoelastic in nature. This results in higher fiber energy absorption and high dissipation of heat at the maximum glass transition temperature, which were inferred from the maximum loss modulus curve values of all the composite materials [30, 31]. Pure epoxy composite resulted in a lower loss modulus value of 0.266 GPa at a temperature of 80 °C while the hybrid composites with 30 wt.% of fibers had the highest value of loss modulus of 1.45 GPa at 110 °C. This was due to the unrestricted molecular motion at high temperature in the former case while equal proportions of ramie and flax restricted the movement of molecules within the composite, resulting in maximum dissipation of energy and minimum degree of freedom to the molecules in the latter case [26]. Meanwhile, for ramie and flax individual fiber composites the loss modulus was found to be 0.9 and 0.63 GPa at 95 and 91 °C respectively. It could be seen from the experiments that equal weight proportions of fibers resulted in better strength and low deformation with varying time and temperature.

### 15.3.4 Damping Factor

Damping factor ($\tan\delta$) is the measure of the ratio between loss modulus and storage modulus, which is influenced by the fiber–matrix adhesion and is a function of temperature [25]. It is also termed as loss factor and is measured at a frequency level of 1 Hz. It was studied by many authors that the loss factor decreases within a temperature range of 20 and 200 °C owing to the increase in content of fibers. Results of the experiments show that the loss factor was minimum for hybrid composites with a value of 0.53 while it was observed to be maximum for the pure bio-epoxy composites with a value of 0.72. This was purely due to the incorporation of fibers into the epoxy resin to form composites that could have acted as a blockade for the polymeric chain molecular movement resulting in reduction in the rate of energy loss and degree of freedom of molecules [28]. Hence, the viscous flow at the glassy region is very less due to the elastic deformation and molecular slip down within the composites. As the region shifts to the transition zone, at maximum glass transition temperature the loss factor increases with increase in temperature and when the temperature reaches the rubbery region, the loss factor drastically falls owing to the freedom of molecules to move around without resistance, ultimately ending up with lower loss modulus value for the hybrid composite material. Individual ramie and flax fiber-reinforced composite materials exhibited a loss factor of 0.66 and 0.61 respectively, which were higher than the value of pure epoxy composites [17].

Figure 15.4 shows the storage modulus value of all composite materials with glass and ramie fiber ratio of 0 : 100 and 75 : 25 as determined by a few researchers. It could be observed from the figure that the value of storage modulus increased with the increase in content of fibers over the whole temperature range for all the composites analyzed. Figure 15.5 depicts the loss modulus curves for glass/ramie composites with fiber proportions of 100 : 0 and 25 : 75 over various temperature ranges. For all the weight fraction of glass/ramie composites analyzed, the value of loss modulus experienced an upward trend as the temperature increases. Owing to the higher dissipation of mechanical energy, the loss modulus curves reached its maximum values and then decreases with increase in temperature due to the unrestricted molecular movement in polymer chains. A few authors discussed that this increase in storage modulus was due to the constraint posed by glass fibers upon the

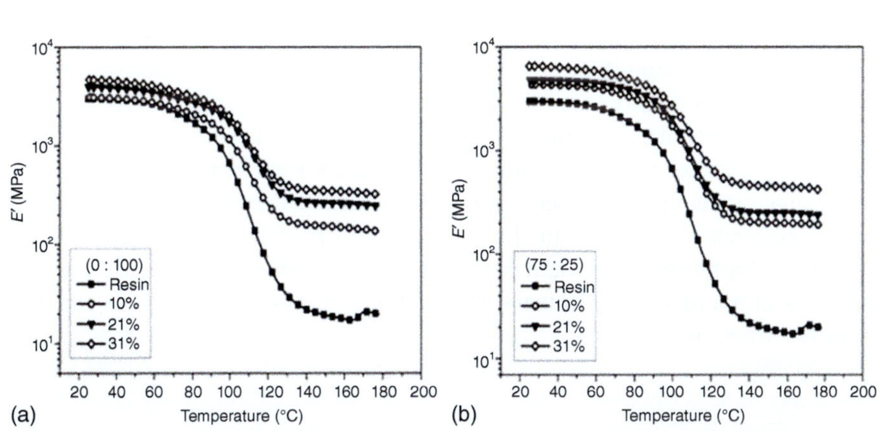


**Figure 15.4** Storage modulus of hybrid composites with fiber ratio (a) 0 : 100 and (b) 75 : 25. Source: Modified from Ornaghi et al. [32].

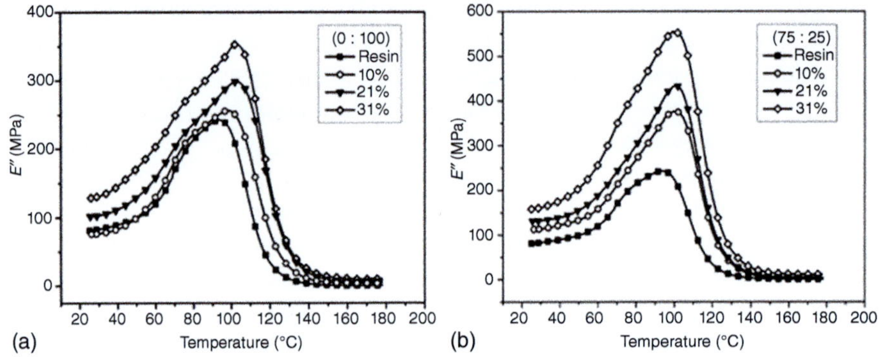


**Figure 15.5** Loss modulus of hybrid composites with fiber ratio (a) 0 : 100 and (b) 75 : 25. Source: Modified from Ornaghi et al. [32].

epoxy matrix, restricting the molecular movement and increasing the rate of stress transfer at the interface region [33]. For the freeness of molecular motion to happen, the interference of the adjacent molecular chains is very important along with the compatibility between molecules so that the value of $E'$ is enhanced. A few authors inferred from the study of ramie fiber composites that when the content of ramie fiber increases it allows the matrix to act on the mechanical restrictions resulting in reversible viscoelastic deformations, which in turn enhanced the stiffness of the composite material entirely [34]. Effectiveness coefficient ($C$) was observed between 40 and 160 °C for all the composites and it reached the lowest value when the content of fibers was 31 wt.%. But this value could not be validated properly by conducting experiments beyond 31 wt.% of fibers and cannot be compared with other fiber proportions owing to the manufacturing difficulties. A few authors experimentally determined the effectiveness coefficient for banana/sisal hybrid polyester composites with a fiber proportion of 1 : 1 and stated that the value of $C$ was low for the fiber content of 40 wt.%. This could be attributed to the tight package of fibers within a volume of composite and acute fiber contacts at higher fiber loadings leading to the least effective transfer of stresses between the fiber and the matrix at its interface [35].

A few experimental studies aimed at the analysis of effect of hybridization of ramie fibers with glass in polyester matrix and the dynamic response of those composites over a specific frequency. Dynamic mechanical properties such as tan $\delta$, $E''$, and $E'$ were measured by varying the contents of hybrid glass and ramie fibers as a function of temperature. The magnitude of glass transition temperature was kept as an indicator for measuring the influence of reinforcements while assessing the storage modulus. Alongside, some other parameters such as relaxation area, peak width at half maximum, and peak height were assessed to evaluate $E'$ and tan $\delta$ for the composites, which reveals the effect of glass transition temperature for all the composites. As the assessing frequency increased, the value of tan $\delta$ also increased when the temperature was maximum, and at this stage the activation energy of 75 wt.% glass fiber proportioned composites were noted to be high. Since this experiment focuses on the investigation of influence of glass/ramie fibers hybridization and the dynamic mechanical characteristics, the storage modulus was found to be greatly influenced by the addition of fibers into the polyester matrix, which increases the stiffness of the composite in the glassy region.

It was also observed from the results that the influence of reinforced glass/ramie fibers over the storage modulus was high at the temperatures above $T_g$ and was low below it. As far as $E'$ and tan $\delta$ were considered, the composites with the fiber ratio of 75 : 25 were characterized by higher relaxation time distribution, which were measured from peak width at half maximum and relaxation time, along with high dissipation of energy measured from peak height beneath the glass transition temperature. It was also observed from the weight fraction of fibers that increase in the content of glass fiber led to agglomeration of fibers, reducing the interaction between the fiber and the matrix while the increase in ramie fiber content enhanced the bio-compatibility between fibers and rendered good interfacial adhesion. Hence, the evaluation of fiber properties was far more important than the dynamic analysis at the interface. To conclude, at 75% of glass fibers, the tan δ peak reached its highest point when the temperature and frequency were higher, resulting in high activation energy [36].

Ramie fibers were characterized by superior tensile strength, which paved the way for them to be used in various applications being reinforced in polymeric matrix. Many experiments were performed on 30 wt.% ramie fibers reinforced in epoxy matrix for analyzing the dynamic mechanical characteristics such as tan $\delta$, $E''$, and $E'$ as a function of temperature by using dynamic mechanical tests. Tests were carried out in nitrogen atmosphere within a temperature range of 20 and 200 °C. Results of the experiments revealed that the addition of ramie fibers into the epoxy matrix enhanced the viscoelastic stiffness of the matrix. It was also noted that the structural damping coefficient of the composites significantly changed when the fiber weight fraction was increased. All these changes were due to the restricted movement of polymer chain molecules within the composites due to the interaction between ramie and epoxy. A few experimenters treated the ramie fibers with diglycidyl ether of bisphenol-A (DGEBA) and triethyl tetramine (TETA) and reinforced the modified fibers in epoxy matrix. The treated ramie fiber-reinforced epoxy composites, when subjected to DMA for assessing $E'$ and tan $\delta$, showcased increased properties particularly when aligned and continuous orientation of ramie fibers were used for all volume fractions. These quasi-static stiffness results were on par with the results obtained for the same composites during flexural tests. Owing to the addition of treated ramie fibers into the epoxy matrix, it becomes soft at higher temperatures and so the shift of $E'$, $E''$, and tan $\delta$ toward the lower values was observed at higher glass transition temperatures. All these lower values were attributed to the loosening of epoxy chain molecules and lower interfacial adhesion between ramie and epoxy, rendering an unrestricted polymeric chain movement. SEM images showcase the weak interfacial adhesion between the ramie fibers and the epoxy matrix, which results in low shear stress at the interface and least effective load transfer [36]. Figure 15.6 depicts the SEM micrograph of ramie fiber–epoxy composites with different magnification, showing debonding of fiber and matrix at the interface in detail.

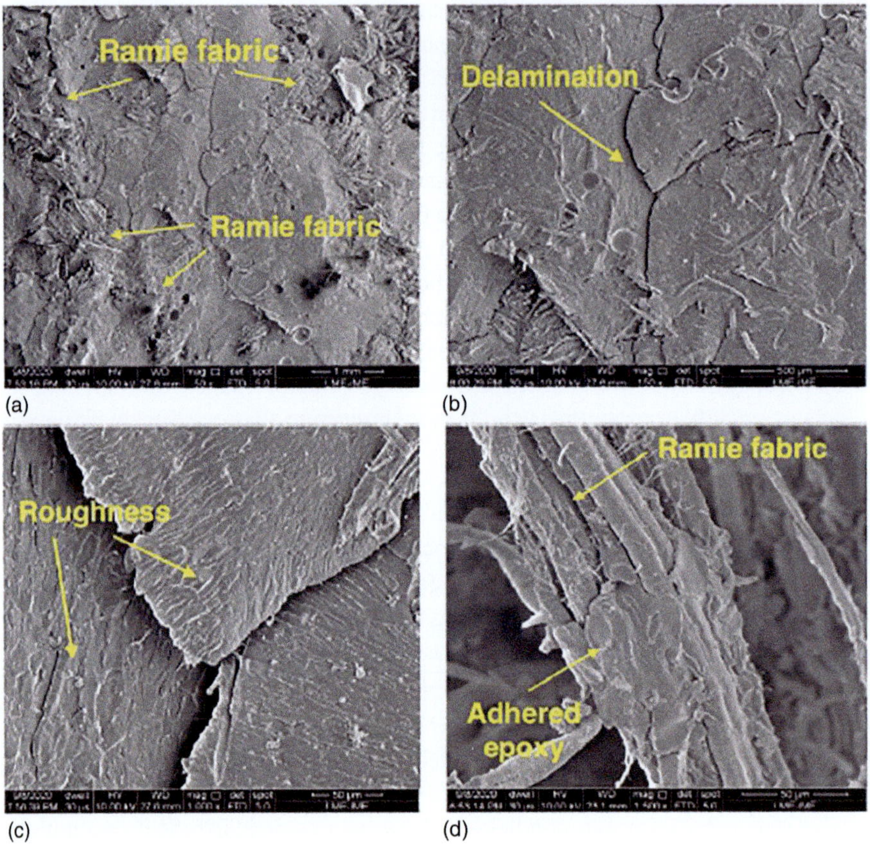


**Figure 15.6** SEM fractography of ramie fiber-reinforced epoxy composites; (a) 50 ×, (b) 150 ×, (c) 1000 × and (d) 1500 ×. Detailing the fiber–matrix debonding. Source: Margem et al. [33].

## 15.4 Conclusion

Analysis of ramie fiber reinforced with various combinations in terms of both natural and artificial materials is made. The specific analysis taken for consideration is the mechanical and dynamic mechanical characteristic evaluation. The evaluation was made for various mechanical strengths such as tensile, impact, flexural, and so on. Similarly, the dynamic mechanical strength was assessed using the influence of temperature, storage modulus, viscous modulus, and damping factor. The hybrid combination and evaluation reveal that these combinations have potential strength, which in turn has the capability to replace metals to some extent. Further research with novel lightweight material could contribute much effectively to the replacement of metals. Ramie fibers were also hybridized with other natural fibers and it resulted in producing low-cost and environment-friendly composites. This is the primary motive of utilization of ramie fiber composites in the agricultural industries as green materials. By the hybridization of synthetic fibers such as carbon with ramie, hardness and impact strength of the composites could be enhanced and the ability to be heat-treated may also improve. Improvement of adhesion between the fiber and the matrix at the interface should be achieved by following novel methodologies, which may also help in improving the properties of ramie-based composites. These composites find majority of its applications in wall cladding, door panels, home appliance outer shell sheathing, construction of room partitions, and electrical housing. It could be concluded from the above facts that the use of ramie fiber composites as individual fiber reinforced or hybrid composites is inevitable in bringing out effective and potential applications as novel materials or substitutes for artificial fibers-reinforced composites, contributing majorly to the economy of the nation ultimately.

## References


1 Ramesh, M., Deepa, C., Aswin, U.S. et al. (2017). Effect of alkalization on mechanical and moisture absorption properties of *Azadirachta indica* (Neem tree) fiber reinforced green composites. *Trans. Indian Inst. Met.* 70 (1): 187–199.

2 Bhoopathi, R., Deepa, C., Sasikala, G., and Ramesh, M. (2015). Experimental investigation on mechanical properties of hemp-banana-glass fiber reinforced composites. *Appl. Mech. Mater.* 766: 167–172.

3 Ramesh, M. and Sudharsan, P. (2018). Experimental investigation of mechanical and morphological properties of flax-glass fiber reinforced hybrid composite using finite element analysis. *Silicon* 10 (3): 747–757.

4 Binita, B.K., Nabaneeta, G., and Sanghamitra, K. (2013). Properties of ramie and its blends. *Int. J. Eng. Res. Gen. Sci.* 1: 1–6.

5 Irawan, A.P., Soemardi, T.P., Widjajalaksmi, K., and Reksoprodjo, A.H.S. (2011). Tensile and flexural strength of ramie fibre reinforced epoxy composites for Socket prosthesis application. *Int. J. Mech. Mater. Eng.* 6: 46–50.

6 Xu, C., Gu, Y., Yang, Z. et al. (2016). Mechanical properties of surface-treated ramie fibre fabric/epoxy resin composite fabricated by vacuum-assisted resin infusion molding with hot compaction. *J. Compos. Mater.*, http://jcm.sagepub.com/content/early/2015/06/19/0021998315590259.abstract.

7 Zhang, Y., Wen, B., Cao, L. et al. (2015). Preparation and properties of unmodified ramie fibre reinforced polypropylene composites. *J. Wuhan Univ. Technol.-Mater. Sci. Ed.* 30: 198–202.

8 Paiva Junior, C.Z., de Carvalho, L.H., Fonseca, V.M. et al. (2004). Analysis of the tensile strength of polyester/hybrid ramie–cotton fabric composites. *Polym. Test.* 23: 131–135.

9 Krasowska, K., Brzeska, J., Rutkowska, M. et al. (2010). Environmental degradation of ramie fibre reinforced biocomposites. *Pol. J. Environ. Stud.* 19: 937–945.

10 He, L.P., Tian, Y., and Wang, L.L. (2008). Study on Ramie fiber reinforced polypropylene composites (RF-PP) and its mechanical properties. *Adv. Mater. Res.* 41–42: 313–316.

11 Wang, H., Xian, G., Li, H., and Sui, L. (2014). Durability study of a ramie-fibre reinforced phenolic composite subjected to water immersion. *Fibers Polym.* 15: 1029–1034.

12 Giridharan, R. (2019). Preparation and property evaluation of glass/ramie fibers reinforced epoxy hybrid composites. *Composites Part B* https://doi.org/10.1016/j.compositesb.2018.12.049.

13 Lokesh, M., Xavier, J., Rodney, K.D. et al. (2019). Mechanical characterisation of epoxy polymer composite reinforced with ramie and synthetic fiber. *Proceedings of International Conference on Recent Trends in Computing, Communication & Networking Technologies (ICRTCCNT)* (19 August 2019). https://ssrn.com/abstract=3439194 or http://dx.doi.org/10.2139/ssrn.3439194.

14 Khalili, P., Blinzler, B., Kádár, R. et al. (2020). Ramie fabric Elium® composites with flame retardant coating: flammability, smoke, viscoelastic and mechanical properties. *Composites Part A* https://doi.org/10.1016/j.compositesa.2020.105986.

15 Dang, C.-Y., Nie, H.-J., Jia, R. et al. (2019). Effect of ramie fibers on the mechanical and shrinkage properties of spray-free polypropylene composite. *Mater. Res. Express* https://doi.org/10.1088/2053-1591/ab5e4d.

16 Kumar, S., Zindani, D., and Bhowmik, S. (2020). Investigation of mechanical and viscoelastic properties of flax-and ramie-reinforced green composites for orthopedic implants. *J. Mater. Eng. Perform.* 29: 3161–3171. https://doi.org/10.1007/s11665-020-04845-3.

17 Han, Q., Zhao, L., Lin, P. et al. (2020). Poly(butylene succinate) biocomposite modified by amino functionalized ramie fiber fabric towards exceptional mechanical performance and biodegradability. *React. Funct. Polym.* 146: 104443.

18 Santhi, K.A., Srinivas, C., and Kumar, R.A. (2020). Experimental investigation of mechanical properties of Jute-Ramie fibres reinforced with epoxy hybrid composites. *Mater. Today: Proc.* https://doi.org/10.1016/j.matpr.2020.04.368.

19 Menard, K.P. (1999). *Dynamic Mechanical Analysis: A Practical Introduction*, 13–29. CRC Press LLC.

20 Ramesh, M. (2019). Flax (*Linum usitatissimum* L.) fibre reinforced polymer composite materials: a review on preparation, properties and prospects. *Prog. Mater Sci.* 102: 109–166.

21 Ornaghi, H.L. Jr., Pistor, V., and Zattera, A. (2012). Effect of the epoxycyclohexyl polyhedral oligomeric silsesquioxane content on the dynamic fragility of an epoxy resin. *J. Non-Cryst. Solids* 358: 427–432.

22 Pistor, V., Ornaghi, F.G., Ornaghi, H.L. Jr., and Zattera, A.J. (2012). Dynamic mechanical characterization of epoxy/epoxycyclohexyl-POSS nanocomposites. *Mater. Sci. Eng., A* 532: 339–345.

23 Ramesh, M., Rajesh Kumar, L., Khan, A., and Asiri, A.M. (2020). Self-healing polymer composites and its chemistry. In: *Self-Healing Composite Materials*, 415–427. Woodhead Publishing – Elsevier. https://doi.org/10.1016/B978-0-12-817354-1.00022-3.

24 Saba, N., Jawaid, M., Alothman, O.Y., and Paridah, M.T. (2016). A review on dynamic mechanical properties of natural fibre reinforced polymer composites. *Constr. Build. Mater.* 106: 149–159.

**25** Pandian, C.A. and Jailani, H.S. (2018). Investigation of viscoelastic attributes and vibrational characteristics of natural fabrics-incorporated hybrid laminate beams. *Polym. Bull.* 75 (5): 1997–2014.

**26** Rajesh, M., Jeyaraj, P., and Rajini, N. (2016). Mechanical, dynamic mechanical and vibration behavior of nanoclay dispersed natural fiber hybrid intra-ply woven fabric composite. In: *Nanoclay Reinforced Polymer Composites* (eds. M. Jawaid, A. el Kacem Qaiss and R. Bouhfid), 281–296. Singapore: Springer.

**27** Chaudhary, V., Bajpai, P.K., and Maheshwari, S. (2018). An investigation on wear and dynamic mechanical behavior of jute/hemp/flax reinforced composites and its hybrids for tribological applications. *Fibers Polym.* 19 (2): 403–415.

**28** Jawaid, M., Khalil, H.A., Hassan, A. et al. (2013). Effect of jute fibre loading on tensile and dynamic mechanical properties of oil palm epoxy composites. *Composites Part B* 45 (1): 619–624.

**29** Saw, S.K., Sarkhel, G., and Choudhury, A. (2011). Dynamic mechanical analysis of randomly oriented short bagasse/coir hybrid fibre reinforced epoxy novolac composites. *Fibers Polym.* 12 (4): 506–513.

**30** Gupta, M.K. (2018). Thermal and dynamic mechanical analysis of hybrid jute/sisal fibre reinforced epoxy composite. *Proc. Inst. Mech. Eng., Part L* 232 (9): 743–748.

**31** Ramesh, M. and Rajesh Kumar, L. (2020). Bioadhesives. In: *Green Adhesives* (eds. R.B. Inamuddin, M.I. Ahamed and A.M. Asiri), 145–161. Wiley-Scrivener Publisher. https://doi.org/10.1002/9781119655053.ch7.

**32** Ornaghi, H.L. Jr., Bolner, A.S., Fiorio, R. et al. (2010). Mechanical and dynamic mechanical analysis of hybrid composites molded by resin transfer molding. *J. Appl. Polym. Sci.* 118: 887–896.

**33** Margem, F.M., Monteiro, S.N., Neto, J.B. et al. (2010). The dynamic mechanical behavior of epoxy matrix composites reinforced with ramie fibers. *Rev. Mater.* 15: 167–175.

**34** Idicula, M., Malhotra, S.K., Joseph, K., and Thomas, S. (2005). Dynamic mechanical analysis of randomly oriented intimately mixed short banana/sisal hybrid fiber reinforced polyester composites. *Compos. Sci. Technol.* 65: 1077–1087.

**35** Romanzini, D., Ornaghi, H.L. Jr., Amico, S.C., and Zattera, A.J. (2012). Influence of fiber hybridization on the dynamic mechanical properties of glass/ramie fiber-reinforced polyester composites. *J. Reinf. Plast. Compos.* 31 (23): 1652–1661.

**36** Margem, F.M., Monteiro, S.N., Bravo Neto, J. et al. (2010). The dynamic-mechanical behavior of epoxy matrix composites reinforced with ramie fibers. *Matéria (Rio de Janeiro)* 15 (2): 164–171.

# 16

# Fracture Toughness of the Natural Fiber-Reinforced Composites: A Review

*Haasith Chittimenu[1], Monesh Pasupureddy[1], Chandrasekar Muthukumar[1], Senthilkumar Krishnasamy[2], Senthil Muthu Kumar Thiagamani[3], and Suchart Siengchin[4]*

[1]*Hindustan Institute of Technology & Science, School of Aeronautical Sciences, Rajiv Gandhi Salai, Padur, Kelambakkam, Chennai 603103, Tamilnadu, India*
[2]*King Mongkut's University of Technology North Bangkok, Center of Innovation in Design and Engineering for Manufacturing (CoI-DEM), 1518 Pracharat 1, Wongsawang, Bangsue, Bangkok, 10800, Thailand*
[3]*Kalasalingam Academy of Research and Education, Department of Mechanical Engineering, Anand Nagar, Krishnankoil 626126, Tamil Nadu, India*
[4]*King Mongkut's University of Technology North Bangkok, The Sirindhorn International Thai-German Graduate School of Engineering (TGGS), Department of Materials and Production Engineering, 1518 Wongsawang Road, Bangsue, Bangkok, 10800, Thailand*

## 16.1 Introduction

Fiber-reinforced composites that are used in high-performance applications are based on glass and carbon fibers. Despite their excellent mechanical properties and superior resistance to environmental aging, these fibers have dangerous effects on human health and the ecosystem. Thus, there arises a need for biodegradable materials such as natural fibers. Natural fibers are considered as a key replacement for synthetic fibers in structural and nonstructural applications [1–4]. The automobile industry mainly offers many applications for natural fiber-reinforced composites (NFCs), where they are used for interior parts [5–7].

Natural fibers are suitable for application in lightweight structures due to their superior stiffness and better strength-to-weight ratio [8, 9]. Other such benefits of plant fibers are the relatively low cost compared to that of synthetic fibers and the absence of threat to the environment after their disposal [10–12]. There are various types of natural fibers used as reinforcements in composites, some of which are listed here:

- Date palm fiber
- Basalt fiber
- Bamboo fiber
- Coconut spathe fiber
- Sugar palm fiber
- Kenaf fiber
- Banana fiber
- Jute fiber
- Hemp fiber
- Sisal fiber
- Pineapple leaf fiber
- Cotton fiber
- Grass fiber
- Coir fiber

*Mechanical and Dynamic Properties of Biocomposites*, First Edition. Edited by Senthilkumar Krishnasamy, Rajini Nagarajan, Senthil Muthu Kumar Thiagamani, and Suchart Siengchin.

Any composite intended for load-bearing applications should be able to withstand the loads applied on it and display excellent performance under various conditions. During the functioning of the composite, it may get indented or impacted with foreign objects causing surface cracks, which may extend into the inner layers of the composite with prolonged usage, leading to material failure. This crack propagation rate should be within permissible limits to avoid catastrophic failure. Therefore, the determination of fracture toughness required for understanding the damage tolerance and crack resistance capabilities of a material is needed.

### 16.1.1 Fracture Toughness Tests

The relation between fracture energy, crack size, and strength of the material is the deciding factor for the crack resistance of a composite material. There are three common fracture toughness test methods based on the mode of delamination, as shown in Table 16.1. The crack opening based on the discussed modes is presented in Figure 16.1.

There are many standards that guide in conducting fracture toughness tests for the natural fiber composites stated by the American Society for Testing and Materials (ASTM) as shown in Table 16.2.

The commonly used test to assess the crack propagation resistance of the composites is shown in Figure 16.2.

**Table 16.1** Fracture modes in composites and their applications.

| Mode | Loading | Remarks |
|---|---|---|
| I | Tension | Simple and commonly used method |
| II | Shear | Predicts delamination in the composite laminate |
| III | Scissoring-shear | Mixed mode and very reliable method, which replicates the real loading scenario in the composite laminate |

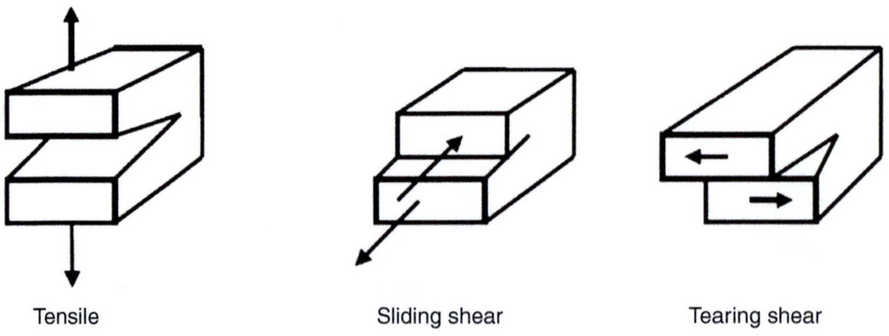


**Figure 16.1** Common fracture modes. Source: Prasad et al. [13].

**Table 16.2** ASTM standards to measure fracture toughness in composites based on the loading.

| Material | Description | ASTM standard |
|---|---|---|
| Any | Standard method | ASTM E1820 |
| | Standard method | ASTM E1823 |
| Plastics | Plane-strain fracture toughness and strain energy release rate | ASTM D5045 |
| Polymer-based laminates and pultruded composites | Translaminar fracture toughness | ASTM E1922 |
| Unidirectional fiber-reinforced polymer composites | Mode I | ASTM D5528 |
| | Mode II | ASTM D7905 |
| | Mixed mode | ASTM D6671 |

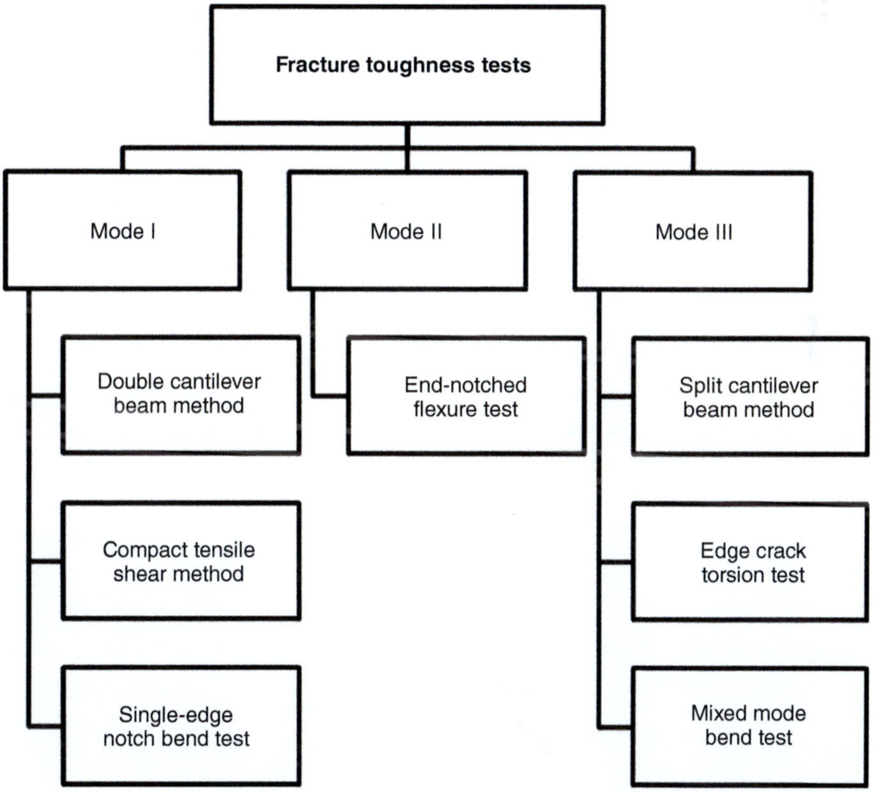


**Figure 16.2** Classification of fracture toughness test methodologies based on loading.

### 16.1.2 Mode-I Loading

#### 16.1.2.1 Double Cantilever Beam Method (DCB)

Double cantilever beam (DCB) test is usually done on the composite containing cracks in the midplane as per the ASTM D5528-13. Pre-crack is created with the help of a thin nonadhesive material inserted during fabrication of the composite. The composite film to be tested can have a dimension of 125 mm × 25 mm × 13 μm. Cross-head displacement speed can range from 1 to 5 mm/min [13]. The hinges are adhesively bonded to the specimen, and even bonded blocks can be used as shown in Figure 16.3a,b. The free ends of the hinges are gripped in standard tensile wedge grips.

The interlaminar fracture toughness ($G_{IC}$) under mode I loading can be obtained from Eq. (16.1):

$$G_{IC} = \frac{3P\delta}{2b\,(a + |\Delta|)}\ \mathrm{J/m^2} \tag{16.1}$$

where $P$ is the load endured by the specimen before failure, $\delta$ is the displacement of the load point, $b$ is the width of the tested specimen, $a$ represents the length of delamination, and $\Delta$ represents a standard correction factor.

#### 16.1.2.2 Compact Tensile Method (CT)

A composite specimen with pre-crack and two holes as illustrated in Figure 16.4 is pulled apart by the tensile clevises equally on the either side of the hole.

The width to thickness ratio of the tested specimen should be 2 : 1. The initial crack length to width ratio can be between 0.45 and 0.55. The initial crack length, which is two times the slot width, is usually generated by carefully tapping a thin cutter into the end of the slot until reaching the crack length needed [15].

#### 16.1.2.3 Single-Edge Notch Bend Test (SENB)

Flexural test is conducted on the specimens with the initial crack facing opposite to the loading direction, and the supports are similar to that in flexural testing, as shown in Figure 16.5. The allowed range for the cross-head speed is 0.5–20 mm/min, and a sharp notch at the root can be ensured using a blade.

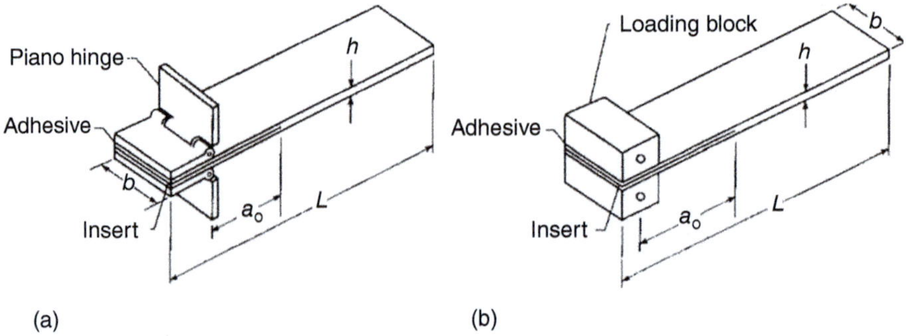


**Figure 16.3** DCB test setup. (a) With piano hinges and (b) with loading blocks. Source: ASTM [14]. © 2001, ASTM International.

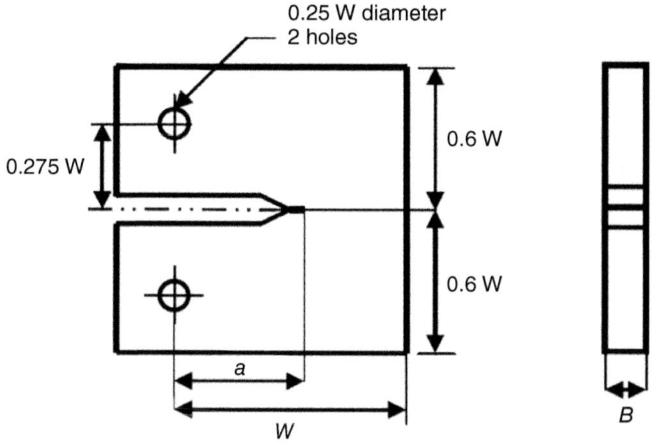


**Figure 16.4** Specimens dimensions for compact tension fracture toughness test. Source: Prasad et al. [13].

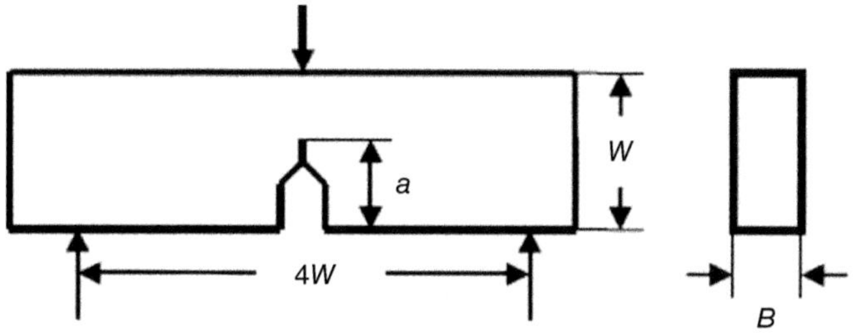


**Figure 16.5** Test setup of the single-edge notch bend (SENB) test. Source: Prasad et al. [13].

### 16.1.3 Mode-II Loading

#### 16.1.3.1 End-Notched Flexure Test (ENF)

The end-notched flexure (ENF) test used to measure mode II fracture toughness ($G_{IIC}$) is as per the ASTM D7905-14. In this test, a composite beam with a transverse midplane crack is subjected to flexural load, as shown in Figure 16.6.

### 16.1.4 Mode-III Loading

#### 16.1.4.1 Split Cantilever Beam Method (SCB)

Split cantilever beam (SCB) test with the specimen dimension shown in Figure 16.7 is used to determine the mode-III fracture energy ($G_{IIIC}$).

**Figure 16.6** End-notched flexure (ENF) test setup. Source: Prasad et al. [13].

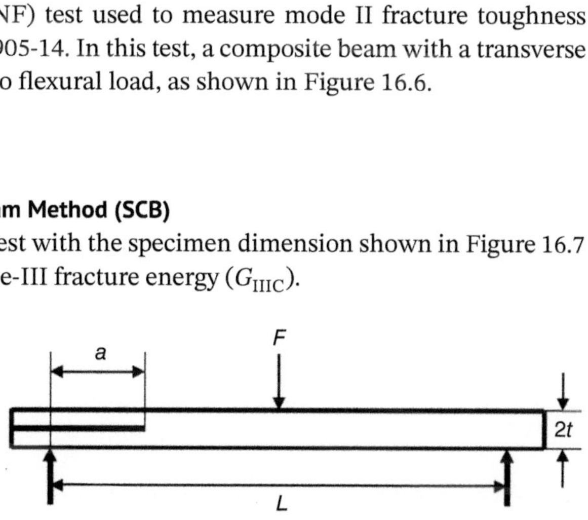

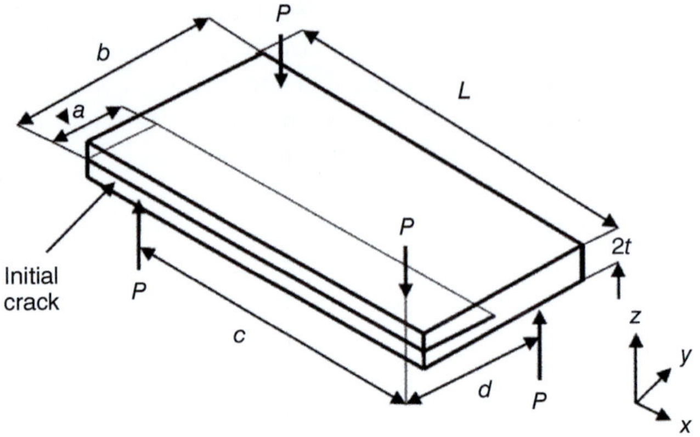


Figure 16.7 Specimen dimension for the SCB test. Source: Prasad et al. [13].

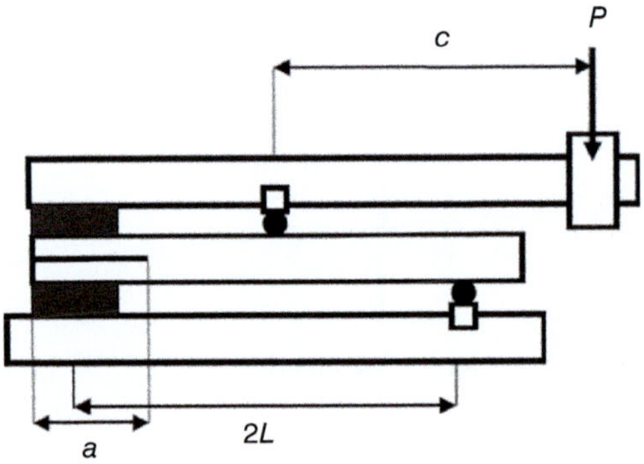


Figure 16.8 MMB test setup. Source: Prasad et al. [13].

#### 16.1.4.2 Edge Crack Torsion Test (ECT)

Delamination in the composite laminate can be characterized using the ECT test by subjecting the composite with pre-crack in the midplane to torsion.

#### 16.1.4.3 Mixed Mode Bend Test (MMB)

Mixed mode bend (MMB) test performed according to the ASTM D6671 involves application of mode I in DCB and mode II in ENF simultaneously (Figure 16.8). A composite laminate with midplane crack in the specimen length is attached to the fixture base on one end and the other two points are loaded with the hinges. The position of the translating loading saddle has to be adjusted to create variation in the ratios of mode I to mode II loading. Mixed mode fracture toughness ($G_c$) can be determined in accordance with the equations in ASTM D6671 [16].

## 16.2 Factors Affecting the Fracture Energy of the Biocomposites

### 16.2.1 Fiber Parameters

Silva et al. investigated the influence of fiber type, fiber architecture, and displacement rate on $G_{IC}$ of the sisal and coconut-based castor oil polyurethane composites using the compact tensile (CT) method. Sisal and coconut fibers were used in chopped form while the sisal fabric was also employed as the fiber reinforcement. It can be observed from their results that the composite with the chopped sisal fiber possessed better $G_{IC}$. However, the composite with woven sisal fabric displayed the highest $G_{IC}$. The displacement rate employed in the test was also found to affect the $G_{IC}$ [17].

Ismail et al. investigated the effect of fiber orientation on the $K_{IC}$ of the woven kenaf/polyester composite using the single-edge notch bend (SENB) test. They indicated that varying the fiber orientation in the plies of the laminate had a negative effect on the fracture energy. The composite with various ply orientations displayed lower $K_{IC}$ values than the composite with unidirectional fibers. The composite with unidirectional fibers showed fiber bridging in the wake of the cracks as opposed to the absence of fiber bridging in the woven fiber-based composite with various ply orientation angles [18].

Liu and Hughes examined the stacking sequence, effect of weave pattern, and fiber yarn linear density on the woven flax/epoxy composite using the CT method. Their results indicate that the presence of woven fiber in the matrix enhances the $K_{IC}$. It is also evident that crack propagation in the warp and weft direction also affects the $K_{IC}$ value. Notably, weave pattern had the least influence on the $K_{IC}$ value of the composite [19].

### 16.2.2 Hybridization

Zhang et al. investigated the delamination resistance termed as R-curve ($K_{IC}$ plotted against the crack growth length) of the unidirectional flax/glass-based hybrid composites from the SENB test. They found that hybrid composites have better fracture toughness than pure flax and glass-based composites. The improved fracture toughness of the hybrid composite is due to the fiber bridging and fiber entanglement in the wake of the delamination (Figure 16.9) [20].

Glass/epoxy composites coated with cellulose particulates at 5, 7.5, and 10 wt.% were subjected to the DCB and the ENF tests. Their results indicate that 5 wt.% coated specimens possessed better $G_{IC}$ from the DCB test while the composites with 10 wt.% cellulose coating had superior $G_{IIC}$ values than the glass/epoxy composites [21].

The influence of 2–4 wt.% rice husk ash (RHA) on the mode-I fracture toughness of RHA/$Al_2O_3$/Al-Mg-Si hybrid composite was investigated. Optimum $K_{IC}$ value was obtained for the hybrid composite containing 2 wt.% RHA while a further increase of RHA led to a decrease in the $K_{IC}$ [22].

### 16.2.3 Fiber Treatment

Abdullah et al. immersed coconut spathe fiber in alkali, silane, and combined alkali–silane and studied the effect of fiber treatment on the $K_{IC}$ of the coconut spathe fiber/epoxy composites. Composites with treated fibers showed lower $K_{IC}$ values than their counterparts with the untreated fibers. In particular, $K_{IC}$ dropped to about 60% in magnitude for the alkaline treatment, while for the other treatments, there was only a slight drop in the values. The drastic decrease in $K_{IC}$ for composites

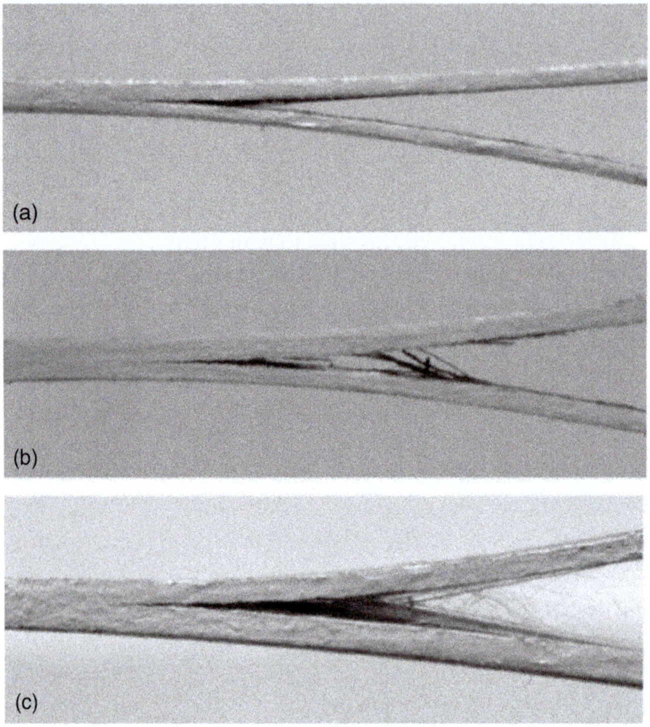


**Figure 16.9** Mode-I failure of composite from the DCB test. (a) Glass/phenolic composite, (b) flax/phenolic composite, and (c) glass/flax/phenolic hybrid composite. Source: Zhang et al. [20].

with alkali-treated fibers was due to the improved wettability between the fiber and the matrix, which reduced the fiber pullout and fiber–matrix debonding such that the energy absorption capacity of the composite decreased [23]. Pickering et al. also reported the trend of decreasing $K_{IC}$ due to fiber treatment with alkali on short hemp/poly-lactic acid (PLA) composites. According to their results, $K_{IC}$ of the composites with the treated fibers showed inferior values compared to the composite with raw fibers. The improved wettability of the alkali-treated fiber with the matrix allowed the pre-notched crack to easily propagate under the load such that composite specimens showed lower $K_{IC}$ values. Another observation from their study was that increase in fiber loading was not beneficial in terms of the fracture toughness for both the former and the latter [24].

Li et al. reported improvements in $G_{IC}$ and $G_{IIC}$ for the woven sisal–vinyl ester composite due to the fiber treatments with silane and potassium permanganate–dicumyl peroxide solution. Despite the improved interfacial bonding between the fiber and the matrix due to the fiber treatment, the fiber crack bridging offered by the woven fiber was effective in arresting the crack growth, and hence the higher fracture toughness values [25]. In their study, Pinto et al. reported that 1 wt.% aqueous silane treatment of jute resulted in a slight improvement in the $G_{IC}$ values for the composite obtained from the DCB test. The reason for the increase in $G_{IC}$ was the presence of fiber–matrix debonding, which allows increased energy absorption. It also indicates that the chosen silane concentration was not effective in improving the interfacial adhesion, thus allowing the composite to absorb more energy leading to slightly higher fracture toughness [26].

### 16.2.4 Aging

Alomayri et al. exposed cotton/geopolymer composites to water aging for 133 days. Their results revealed a decrease in the $K_{IC}$ for the exposed composite specimens. As the fiber loading was increased, aged composites showed lower $K_{IC}$ than the dry specimens; however, fracture toughness retention was better at higher fiber loading. Matrix cracking and fiber–matrix debonding, which are typical failures due to the absorption of water molecules into the composite, were evident from the micrographs of the failed specimens [27].

Alamri and Low reported that the addition of nanoclay helped in slightly higher retention of the fracture toughness on the recycled cellulose paper sheet/epoxy nanocomposites. Composites without nanoclay showed a 60% drop in $K_{IC,}$ while composites incorporated with the nanoclay between 1 and 5 wt.% displayed a 50% drop in value. A slight improvement for nanocomposites is due to the presence of nanoclay, which functions in the following ways: (i) it acts as a water barrier and reduces the moisture absorption into the composite and (ii) arrests micro-crack propagation into the matrix [10].

Kenaf/glass-based polyester composites were immersed in seawater, distilled water, and rainwater between 1 and 4 weeks, and the SENB test was conducted on the aged composites. In general, aged specimens showed a decrease in the $K_{IC}$ as the immersion time was prolonged. Moisture absorbed into the composite, as evident from the weight gain, caused fiber damage as well as matrix plasticization leading to lower fracture toughness [28].

According to Almansour et al. [29], the hybrid composite with flax/basalt layers showed better $G_{IC}$ under water aging than the flax/vinyl ester composite as determined from the DCB test. The presence of basalt fabric on the top and bottom acted as a water barrier and improved the $G_{IC}$. Flax/vinyl ester without basalt underperformed when exposed to water aging. Flax/basalt hybrid vinyl ester composites displayed better performance under mode-II loading assessed from the ENF test. In another study, Almansour et al. demonstrated that $G_{IIC}$ was higher for flax-basalt based hybrid composites than the pure flax/vinyl ester composite [30].

Islam et al. investigated the influence of water aging (25 °C for three months) and hygrothermal aging (50 °C for three months) on the $K_{IC}$ of the hemp/PLA composites through the SENB test. They found that both the aging conditions had a detrimental effect on the $K_{IC}$. However, the effect was more pronounced in the case of hygrothermal aging. This is because of the increased moisture absorption by the hemp fibers at elevated temperatures than at room temperature. They further highlighted that this limitation could be overcome by using alkali-treated hemp fibers, which showed better retention in $K_{IC}$ [31]. Islam et al. further extended their study to accelerated weathering. They exposed hemp/PLA composites to

UV irradiation and water spray for 250, 500, 750, and 1000 hours respectively. PLA, which was resistant to hygrothermal aging in their previous study, showed a decline in $K_{IC}$ with an increase in weathering time. Despite the decrease in fracture toughness, composites were found to perform better under weathering over the neat PLA [32].

## 16.3 Conclusion

Based on the literature review, the following conclusions were reached:

- Most composites encompass a brittle matrix with a high elastic modulus fiber, which are destined for structural-based applications. Thus, the fracture toughness can be evaluated for these materials by employing the linear elastic fracture mechanics (LEFMs) approach.
- Stress intensity factor (SIF) and fracture energy are the primary terms used to evaluate the fracture resistance of fiber-reinforced composites. In the case of the ductile matrix, the J-integral procedure could be applied.
- Crack resistance and fracture energy of the composites can be tailored by fiber parameters such as fiber type, fiber architecture, fiber hybridization, and fiber ply orientations. Use of long fibers and mat type yielded better crack resistance in the biocomposites while changing the ply orientation led to decline in the SIF. Reduction in SIF was reported for the composites reinforced with the treated fibers. Composites immersion in distilled water, seawater, and hygrothermal conditions showed lower crack resistance. Specifically, both the seawater and hygrothermal conditions had more degradation effects compared to the composites immersed in distilled water.

## Acknowledgments

We hereby acknowledge and sincerely appreciate the unalloyed support from the managements of the following institutions: Hindustan Institute of Technology & Science, Tamilnadu, India, as well as the King Mongkut's University of Technology North Bangkok (KMUTNB), Thailand.

## References

1 Low, I.M. et al. (2007). Mechanical and fracture properties of cellulose-fibre-reinforced epoxy laminates. *Composites Part A* 38 (3): 963–974.
2 Gupta, M.K. and Srivastava, R.K. (2015). Effect of sisal fibre loading on dynamic mechanical analysis and water absorption behaviour of jute fibre epoxy composite. *Mater. Today: Proc.* 2 (4–5): 2909–2917.
3 Krishnasamy, S. et al. (2019). Recent advances in thermal properties of hybrid cellulosic fiber reinforced polymer composites. *Int. J. Biol. Macromol.* 141: 1–13.

**4** Thiagamani, S.M.K. et al. (2019). Investigation into mechanical, absorption and swelling behaviour of hemp/sisal fibre reinforced bioepoxy hybrid composites: effects of stacking sequences. *Int. J. Biol. Macromol.* 140: 637–646.

**5** Zarges, J.-C., Minkley, D., Feldmann, M., and Heim, H.-P. (2017). Fracture toughness of injection molded, man-made cellulose fiber reinforced polypropylene. *Composites Part A* 98: 147–158.

**6** Chandrasekar, M., Siva, I., Kumar, T.S.M. et al. (2020). Influence of fibre inter-ply orientation on the mechanical and free vibration properties of banana fibre reinforced polyester composite laminates. *J. Polym. Environ.* 28 (11): 1–12.

**7** Chandrasekar, M. et al. (2019). Flax and sugar palm reinforced epoxy composites: effect of hybridization on physical, mechanical, morphological and dynamic mechanical properties. *Mater. Res. Express* 6 (10): 105331.

**8** Prasad, V., Sekar, K., Varghese, S., and Joseph, M.A. (2019). Enhancing Mode I and Mode II interlaminar fracture toughness of flax fibre reinforced epoxy composites with nano $TiO_2$. *Composites Part A* 124: 105505.

**9** Senthil Muthu Kumar, T., Senthilkumar, K., Chandrasekar, M. et al. (2019). Characterization, thermal and dynamic mechanical properties of poly(propylene carbonate) lignocellulosic *Cocos nucifera* shell particulate biocomposites. *Mater. Res. Express* 6 (9): 1–10.

**10** Alamri, H. and Low, I.M. (2013). Effect of water absorption on the mechanical properties of nanoclay filled recycled cellulose fibre reinforced epoxy hybrid nanocomposites. *Composites Part A* 44: 23–31.

**11** Chandrasekar, M., Ishak, M.R., Sapuan, S.M. et al. (2017). A review on the characterisation of natural fibres and their composites after alkali treatment and water absorption. *Plast. Rubber Compos.* 46 (3): 119–136.

**12** Senthilkumar, K. et al. (2018). Mechanical properties evaluation of sisal fibre reinforced polymer composites: a review. *Constr. Build. Mater.* 174: 713–729.

**13** Prasad, M.S.S., Venkatesha, C.S., and Jayaraju, T. (2011). Experimental methods of determining fracture toughness of fiber reinforced polymer composites under various loading conditions. *J. Miner. Mater. Charact. Eng.* 10 (13): 1263.

**14** ASTM D5528-01 (2001). *Standard test method for Mode I interlaminar fracture toughness of unidirectional fiber-reinforced polymer matrix composites.* Annu. B. ASTM Stand., vol. 15.

**15** Adams, D.A. Compact tension fracture toughness testing [Online]. Available: https://www.compositesworld.com/articles/compact-tension-fracture-toughness-testing (Accessed: 05 July 2020).

**16** Adams, D.O. Fracture mechanics testing of composites [Online]. Available: https://www.compositesworld.com/articles/fracture-mechanics-testing-of-composites (Accessed: 05 July 2020).

**17** Silva, R.V., Spinelli, D., Bose Filho, W.W. et al. (2006). Fracture toughness of natural fibers/castor oil polyurethane composites. *Compos. Sci. Technol.* 66 (10): 1328–1335.

**18** Ismail, A.E. et al. (2016). Fracture toughness of woven kenaf fibre reinforced composites. *IOP Conf. Ser. Mater. Sci. Eng.* 160 (1): 12020.

**19** Liu, Q. and Hughes, M. (2008). The fracture behaviour and toughness of woven flax fibre reinforced epoxy composites. *Composites Part A* 39 (10): 1644–1652.

**20** Zhang, Y., Li, Y., Ma, H., and Yu, T. (2013). Tensile and interfacial properties of unidirectional flax/glass fiber reinforced hybrid composites. *Compos. Sci. Technol.* 88: 172–177.

**21** Uppin, V.S., Ashok, I.S., and Gouda, P.S.S. (2016). Interlaminar fracture toughness in glass-cellulose reinforced epoxy hybrid composites. *IOP Conf. Ser. Mater. Sci. Eng.* 149 (1): 12113.

**22** Alaneme, K.K., Akintunde, I.B., Olubambi, P.A., and Adewale, T.M. (2013). Fabrication characteristics and mechanical behaviour of rice husk ash–alumina reinforced Al-Mg-Si alloy matrix hybrid composites. *J. Mater. Res. Technol.* 2 (1): 60–67.

**23** Abdullah, A.H., Abdul Mutalib, F.F., and Mat, M.F. (2016). Tensile and fracture toughness properties of coconut spathe fibre reinforced epoxy composites: effect of chemical treatments. *Adv. Mater. Res.* 1133: 603–607.

**24** Pickering, K.L., Sawpan, M.A., Jayaraman, J., and Fernyhough, A. (2011). Influence of loading rate, alkali fibre treatment and crystallinity on fracture toughness of random short hemp fibre reinforced polylactide bio-composites. *Composites Part A* 42 (9): 1148–1156.

**25** Li, Y., Mai, Y.-W., and Ye, L. (2005). Effects of fibre surface treatment on fracture-mechanical properties of sisal-fibre composites. *Compos. Interfaces* 12 (1–2): 141–163.

**26** Pinto, M.A., Chalivendra, V.B., Kim, Y.K., and Lewis, A.F. (2013). Effect of surface treatment and Z-axis reinforcement on the interlaminar fracture of jute/epoxy laminated composites. *Eng. Fract. Mech.* 114: 104–114.

**27** Alomayri, T., Assaedi, H., Shaikh, F.U.A., and Low, I.M. (2014). Effect of water absorption on the mechanical properties of cotton fabric-reinforced geopolymer composites. *J. Asian Ceram. Soc.* 2 (3): 223–230.

**28** Salleh, Z., Taib, Y.M., Hyie, K.M. et al. (2012). Fracture toughness investigation on long kenaf/woven glass hybrid composite due to water absorption effect. *Procedia Eng.* 41: 1667–1673.

**29** Almansour, F.A., Dhakal, H.N., and Zhang, Z.Y. (2017). Effect of water absorption on Mode I interlaminar fracture toughness of flax/basalt reinforced vinyl ester hybrid composites. *Compos. Struct.* 168: 813–825.

**30** Almansour, F.A., Dhakal, H.N., and Zhang, Z.Y. (2018). Investigation into Mode II interlaminar fracture toughness characteristics of flax/basalt reinforced vinyl ester hybrid composites. *Compos. Sci. Technol.* 154: 117–127.

**31** Islam, M.S., Pickering, K.L., and Foreman, N.J. (2010). Influence of hygrothermal ageing on the physico-mechanical properties of alkali treated industrial hemp fibre reinforced polylactic acid composites. *J. Polym. Environ.* 18 (4): 696–704.

**32** Islam, M.S., Pickering, K.L., and Foreman, N.J. (2010). Influence of accelerated ageing on the physico-mechanical properties of alkali-treated industrial hemp fibre reinforced poly(lactic acid)(PLA) composites. *Polym. Degrad. Stab.* 95 (1): 59–65.

# 17

# Dynamic Mechanical Behavior of Hybrid Flax/Basalt Fiber Polymer Composites

*Arun Prasath Kanagaraj[1], Amuthakkannan Pandian[2], Veerasimman Arumugaprabu[1], Rajendran Deepak Joel Johnson[3], Vigneswaran Shanmugam[1,3], and Vairavan Manikandan[4]*

[1] *Kalasalingam Academy of Research and Education, Department of Mechanical Engineering, Anand Nagar, Krishnankoil, Tamil Nadu 626126, India*
[2] *PSR Engineering College, Department of Mechanical Engineering, Sivakasi, Tamil Nadu 626140, India*
[3] *Saveetha School of Engineering, Saveetha Institute of Medical and Technical Sciences, Department of Mechanical Engineering, Chennai, Tamilnadu, India*
[4] *PSN College of Engineering and Technology, Department of Mechanical and Automation, Tirunelveli, Tamil Nadu 627152, India*

## 17.1 Introduction

Recently most of the researchers increase their research area by the usage of natural fiber-based applications, due to its environmental sustainability and their environment-friendly nature [1]. In this context, the natural fibers are mainly established in the fields of automobile and heavy building structures, aerospace, marine, electronics, and structural-based applications [2]. Natural fibers contain lignocellulose, which helps to achieve better strength for the composite materials [3]. The natural fiber has many advantages such as easy availability, less cost, low density, biodegradability, acceptable mechanical properties, and low energy consumption. Nevertheless, natural fiber has some drawbacks in terms of absorbing the moisture, which would reduce the strength of the composite materials. Further, they tend to deteriorate the fiber–matrix adhesion characteristics in polymeric matrix composites [4]. To satisfy the gap, the new material is introduced in the composite composition called synthetic fiber or man-made fibers to fulfill the gap in the strength and adhesiveness of the composite [5]. Hybridization of composite produces better properties [6–8]. Incorporation of natural fiber with synthetic fiber improves the stiffness and results in high strength-to-weight ratio of the composites.

Dynamic mechanical analysis (DMA) is used to measure the applied stress and strain response of the polymers. DMA is one of the precise methods to evaluate and characterize the polymers in the state of elastic and plastic behavior. The temperature and elastic strain variation allow us to determine the complex modulus of

*Mechanical and Dynamic Properties of Biocomposites*, First Edition. Edited by Senthilkumar Krishnasamy, Rajini Nagarajan, Senthil Muthu Kumar Thiagamani, and Suchart Siengchin.

the materials. In general, the complex modulus helps to locate the glass transition temperature ($T_g$) of the material. $T_g$ is one of the significant characteristics of the polymers to identify their transition rate between hard and soft regions during the polymerization [9].

The combination of hemp and carbon fiber-reinforced composites produced better results in the dynamic mechanical property. Increasing the content of the hemp fiber improves the bonding and allows the polymer molecules to flow, which improves the molecular dynamics of the polymer chain and enables the composites to absorb more energy [10]. Always the smooth flow of molecules within the fibers produces better results for a pineapple leaf and glass fiber composite, due to loss of heat to achieve better dynamic mechanical properties [11]. Kevlar reinforced with bucky paper exhibits improved elastic properties with good interfacial bonding. Mostly, Kevlar-based composites produce good stability with high thermal resistance during dynamic conditions [12].

The toughened Elium thermosetting polymer improves the interlocking between the layers of flax fiber in the composite. The high crystalline structure of flax fiber strengthens the interlocking between the adjacent layers of the fibers and improves the polymer motions within the composites for improved dynamic responses [13]. Siakeng et al. reported that the pineapple leaf fiber/coir fiber/polylactic acid composites exhibited higher storage and loss modulus than the pure fiber-reinforced composites [14].

In the glassy region of the polymers, higher storage modulus is observed in coir pith/pine apple leaf/polylactic acid-based composites. Below the glass transition temperature, there is no significant change in the storage modulus of hybrid composites observed [15]. Doddi et al. reported that hybridizing the basalt/pineapple leaf fiber/epoxy matrix composites showed enhanced results in DMA than the pure fiber-reinforced composites such as basalt/epoxy and pineapple leaf fiber/epoxy matrix composites [16]. Shen et al. introduced a new method called composite delaminate detection, which was based on the use of the DMA test. They reported that the delamination of the composites could be easily found by employing the composite delaminate detection technique [17]. Researchers compared the DMA characteristics of untreated and treated basalt fiber-reinforced polymer matrix composites. Analysis of the results revealed that the untreated fiber-reinforced composites showed improved storage modulus, loss modulus, and damping factor than the treated fiber composites [18]. Senthilkumar et al. studied the DMA properties of pure sisal/bioepoxy, pure hemp/bioepoxy, and their hybrid combinations. It was reported that the stiffness and their thermal stability were found to be increased for specific hybrid configurations.

Furthermore, they suggested these sisal/hemp/bioepoxy composites for structural-based applications [19]. From the overall study, the hybridization of natural and synthetic fibers showed improved thermal properties. The increment of natural fiber content in the composite showed more relaxation of polymer motion and resulted in better thermal properties.

In this chapter, the importance of using natural fiber and the advantages of the hybridization of fibers in DMA are investigated. The hybridization of basalt and flax-based composites was developed, and their dynamic responses were studied in the present work.

## 17.2 Materials and Methods

### 17.2.1 Materials

Flax fiber was purchased from Varghese fibers Kerala, India, and Basalt fiber was imported from Austria. The general-purpose polyester resin (grade VBR 2303) and curing agents such as methyl ethyl ketone peroxide (accelerator) and cobalt naphthenate (catalyst) were purchased from Vasavibala Resins Pvt. Ltd., Chennai, India. The essential properties of flax, basalt, and polyester resin are given in Tables 17.1 and 17.2, respectively.

### 17.2.2 Fabrication of Composites

Initially, the fibers were cut to a length of 40 mm and washed with distilled water. Further, the fibers were dried in air. The pure flax, pure basalt, and their hybrid combinations were fabricated by using a compression molding machine under a pressure of 80 bar. The fibers were taken according to the type of configurations, and the resin mixture was poured over the fiber surfaces. After pouring the resin mixture, the mold was closed and compressed in a compression molding machine for the next 24 hours at room temperature. The pure and hybrid composites were fabricated with different fiber volume fractions such as (flax:basalt) 20 : 80, 50 : 50, and 80 : 20, respectively.

### 17.2.3 Dynamic Mechanical Analysis

The DMA was carried out on a Hitachi DMA 7100 machine under 1 Hz frequency with a fixed load of 2 N. The dynamic analysis is carried out at a room temperature of 120 °C with the heating rate of 3 °C/min. The samples were prepared according to the ASTM D4065-01 dimensions as $60 \times 12 \times 3\ mm^3$. The samples were machined in the abrasive water jet machine, and the average results of three trials are chosen for the study.

**Table 17.1** Properties of flax and basalt fiber.

| Name of the Fiber | Tensile strength (GPa) | Young's modulus (GPa) | Diameter (μm) | Density (g/cm³) | Specific modulus | Specific strength |
|---|---|---|---|---|---|---|
| Flax fiber | 6.5–8 | 60 | 0.02 | 1.4–1.5 | ~41 | ~480 |
| Basalt fiber | 2.8–3.1 | 85–87 | 0.018 | 2.63–2.8 | ~32.1 | ~1.09 |

**Table 17.2** Properties of polyester resin.

| Type of matrix | Tensile strength (GPa) | Young's modulus (GPa) | Curing shrinkage | Density (g/cm³) | Heat distortion temperature (°C) | Viscosity (around 25 °C) |
|---|---|---|---|---|---|---|
| Polyester | 0.04–0.090 | 2–4.5 | 4–8% | 1.2–1.5 | 54 | 250–300 cP |

## 17.3 Result and Discussion

### 17.3.1 Damping Factor (Tan $\delta$) Response of Basalt/Flax Fiber Composite

The damping factor (tan $\delta$) is the ratio of loss modulus to storage modulus. Figure 17.1 shows the relationship between the damping factor and temperature responses of basalt/flax fiber-reinforced hybrid composites. The 80 : 20 (flax:basalt) hybrid composites showed the highest damping factor. It represents a higher degree of polymerization with more molecular mobility [18]. Further, it was followed by the 20 : 80 and 50 : 50 hybrid combinations. Major changes in the tan $\delta$ values were seen between the temperatures of 100 and 150 °C [19]. The mobility of the polymer molecules was decreased for composites having 50 wt.% of flax/50 wt.% of basalt fiber, which produced a lower damping factor. Thus, the 50 : 50 hybrid composites showed the least damping factor. It was ascribed to the increased fiber–matrix bonding characteristics than in the rest of the composite configurations [20].

### 17.3.2 Storage Modulus (E′) Response of Basalt/Flax Fiber Composite

Figure 17.2 shows the storage modulus of flax/basalt fiber-reinforced hybrid composites. It is observed that the temperature increases to a maximum level of glassy regions, and they fall toward the rubbery regions due to recrystallization. In the lesser temperature ranges, the 80 : 20 (basalt:flax) fiber-reinforced composites exhibited the least storage modulus followed by the 20 : 80 and 50 : 50 combinations. On increasing the temperature, the above-said pattern was slightly changed. When reaching the glass transition temperature, the 50 : 50 maintained the highest storage

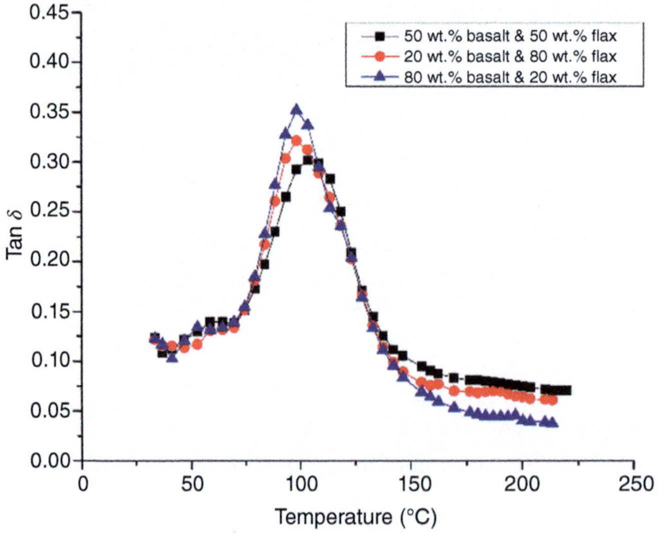


**Figure 17.1** Tan $\delta$ of flax/basalt fiber-reinforced polyester matrix composites.

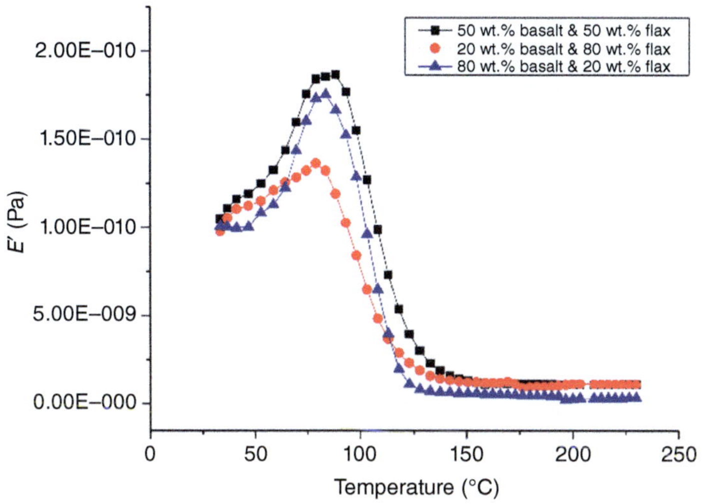


**Figure 17.2** Storage modulus of flax/basalt fiber-reinforced polyester matrix composites.

modulus, followed by the 80 : 20 and 20 : 80. However, the 50 : 50 and the 80 : 20 did not show any significant differences in the glass transition temperature ranges. On further increasing the temperature, the 20 : 80 loaded hybrid composites exhibited improved storage modulus, followed by 50 : 50 and 80 : 20.

### 17.3.3 Loss Modulus Performance of Basalt/Flax Fiber Composites

The loss modulus of the material measures the viscoelastic response of the polymer and dissipation of energy in the form of heat. Figure 17.3 shows the loss modulus function and the corresponding responses with respect to temperature for the chosen composite. Before reaching the glass transition temperature, the 80 : 20 (basalt:flax) composites exhibited the least loss modulus. When increasing the temperature, the 50 : 50 and 80 : 20 hybrid composites performed well; these composites showed a higher loss modulus than the 20 : 80 (basalt:flax) type of composites. When comparing the storage modulus and loss modulus of 50 : 50 hybrid composites, both followed a similar behavior. It was ascribed to the increased fiber–matrix bonding characteristics. However, 80 : 20 hybrid composites also showed higher loss modulus at the glass transition temperature. On further increasing the temperature, the loss modulus of all the hybrid combinations was found to drop. At higher temperatures, the 50 : 50 hybrid configuration consistently maintained its performance.

## 17.4 Conclusions

- This chapter shows the performance of storage modulus, loss modulus, and damping factor for flax/basalt fiber-reinforced composites under dynamic environment. Among the composites, the 50 : 50 configuration provided better fiber/matrix than the rest of the configurations.

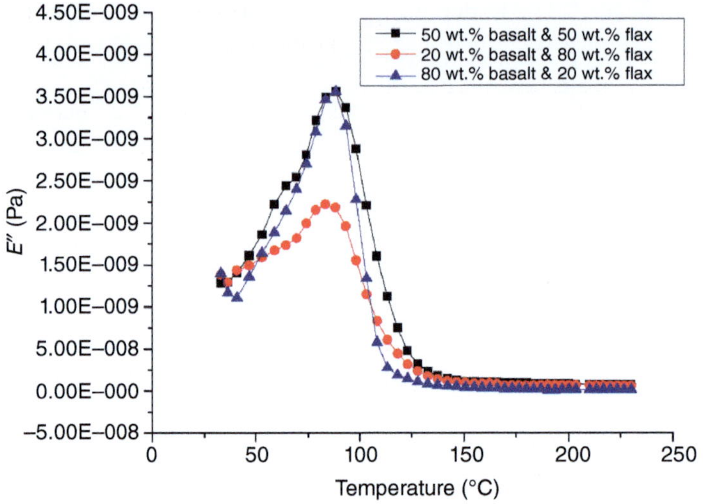


**Figure 17.3** Loss modulus of flax/basalt fiber-reinforced polyester matrix composites.

- Regarding the tan $\delta$, the 80 : 50 (basalt:flax) of hybrid composites showed the highest values, followed by 20 : 80 and 50 : 50.
- In the case of storage modulus, 50 : 50 (basalt:flax) composites exhibited the highest values at the glass transition temperature, followed by 80 : 20 and 20 : 80. Nevertheless, the 20 : 80 hybrid combinations showed the maximum storage values in the rubbery plateau region.
- In loss modulus, the 50 : 50 and 80 : 20 (basalt:flax) composites showed improved values at the glass transition temperature. On further increasing the temperature, the loss modulus of all the hybrid composites was found to drop.

## Acknowledgments

The authors wish to thank the Center for Composite Materials, Department of Mechanical Engineering, Kalasalingam Academy of Research and Education, Krishnankoil, Tamilnadu 626126, India for their kind permission to carry out the preparation and testing of the composites.

## References

1 Fiore, V. and Calabrese, L. (2019). Effect of stacking sequence and sodium bicarbonate treatment on quasi-static and dynamic mechanical properties of flax/jute epoxy-based composites. *Materials* 12 (9): 1363.
2 Shishevan, F.A. and Akbulut, H. (2019). Effects of thermal shock cycling on mechanical and thermal properties of carbon/basalt fiber-reinforced intraply hybrid composites. *Iran. J. Sci. Technol., Trans. Mech. Eng.* 43 (1): 441–449.

3 Mochane, M.J., Mokhena, T.C., Mokhothu, T.H. et al. (2019). Recent progress on natural fiber hybrid composites for advanced applications: a review. *eXPRESS Polym. Lett.* 13: 159–198.

4 Wu, C., Yang, K., Gu, Y. et al. (2019). Mechanical properties and impact performance of silk-epoxy resin composites modulated by flax fibers. *Composites Part A* 117: 357–368.

5 Han, L., Ma, F., Chen, S., and Pu, Y. (2019). Effect of short basalt fibers on durability, mechanical properties, and thermal properties of polylactic acid composites. *Polym. Renewable Resour.* 10 (1–3): 45–59.

6 Karthik Babu, N.B., Muthukumaran, S., Ramesh, T., and Arokiasamy, S. (2019). Effect of agro-waste microcoir pith and nano-alumina reinforcement on thermal degradation and dynamic mechanical behavior of polyester composites. *J. Nat. Fibers* 6: 1–13.

7 Fombuena, V., Petrucci, R., Dominici, F. et al. (2019). Malcinized linseed oil as epoxy resin hardener for composites with high bio content obtained from linen byproducts. *Polymers* 11 (2): 301.

8 Obada, D.O., Kuburi, L.S., Dodoo-Arhin, D. et al. (2019). Synthesis, characterization and industrial application. In: *Fillers*, 1–11. IntechOpen https://doi.org/10.5772/intechopen.82889.

9 Verma, S.K., Gupta, A., Patel, V.K. et al. (2019). The potential of natural fibers for automotive sector. In: *Automotive Tribology* (eds. J.K. Katiyar, S. Bhattacharya, V.K. Patel and V. Kumar), 31–49. Singapore: Springer.

10 Poyyathappan, K., Bhaskar, G.B., Rajesh, S., and Pazhanivel, K. (2019). Experimental analysis and comparative mechanical testing on glass-carbon hybrid composites. *Mater. Today: Proc.* 16: 612–620.

11 Zin, M.H., Abdan, K., Mazlan, N. et al. (2019). Automated spray up process for Pineapple Leaf Fiber hybrid bio composites. *Composites Part B* 177: 107–306.

12 Sharma, S., Dhakate, S.R., Majumdar, A., and Singh, B.P. (2019). Improved static and dynamic mechanical properties of multiscale bucky paper interleaved Kevlar fiber composites. *Carbon* 152: 631–642.

13 Khalili, P., Blinzler, B., Kádár, R. et al. (2019). Flammability, smoke, mechanical behaviours and morphology of flame retarded natural fiber/elium. *Compos. Mater.* 12 (17): 2648.

14 Siakeng, R., Jawaid, M., Ariffin, H., and Sapuan, S.M. (2019). Mechanical, dynamic, and thermomechanical properties of coir/pineapple leaf fiber reinforced polylactic acid hybrid biocomposites. *Polym. Compos.* 40 (5): 2000–2011.

15 Fortunato, G., Anghileri, L., Griffini, G., and Turri, S. (2019). Simultaneous recovery of matrix and fiber in carbon reinforced composites through a Diels–Alder solvolysis process. *Polymers* 11 (6): 1007.

16 Doddi, P.R.V., Chanamala, R., and Dora, S.P. (2019). Dynamic mechanical properties of epoxy based PALF/basalt hybrid composite laminates. *Mater. Res. Express* 6 (10): 105343.

17 Shen, Y., Tan, J., Fernandes, L. et al. (2019). Dynamic mechanical analysis on delaminated flax fiber reinforced composites. *Materials* 12 (16): 2559.

**18** Amuthakkannan, P. and Manikandan, V. (2018). Free vibration and dynamic mechanical properties of basalt fiber reinforced polymer composites. *Indian J. Eng. Mater. Sci.* 25 (10): 12–22.

**19** Krishnasamy, S., Thiagamani, S.M.K., Muthukumar, C. et al. (2019). Effects of stacking sequences on static, dynamic mechanical and thermal properties of completely biodegradable green epoxy hybrid composites. *Mater. Res. Express* 6 (10): 105351.

**20** Senthilkumar, K., Chandrasekar, M., Rajini, N. et al. (2019). Characterization, thermal and dynamic mechanical properties of poly(propylene carbonate) lignocellulosic *Cocos nucifera* shell particulate bio composites. *Mater. Res. Express* 6 (9): 096426.

# Index



*Mechanical and Dynamic Properties of Biocomposites*, First Edition. Edited by Senthilkumar Krishnasamy, Rajini Nagarajan, Senthil Muthu Kumar Thiagamani, and Suchart Siengchin.

## g



## h



## i

***o***